Rapid One-of-a-kind Product Development

Shane (S.Q.) Xie · Yiliu Tu

Rapid One-of-a-kind Product Development

Strategies, Algorithms and Tools

Springer

Associate Professor Shane (S.Q.) Xie
University of Auckland
Department of Mechanical Engineering
Private Bag 92019
1142 Auckland
New Zealand
s.xie@auckland.ac.nz

Professor Yiliu Tu
University of Calgary
Department of Mechanical and
Manufacturing Engineering
2500 University Drive NW
Calgary, AB T2N 1N4
Canada
paultu@ucalgary.ca

ISBN 978-1-4471-5710-6 ISBN 978-1-84996-341-1 (eBook)
DOI 10.1007/978-1-84996-341-1
Springer London Dordrecht Heidelberg New York

British Library Cataloguing in Publication Data
A catalogue record for this book is available from the British Library

Cover design: eStudioCalamar, Girona/Berlin

Printed on acid-free paper

Springer is part of Springer Science+Business Media (www.springer.com)

Preface

The globalisation of business and industry and the worldwide competitive economy are forcing one-of-a-kind production (OKP) enterprises to fully utilise the best available equipment and techniques. The objective is to have efficient control of the organisational structure in order to produce high quality products at lower prices within a shorter period of time. With the development of modern technologies and theories, a new generation of OKP systems is urgently required for rapid product development (RPD) by OKP enterprises to meet today's increasingly global competition.

RPD has been recognised as playing a key role in improving the competitiveness of OKP companies in the global market. With rapid development of information technology, network technology, modern management technology and other related advanced technologies, tremendous efforts have been made in developing new systems, algorithms and tools for the purpose of RPD through collaboration over the Internet, e.g. Internet-based systems for product design and manufacturing, Internet-based design for X (DFX) and Internet-based integrated systems. These Internet-based systems can be used to rapidly produce high quality products with low costs, bringing tremendous profit for manufacturing companies and allowing them to stay competitive in the global market. Such systems are urgently required in industry. However, the research and development of Internet-based RPD (IRPD) systems for producing OKP products are still at an early stage and further research and development in this area are becoming more and more urgent.

This book systematically reviews the historical background of the innovative technologies for OKP product development (PD). Through systematic overview of the existing systems and recent approaches of computer aided design and manufacturing systems, the problems that emerged from recent approaches have been identified. They include distributed system structure, cost optimisation algorithm, product data modelling, production information flow modelling, DFX, computer aided design (CAD)/computer aided process planning (CAPP)/computer aided manufacturing (CAM) integration technology and process modelling and distributed knowledge management. To overcome these problems and to develop a new generation of OKP systems, research and development of a PD platform has been carried out.

The main findings from this research work are outlined. These include the introduction of rapid OKP PD, a comprehensive review of recent research developments in OKP, new product modelling technologies, definition of the new PD platform, an integrated information framework for rapid development of sheet metal parts, the global data structure for global manufacturing, a CAD/CAPP/CAM integration system, a WWW-based information management system, case studies using the platform for rapid sheet metal and mould PD, and the definition and application of an Internet-based cost estimation and optimisation framework. These research topics and findings constitute the main contents of the book.

The aim of the book is to provide a snapshot of our recent research outcomes and implementation studies in the field of OKP PD. The 16 chapters are divided into four parts. The first part introduces state-of-the-art OKP PD; the second part presents the system and framework developed; the third part introduces new methods for modelling OKP products; and the final part presents our case studies showing the new algorithms, methods and tools developed for rapid development of high value-added OKP products.

As the title suggests, the first chapter gives an overview of OKP. It briefly introduces the history and objectives of the OKP PD process. This is followed by discussions on the current issues involved in OKP PD process.

The second chapter presents the historical background of Internet-based product design and manufacturing systems for the rapid development of OKP products. By reviewing existing OKP systems and recent approaches of Internet-based design and manufacturing systems, the requirements for the next generation of OKP systems and the current techniques that can be used to implement Internet-based product design and manufacturing systems are discussed. The problems that emerged from recent developments are reviewed and solutions are proposed. The future trends of Internet-based collaborative design, decision support, manufacturing support, supply chain management, workflow management, Internet techniques for product design and manufacturing, product modelling, STEP-based data environment, concurrent engineering, etc., are also discussed. The reviewed state-of-the-art approaches are used directly or indirectly as references for the development of new generation OKP systems. A reference system structure for building an Internet-based integrated PD system is then proposed to facilitate rapid development of OKP products.

Targeting product modelling, Chapter 3 presents a comprehensive review of recent developments in product modelling technology. Four types of product modelling methodology are discussed in detail. Two object-oriented product modelling methods, including STEP-based product modelling and UML-based product modelling are reviewed and compared. The research gaps and issues are identified. To implement the product modelling in the current integrated manufacturing environment, a generic product modelling framework is proposed. Future research trends in product modelling are also discussed.

To further develop the reference system structure of the Internet-based integrated PD system into a working PD system, Chapter 4 proposes an integrated PD system for rapid development of OKP products. The structure of the system is discussed and case studies are carried out to test the idea of an Internet-based integrated system.

In sheet metal processing and manufacturing there are many small- or medium-sized enterprises (SMEs). These shops have to make every effort to shorten PD lead time, improve production efficiency, approach high quality standards, while cutting down costs at the same time. Chapter 5 presents a compound cutting and punching production method supported by an integrated CAD/CAPP/CAM system. The solutions include an integrated data integration platform based on Pro/INTRALINK and STEP, and a knowledge-based real-time CAPP (RTCAPP) system for compound sheet metal cutting and punching.

The requirements for CAPP systems have changed in the current integrated manufacturing environment. Requirements have increased in terms of integration, openness, real time and distribution of process planning. Systems must now be open and dynamic with the ability to adapt and accept radical unpredictable changes in structures and industrial practices. Chapter 6 proposes an agent-based process planning system (ABPPS) for optimal OKP PD. Agents defined to carry out tasks include an unfolding agent, a feature recognition agent, a task agent, a nesting agent, a path planning agent, a bending agent, a machining method selection agent, a machine selection agent and a fixture/jig selection agent. This chapter provides a good example of how an Internet-based PD system using modern agent technology is developed.

To model OKP products, Chapter 7 proposes a generic product modelling framework (GPMF) that aims to provide an infrastructure for modelling various types of product information. The outcome of the GPMF is a set of data models defined to model product information at different stages of the development process. This chapter discusses the structure of the GPMF and its main components.

Chapter 8 investigates the EXPRESS data model (EDM) that is developed based on the STEP-based modelling environment and the "five-phase" modelling method. The structure of the EDM is discussed. The elements of the EDM – schemas and the relationships between these schemas – are presented using both EXPRESS language and EXPRESS-G diagrams.

Chapter 9 further develops the product modelling methods into a GPMF for efficient information exchange and sharing. This framework consists of four functional components: the EDM, a STEP-based modelling environment, a "five-phase" modelling method, and three EDM data exchange and sharing methods. The focus of this research is on the modelling methodologies and the definition of schemas for various PD activities such as manufacturing, inspection, etc., and the integration of the schemas with other resources defined within STEP. There are 25 schemas defined to ensure that the proposed GPMF is compatible and can be used to model various types of products. To the best of our knowledge, these aspects have not been studied extensively in the literature.

Information integration is an important issue to address in supporting integrated and concurrent PD. Chapter 10 explores the definition and structure of an information framework for rapid development of sheet metal parts. This framework aims to build an information bridge to fill the gap among sheet metal part design, process planning and manufacturing systems. A tree-based step-structure information modelling methodology for sheet metal parts has been developed and a case study is described.

In the early stages of a PD process, the estimate of a developing product life cycle cost is the primary information needed to determine the profitability of the product. Chapter 11 presents a cost estimation and optimisation system, which is a part or a subsystem of an integrated PD system. The PD cost under consideration in this chapter is the sum of the costs of product design, production and logistics, which is the main cost of a product. Some other possible costs associated with a PD cycle, e.g. overhead costs, are normally proportional to this main PD cost and in practice, these are estimated by adding a fixed percentage to the PD cost. Unlike some earlier research which focused on cost estimation in a single manufacturing company, this cost estimation system aims for PD cost estimate and optimal control under the complexities of a global manufacturing environment. The proposed cost estimate and optimal control system has been implemented in a sheet metal manufacturing company.

Chapter 12 reports on research work that aims to develop an integrated data structure to support RPD in the Internet environment. The emphasis is placed on integrated data management and the reuse of past PD experience to support a company's aim to shorten its PD cycle. The integrated global data structure model was modelled using EXPRESS from STEP with the consideration of real-time data communication within the Internet environment. In terms of this data structure, a design/manufacturing knowledge base was developed as a major part of the WWW (world wide web)-based PD system. The basic principles and concepts of the knowledge base and the WWW-based knowledge management system are presented. An industrial implementation is also reported.

As efficient management of product information that covers the whole life cycle is critical to the enhancement of corporate competitiveness, Chapter 13 explores the design and development of a WWW-based PD information management system for a cross-nation manufacturing corporation in New Zealand. Since product data are often managed in a distributed computing environment, CORBA is employed to ensure interoperability among distributed information objects. This WWW- based information management system is discussed in this chapter.

Internet-based DFX systems have been recognised as efficient tools for implementing concurrent engineering and playing a key role in RPD. Internet-based DFX or IDFX systems can be applied by manufacturing companies to rapidly produce high quality products with low costs and high profits. However, the implementation of IDFX systems is not an easy task. Chapter 14 presents two typical applications of IDFX, i.e. Internet-based design for manufacture (IDFM) and Internet-based design for cost (IDFC) systems, for rapid and economic tool-/mould-making. The structure and the key models of the systems are discussed.

Process planning has become very important for the OKP industry as global competition to reduce product cost has intensified. An optimal process planner can maximise the utilisation of costly raw material resources, improve machining efficiency, and hence reduce product costs. Chapter 15 explores algorithms for solving the path planning issue. This work uses sheet metal products as examples and investigates the two issues using genetic algorithms. The proposed genetic algorithm approach uses a genetic encoding scheme and a genetic reproduction strategy to reach an optimum solution. Case studies are carried out to test the genetic algorithms. The effectiveness of the genetic algorithm path planning approach is compared with that of the ant colony algorithm.

The future trends in PD technologies are discussed, and include such topics as Internet-based collaborative design, decision support, manufacturing support, supply chain management, Internet techniques for product design and manufacturing, system integration, product modelling, STEP-based data environment, and concurrent engineering. This may be used to guide coming research, or act as a reference for a company to deploy or develop new systems, algorithms and tools for producing OKP products.

This book also contains an Appendix that summarises some of the systems developed for RPD. It provides the source code of the tools developed including the quality function deployment (QFD) interface, CAD/CAPP/CAM system, and database interfacing program. These are excellent examples for users or developers.

I would like to also take this opportunity to express my deep appreciation to those who have contributed to this book. The authors are also grateful to Ms Bomiao Li and Ms Jenny Xu for their assistance in compiling the book. It is our sincere hope that readers will find this book useful to their study and research.

Auckland, New Zealand
September, 2010

Shane (S.Q.) Xie

Acknowledgements

The authors would like to acknowledge funding support from the Foundation for Research, Science and Technology of New Zealand. The project (No. PROJ-13546-IIoF-UoA) is titled "Enabling Technology for Rapid Development of High Value-added Customized Products", and funded under the International Investment Opportunities Fund scheme.

Contents

Part I State of the Art

Part I
State of the Art

In today's global market, more and more manufacturing companies have realised that the ability to quickly develop a customised product in an economic and efficient way is critical for them to survive in today's competitive international market. This is particularly true for those small or medium-sized one-of-a-kind production (OKP) companies. A new generation of OKP systems needs to be developed to help these companies maintain competitiveness in the global marketplace and improve the ability to rapidly combine the strengths of manufacturing partners to meet market needs.

Part I of the book, as its name suggests, aims to provide a comprehensive state-of-the-art review of recent developments in the methods, tools and systems for rapid development of OKP products in order to achieve high profits. This part includes three chapters.

Chapter 1 discusses the main objectives of rapid OKP development in the global manufacturing environment. In this chapter, the background of recent approaches for rapid development of OKP products is reviewed. After systematically reviewing the existing OKP systems and recent developments of new technology and systems, the authors discuss current issues and requirements for developing a new generation of OKP systems for rapidly producing OKP products.

Chapter 2 reviews the historical background of product design and manufacturing systems for rapid development of OKP products. Through over viewing the existing OKP systems and recent approaches to product design and manufacturing systems, the author will discuss current techniques that can be used to implement product design and manufacturing systems to produce OKP products. The problems that emerged from recent developments are summarized in this chapter. The future trends of Internet-based collaborative design, decision support, manufacturing support, supply chain management, workflow management, Internet techniques for product design and manufacturing, product modelling, STEP-based data environment, concurrent engineering, etc. will also be discussed in this chapter. At the end of the chapter, a reference system structure for building an Internet-based integrated product development system is proposed for rapid development of OKP products.

As for OKP companies, how to better manage and record the previous product development knowledge has become a core issue for them to improve the product development process, cut down development cost and reduce lead time. In recent years, considerable effort has been placed on developing new enabling technologies for OKP companies to achieve high quality and productivity, and to quickly respond to the changing market to meet customer requirements. Product modelling is a pivotal activity in the product development (PD) processes. Well-defined product models organize product data, production information and knowledge to satisfy the requirements of the rapidly changing PD environment. Chapter 3 provides a comprehensive review on recent development of product modelling technology. Four types of product modelling methodologies are discussed in detail. Two object-oriented product modelling methods, including STEP-based product modelling and UML-based product modelling are reviewed and compared. The research gaps and issues are identified. To implement the product modelling into current integrated manufacturing environment, a generic product modelling framework is proposed. The future research trend of product modelling is also discussed.

Chapter 1
Background and Motivation

Abstract In today's global market, more and more manufacturing companies have realised that the ability to quickly develop a customised product in an economic and efficient way is critical for them to survive in today's competitive international market. This is particularly true for one-of-a-kind production (OKP) companies. A new generation of OKP systems needs to be developed to help these companies maintain competitiveness in the global marketplace and improve the ability to rapidly combine the strengths of manufacturing partners to meet market needs. This chapter gives a definition of OKP and introduces the main issues and objectives of rapid OKP development in the global manufacturing environment.

1.1 Rapid One-of-a-kind Product Development

Nowadays more and more manufacturing companies have realised that the ability to quickly develop a customised product in an economic and efficient way is critical for them to survive in the keen competitive international market. This is particularly true for those OKP companies. Wortmann (1991) defined various types of OKP companies using a 2D typology. One dimension is determined by a company's production system position strategies, *i.e.*, product-oriented or capacity-oriented. The other dimension is determined by a company's market strategies, *i.e.*, make to stock, assemble to order, make to order, and engineer to order. The research as reported in this book focuses on the rapid product development (RPD) problems in an OKP company which has a product-oriented production system and adopts an engineer-to-order market strategy.

OKP, as predicated by scholars such as Rolstadas (1991), Wortmann (1991) and Hirsch (1992), could be a promising manufacturing model for the factory of the future. At the same time OKP poses challenges to the factory of the future. Tu (1996) characterised the OKP philosophy as:

1. high customisation;
2. successful product development (PD) and production in one go;

3. optimal or rational utilisation of technologies and resources;
4. adaptive production planning and control;
5. continuous customer influence throughout the production;
6. incremental process planning;
7. distributed control and inter-organisational autonomy; and
8. virtual company structure and global manufacturing.

From a practical viewpoint, an OKP company can be loosely understood as an advanced manufacturing company, which provides customised products in a certain manufacturing area (*i.e.*, 'kind' in OKP). It is often a product-focused small or medium sized enterprise (SME). According to its market needs, it produces a very wide range of product variations according to the customer requirements. It has a much wider product domain and more flexible batch size (maybe one) than a product-focused batch production company has. A typical OKP company can be a sheet metal company, injection mould/tool manufacturer, a steel construction company, *etc.* An OKP company has a greater flexibility to adopt market changes than a mass/batch production company, and higher production efficiency than a job shop. However, OKP companies, owing to high customisation, face a large amount of uncertainties and consequently may deal with a lot of rework. PD cost is also normally higher and development lead time longer than product-focused manufacturing companies.

To rapidly develop products with low costs, a new paradigm called rapid product development (RPD) has been proposed for the rapid development of products with low cost, high value addition and acceptable quality. Bullinger *et al.* (2000) defined RPD as an interdisciplinary methodology to combine all influences of an engineering process to an iterative PD. Its research topics focus not only on product, but also on their development process. The rapid development of OKP product is a holistic organisational concept that describes the rapid OKP PD process achieved by combining and integrating various innovative technologies and tools, *e.g.*, rapid manufacturing (http://www.albright1.com/manufacture.shtml), simultaneous engineering, computer supported cooperative work tools (Bullinger *et al.* 1996), and a supportive environment (Tu *et al.* 2001).

In this book, the scope of RPD will focus on the development of customised, individual or engineer-to-order OKP products. The P in OKP means OKP products, *e.g.*, a crystal oscillator, a customised sheet metal product or injection moulding. As shown in Figure 1.1, the RPD processes include all the OKP PD processes starting from the customer's requirements until the products are delivered.

Many OKP companies in New Zealand and overseas are striving to develop new computer aided tools for supporting their PD activities including product design, validation, resource allocation and optimisation, scheduling and machining. The main objectives are:

1. to shorten OKP time to market (from product definition to market launch);
2. to develop OKP products by optimising key factors, *e.g.*, time to market, cost and quality;

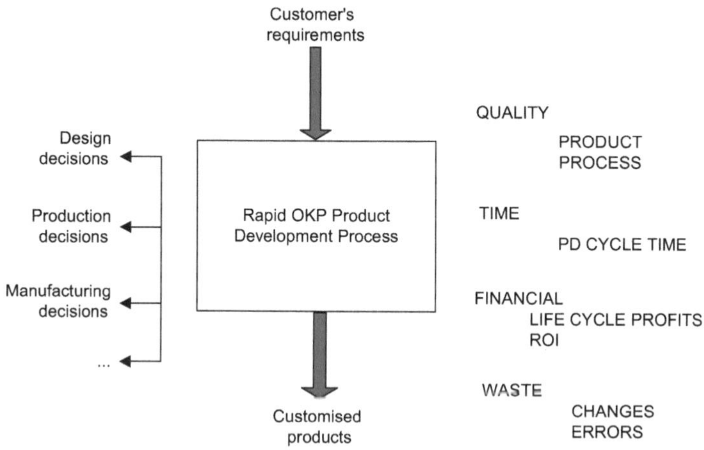

Figure 1.1 PD metrics and perceptions

3. to increase OKP products quality, achieve high profit and decrease waste (changes, reworks and errors);
4. to rapidly respond to the customer's requirements and market changes.

1.1.1 The Performance of RPD

The effectiveness of an OKP company's PD process must be evaluated against some objective criteria – a set of quantifiable parameters/metrics (as shown in Figure 1.1). Generally speaking, the metrics typically fall into the following four broad categories – quality, time, financial, and waste. Hence, the performance of PD can be weighed by following these factors (Floyd 1993).

1.1.1.1 Product Quality

The definition of quality is often used in today's manufacturing environment as "meeting the needs and expectations of customers". In today's market, it is no longer sufficient to define quality simply as "conformance to specifications". Sometimes, the specification may not fully meet the customers' needs which change with time. In the new definition, the operative words are needs, expectations and customers. The definitions of needs and expectations given by Floyd (1993) are briefly quoted in the following paragraph.

Needs are those functions, characteristics, and features that customers expressly want or must have in a product that they are acquiring. Customers are aware of their needs and, if asked, can usually state them with a reasonable degree of clarity.

Expectations are more difficult to determine and identify. Expectations are those things customers automatically assume to be inherent in the product. Often, they do not explicitly state their expectations, making it difficult to interpret using computer technology.

1.1.1.2 Lead Time

There are many ways to define the time to develop an OKP product. Some common terms are time to market, development lead time, PD time, and PD cycle time. These terms are generally used interchangeably – and all are intended to describe the total time it takes to implement the PD process from conception of a product idea to the point at which the product is both released for sale and in full production. In this book, the lead time is understood as the time from starting a product definition to finally delivering the product to the customer.

World-class OKP companies are constantly striving to improve their performance, and thereby to reduce the PD cycle time. Through a combination of innovative RPD process changes and continual improvement with new technology, many manufacturing companies have made dramatic reductions in the PD lead time. For example, during recent decades, companies striving for world-class status have cut cycle times by as much as 50%, as reviewed by Floyd (1993). These companies are continuing to improve.

1.1.1.3 Financial

For many OKP companies, the main reason to develop customised products is the possibility of bringing high profit returns. Product development not only requires a significant amount of investment, but also establishes the costs of producing OKP products. The return on investment (ROI) has always been the focus of PD. In recent years, cost control and containment have become even more important than before.

PD process costs are generally grouped into PD costs and product manufacturing costs. Others include costs associated with activities such as training and relocating resources. PD costs are the total costs incurred through a PD process. These costs are frequently referred to as NRE (non-recurring engineering/development) costs since they only happen once. The cost for an OKP product is mainly PD cost or NRE cost since an OKP product will be developed in one go and rarely repeated in future.

The manufacturing cost (sometimes called the recurring cost) represents the cost of producing one unit of the product. It includes material, labour and manufacturing overhead costs for the product. It is important to note that much of this cost is established during the PD process because of the design of the parts, the manufacturing methods, processes, tooling, *etc.* An OKP product also includes manufacturing cost as a part of its total development cost.

1.1.1.4 Product Development Waste

There are many contributors to waste in the RPD process, but the most costly factors are errors, reworks and changes. These result from a lack of proper preparation and planning, poor implementation, and poor PD processes that allow development to proceed between responsible departments without sufficient information interchange. Designers, for example, may develop the product without taking into account the needs of the manufacturing departments. A flood of design changes may overwhelm the organisation as it tries to modify the product to suit the manufacturing requirements. Although the cost of the changes may be huge, the biggest impact is the delay of market launch while the product is debugged. Another major source of waste results from inadequate product definition. The product designers may be required to rework many man-months of effort when the product definition is changed late in the development cycle.

1.1.2 Product Development Process Evaluation

The first step in improving a PD process is to measure the effectiveness of the one currently in use. This can be done by a comprehensive assessment that includes the use of both metrics and perceptions. The metrics are used as a quantitative assessment of the PD process and the perceptions are used as a qualitative assessment of the process. The combination of the two results in a detailed evaluation. Metrics are used to determine quantitatively how well the process is meeting its objective of yielding timely, competitive, and profitable OKP products. Metrics can provide a basis for comparing the PD performance of a company with its competitors, as well as measuring the ability of the PD organisation to develop products that meet their original plans and objectives.

The effectiveness of a PD process requires its use and support by the customers of the process. If the users of the PD process perceive that the employing process is not one of good quality – *i.e.* it is very bureaucratic and cumbersome – then they will resist using this process. The process will be ignored and the product developers will proceed in their own way.

Management's perception of the relative degree of success demonstrated may be critical to the funding of future OKP PD programmes. If their perception is that product developments are consistently marginal, the probability of obtaining adequate funding for future programmes becomes questionable. Management will have to be convinced, and rightfully so that the sins of the past have been addressed and are on the way of being corrected, before they can invest in development of the product.

Thus, many companies find themselves in a situation where new products are not giving adequate ROI to justify continuing long-term investment and without new products, there is no hope for long-term survival. The solution would be to develop an effective PD process that will yield competitive and timely products.

1.1.3 Benefits of RPD

In the market, successful products usually achieve high marks in each of the categories of performance metrics, *e.g.*, short lead time, low waste and low cost. Many of the metrics affect each other. Quality products, for example, can be sold well in the market, and hence lead to a higher life cycle profit.

Figure 1.2 demonstrates the benefits gained by two manufacturing companies through rapidly delivering quality products to the market. Obviously, company A entered its market early with a quality product, achieved a commanding market share, and generated much more life cycle revenue than its competitor, company B, who entered the market late with a product of equivalent quality.

Life cycle profits are negative during the PD cycle time and only turn positive when sufficient revenues are achieved. Figure 1.2c depicts the corresponding product life cycle profits for companies A and B.

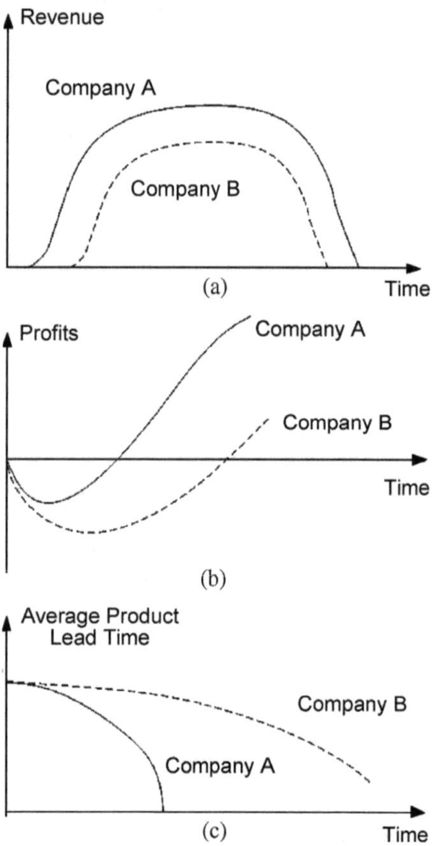

Figure 1.2 Financial and lead time results

1.1.4 Drivers for RPD

There are many factors driving the need for OKP companies to improve their PD process. Some of these factors are within the control of the company and some are not. Some factors within the control or influence of the company are:

1. ROI targets;
2. staff reductions resulting from downsizing activities;
3. excess production capacity.

These factors are specific to the company and its internal environment. They are all related to increasing profitability and hence long-term viability.

In addition to the company internal factors, there are several common external factors that are major drivers forcing the company to move towards RPD. These factors include:

1. aggressive global competition;
2. rapidly changing technologies; and
3. increasingly complex markets.

These factors are briefly discussed in the following subsections.

1.1.4.1 Aggressive Global Competition

Table 1.1 illustrates some key results during the global competition in the auto industry in the 1980s. Twenty auto manufacturers worldwide were evaluated and pertinent metrics determined. The metrics were normalised for a compact car. The chart clearly shows, as most of us are already aware – that the Japanese developed automobiles in far less time and for much fewer development dollars. The Japanese products were also better performers than those of the European high volume manufacturers.

Table 1.1 Global competition in car industry

Activity	Japanese volume producer	US volume producer	European volume producer	European high-end specialists
Engineering hours (millions)	1.7	3.2	3.0	3.0
Percentage to US	53	100	94	94
Lead time, concept to market (months)	45	60	57	63
Percentage to US	75	100	95	105

Source: Product Development Performance: Strategy, Organization, and Management in the World Auto Industry (Figures 4.2 and 4.4) by K.B. Clark and T. Fujimoto. Boston: Harvard Business School Press, 1991.

Although this analysis is for one industry, it is very representative of what is going on in the world. In this increasingly fierce competitive global economy, there are foreign as well as domestic competitors who can do it better, and when they do, they will seize market share. Virtually no industry or marketplace are immune to these competitive challenges.

1.1.4.2 Rapid Changing Technologies

Previously, technology usually remained the same for several generations of the product, but this is not the case today. Today, many technology changes take place within each generation. In some high-tech industries, such as the computer industry and the communication industry, it is not uncommon for technologies to change dramatically every two or three years.

As a result, product life cycles are rapidly declining. This, in turn, reduces life cycle revenues and profits. It is becoming more and more difficult to achieve satisfactory ROI, considering the expense of developing OKP products and the short periods available to obtain financial recovery on that investment. Hence, companies are being forced to develop OKP products at a faster and faster pace. OKP companies that are not making significant strides in reducing their RPD cycle times are struggling to remain competitive.

1.1.4.3 Increasingly Complex Market

Markets continue to become more diverse and complex. Customers, increasingly exposed to the advertising and marketing efforts of global competitors, are becoming more sophisticated and demanding. Competitors, seeking a marketing niche, are decreasing prices and adding product differentiation. This increasingly complex market has put a lot of pressure on OKP manufacturing enterprises to meet the diverse customer needs and demands.

1.1.5 Emerging Research Issues

To improve the performance of OKP manufacturing enterprises and to win the RPD competition, the major research problems for OKP companies to compete in today's global market have been identified (Tu *et al.* 2002).

1.1.5.1 High Customisation

It is obvious that engineer-to-order will result in high customisation. High customisation often leads to diversified customer's requirements and hence great uncertainties. These uncertainties result in numerous reworks, longer PD and production lead times, and high costs.

1.1.5.2 "Once" Successful Approach

The production in OKP companies is different from that in batch or mass production companies. In a batch or mass production company, a prototype of the product will be made first to approve the design and to study the manufacturing processes. In OKP companies, no prototype is made. The product must be made in OKP companies in one go, or a "once" successful approach. It is not economical for a mould/tool maker to make a prototype of the mould first prior to making the final mould. It would be ridiculous for a large oil cargo building company to make a prototype of the ship to approve the design and study the manufacturing processes first, before making another ship as the final product. They all must make their products according to customers' requirements in one go. This once successful approach to product production in OKP companies creates many challenges and problems for the companies:

1. uncertainties in product design, process planning and shop floor scheduling;
2. owing to the aforementioned uncertainties, product design and production must be carried out concurrently, or jointly planned and controlled through the whole PD life cycle;
3. hidden costs or reworks caused by inappropriate or wrong designs, process plan or production schedules;
4. it is difficult (if not impossible) to get a well planned production schedule and inventory plan, and OKP companies often run in a fire fighting situation.

1.1.5.3 Looser or Fatter Production Planning and Control

Owing to the uncertainties mentioned above, OKP companies normally plan their production and allocate production resources in a much looser or fatter way compared to batch and mass production companies. In a batch or mass manufacturing company, production is normally organised along a production line that synchronously moves according to a pre-defined cycle time. In a car manufacturing company, for instance, the cycle time may be 1–2 min. This means that production in this car manufacturing company can be synchronised by this cycle time. Operations in every section throughout the whole plant will be able to be completed in just one to two minutes. In every cycle time, a product (or a car) can be delivered from the company's final assembly line. This would be impossible for an OKP company. In OKP companies, production is asynchronous. The operations are loosely planned according to rough estimates of the processes. This loosely planned asynchronous production results in mixed material flows in an OKP company, and a higher work-in-process inventory and cost is incurred.

1.1.5.4 Continual Customer Influence

In OKP, customers may change their requirements throughout the production procedure, which is different from batch and mass production where the customer's requirements are confirmed or fixed before product design and manufacture. In OKP, owing to limited knowledge on how to produce the required product, the customer may change the requirements after the product has been partially made, *e.g.*, when the customer sees the design of the product, a plastic part made by a rapid prototyping machine, a part of the product, *etc.* In the worst case, a customer may wish to change some of his/her requirements or add some modifications in later stages of the production. It is obvious that the ability to cope with this type of continual customer influence will improve an OKP company's standing in the marketplace.

1.1.5.5 Complicated Product Data and Information Flows

An OKP company may produce a number of products at the same time. Due to the high customisation of these products, each of these products must be managed individually. It is not rare for a part of product A to be wrongly assembled into product B, a wrong order to be placed to the supplier, a wrong inventory to be kept in the company, *etc.* The information and data of product design, process planning, inventory planning and control, and production scheduling are determined for each of the products, and flow simultaneously in the company. On the other hand, no OKP company, particularly small or medium sized companies, would like to equip itself as a super company that can meet all kinds of customer requirements. They often use a lot of partners or subcontractors to compensate for their lack of manufacturing ability. This leads to complicated product data and information flows not only within the OKP company but also across companies.

1.1.5.6 Complicated Logistics Management

This is another problem resulting from the high customisation of OKP. The complexity of logistics management in an OKP company can be viewed internally and externally. Internally, an OKP company normally has complicated material handling and inventory control. Externally, it may have an extensive outsource network that consists of complicated supplier chains and partner/subcontractor chains, and a customer network that consists of complicated delivery chains and sale chains.

1.2 Motivation for the Book

To solve the problems mentioned above, OKP companies have been actively adopting new technologies, particularly computer technologies such as intra/internet communication technology, CAD/CAPP/CAM, rapid prototyping, MRPII/ERP (manufacturing resource planning/enterprise resource planning), PDM (product data management), and computer simulation technology. According to a survey (Whybrew *et al.* 2000), some of these technologies are also widely applied in New

Zealand manufacturing companies as well as overseas. Despite the large efforts that have been made to develop in these areas (a comprehensive review of literature will be given in Chapter 2), these technologies are often used separately rather than in a systematic and integrated way. The applications of these technologies are lacking in system integration and synergy. Our findings also show that most OKP companies find it difficult to adopt commercial software tools for their applications.

To help manufacturing companies, particularly OKP companies, to rationally adopt these computer application technologies to effectively shorten their PD cycle time and reduce PD cost, a research programme has been funded by the Foundation of Science and Technology of New Zealand (NZ) since 1998. This research programme aims to provide these NZ OKP companies with a reference production management system framework, concepts and software tools to support the applications of these computer technologies. The research work reported in this book is a part of this research programme, and mainly focuses on the strategies, algorithms and technology for rapid development of OKP products.

The principal goal of this book is to develop new technology, algorithms, tools and systems to reduce PD lead time and cut down the cost for OKP PD. This goal is to be achieved by the proposal of an Internet-based PD system which integrates modern technologies such as information technology (IT), RPD paradigm, information modelling, distributed multi-objective optimisation, and Internet-based information management. The developed Internet-based RPD or IRPD system is for the strategic purpose of improving the New Zealand manufacturing level in the world, particularly for new PD. The methods, algorithms and tools however, will be useful to many OKP companies in the world. In the book, theories, concepts and methods from the areas of modern control engineering, rapid manufacturing, CAD/CAPP/CAM integration, engineering optimisation, information technology and knowledge-based technology are developed and applied to industry. The research has used two typical New Zealand OKP companies, *i.e.*, customised sheet metal manufacturers and tool/mould makers, as pilot companies or test beds to validate the proposed system and other major research findings. Following similar principles, it can be envisaged that a number of high-tech RPD tools can be developed to help other manufacturing enterprises in New Zealand to better compete in the global market.

1.3 Book Organisation

The presentation of the book strives for a balance between modularity and integrity of the chapters. The chapters are carefully structured in a self-contained fashion in order to address specific issues involved in the rapid development of OKP products. Each of these chapters starts with an overview of the technique and research background of the major objectives of the research, which form the main points of the new system structure, concept and method. The book will provide answers to most of the main research activities that are ongoing around the world in the OKP research field. The main contents of the book are summarised as follows.

1. Investigation of current hurdles that hamper RPD and proposal of a reference architecture for building an Internet-based integrated PD system. Nowadays, the operation of an OKP company tends to build on the basis of cooperation between the company and its partners in its manufacturing network. It is often driven by customer focus and volatile markets. In short, OKP companies operate in a global competitive and collaborative environment. To rapidly develop a new product in this global environment, OKP companies need to update their traditional manufacturing systems and traditional PD processes to meet the requirements of this global environment. The hurdles blocking the way to this upgrading process need to be identified and further studied. To meet the requirements of the global environment and to facilitate the complicated communications between an OKP company and its partners, and between the company and its customers, a reference structure of an Internet-based integrated PD system needs to be developed.

2. The development of a system architecture proposed for OKP PD. To test the feasibility of the proposed reference system architecture, the proposed system architecture is implemented to speed up sheet metal PD. Sheet metal manufacture is a major part of the New Zealand manufacturing sector.

3. An Internet-based information platform or framework for RPD. To rapidly develop a sheet metal part/product, an Internet-based information platform needs to be developed as a major part of the Internet-based rapid sheet metal part/product development system.

4. A new compound machining method to speed up sheet metal PD. A compound cutting and punching production method supported by an integrated CAD/CAPP/CAM system in sheet metal manufacturing needs to be developed to speed up sheet metal product design and manufacture. Many existing commercial CAD/CAM systems are not suitable for this manufacturing method, especially in concurrent and global design and manufacturing environments. Some problems have to be solved before these CAD/CAM systems can be employed and integrated for this compound manufacturing method.

5. Cost estimation and optimisation for rapid development of sheet metal part/product. In order to maintain a strong competitive ability, sheet metal manufacturing companies have to look for the cheapest way to develop their products. This is a cost estimation and optimisation process.

6. An integrated product data model for PD life cycle and application for rapid tool/mould making. As mentioned before, OKP companies tend to design, manufacture and maintain a product through global collaboration in a team. The team members may include engineering, manufacturing and service firms and they are often geographically dispersed. They work together to design, manufacture and support the product. For this reason, an integrated information model has to be developed to support the definition and exchange of product data unambiguously and consistently between team members during the entire course of the PD process. Developing and manufacturing a product through this type of global collaboration is termed virtual manufacturing, and a company that adopts the virtual manufacturing strategy is called a virtual manufacturing company. It is obvious that the virtual manufacturing concept is particularly suitable for rapid OKP PD since the virtual manufacturing strategy can improve the company's flexibility and capacity to handle much wider and diversified customer requirements. Tool/mould making is a typical one-of-a-kind production and is also another major part of the New Zealand manufacturing sector. To apply the research findings and to widely support New Zealand manufacturing companies, the proposed data model is tested through a mould development process instead of sheet metal part/product development.

7. Development of an Internet-based information management system. It is obvious that an Internet-based information management system needs to be developed as an important part of the Internet-based OKP PD system to manage the various databases which can be directly accessed via Internets.

8. An Internet-based DFX (IDFX) system for rapidly tool/mould manufacture. DFX, or design for X, is a broader concept to encapsulate various concurrent engineering approaches for RPD, such as design for manufacture (DFM), design for assembly (DFA), *etc.* To test the suitability and feasibility of adopting this concept into the Internet-based rapid OKP PD system, an Internet-based DFX system or IDFX system needs to be developed and tested in a typical company.

References

Bullinger H-J, Warschat J, Warner K (1996) Management of complex projects as cooperative task. In: Koubek RJ, Karwowski W (eds) Manufacturing agility and hybrid automation: I. Proceedings of the 50th International Conference on Human Aspects of Advanced manufacturing: Agility and Hybrid Automation: Maui, HI, USA, August, 1996, pp 383–386

Bullinger H-J, Warschat J, Fischer D (2000) Rapid product development – an overview. International Journal of Computer in Industry 42:99–108

Floyd DT (1993) Winning the new product development battle. The Institute of Electrical and Electronics Engineers, New York

Hirsch BE (1992) Future research in one-of-a-kind production. In: Hirsch BE, Thoben K-D (eds) One-of-a-kind production: new approaches. Elsevier Science Publishers B.V., North Holland, pp 87–94

Rolstadas A (1991) ESPRIT basic research action 3143–FOF production theory. International Journal of Computer in Industry 16:129–139

Tu YL (1996) A framework for production planning and control in a virtual OKP company. In: Technical Papers of North American Manufacturing Research Institution of SME 1996, pp 121–126

Tu YL, Xie SQ, Zhou ZD (2001) An information modeling framework to concurrent product design and manufacturing. In: Proceedings of IAM'2001, American University in Dubai, U.A.E., pp 564–571

Tu YL, Xie SQ (2002) Rapid one-of-a-kind product development in a global and virtual manufacturing environment. In: Proceedings of AMPS 2002, September 2002, The Netherlands

Wortmann IC (1991) Towards one-of-a-kind production: the future of European industry. In: Eloranta E (ed.) Advances in production management systems. Elsevier, Amsterdam, pp 41–49

Whybrew K, Raine JK, Shaw A, Aitchison DR (2000) Summary of results of the survey: Rapid product development for world class manufacturing. Technical Report 58. Dept of Mechanical Engineering, University of Canterbury

Chapter 2
Review of Rapid OKP Product Development

Abstract This chapter provides a comprehensive review of the historical background of product design and manufacturing systems for the rapid development of OKP products. Through the overview of existing OKP systems and recent approaches to product design and manufacturing systems, the author will discuss current techniques that can be used to implement product design and manufacturing systems for producing OKP products. The problems that have emerged from recent developments are reviewed and sorted in this chapter. The future trends of collaborative design, decision support, manufacturing support, supply chain management, workflow management, Internet techniques for product design and manufacturing, product modelling, STEP-based data environment, concurrent engineering, *etc.*, will also be discussed in this chapter. The reviewed state-of-the-art works have been used directly or indirectly as references for this book. At the end of the chapter, a reference system structure for building an Internet-based integrated product development (PD) system is proposed for the rapid development of OKP products.

2.1 Introduction

In today's competitive market, OKP manufacturing enterprises are facing more severe competition in the market than ever before due to the globalisation of the market, short life cycle of products, increasing product diversity, high demand for quality and short delivery times. To survive and thrive, these enterprises have to work hard to achieve high quality and productivity, reduce cost, and respond quickly and effectively to a market that is becoming more global, dynamic and customer-driven.

OKP companies, owing to the high customisation, face a large amount of uncertainties and consequently must deal with many reworks. PD cost is also normally higher and development lead time longer than those of product-focused manufacturing companies. To respond rapidly to market pressure, many OKP companies have adopted a virtual manufacturing concept, which can be simply understood as a global collaboration business through sub-contracting and technology transfer between a company and its partners. This virtual manufacturing concept enables a company to quickly and often economically meet its customers' needs by combining its own manufacturing strengths with those of its partners. However, these virtual companies usually require technical and theoretical support in order to manage and optimise PD processes through a whole PD cycle. These processes include customer requirement interpretation to decide how to address the customers' needs, product design to address the customers' needs, conceptual and functional prototyping to prove the effectiveness of the designs, manufacturing process planning (which includes subcontract planning) to determine how the product is to be made, design manufacture of tools, pre-production trial, production, and finally delivering the product to the customer or launching the product into the market. OKP companies particularly wish to meet the customers' needs in one-go and minimise the costs spent on intermediate tests and mock-up manufacture.

To shorten the lead time and to reduce the cost of PD for OKP companies, in particular for virtual OKP companies, the combination and integration of existing innovative technologies and the development of new technologies have become essential. The achievement of the goals of RPD is becoming more and more difficult as the PD environment has dramatically changed and the factors that influence the PD process have become more and more complicated. The goals of RPD cannot be achieved by focusing efforts only on a single stage of the PD process, contributions from the entire PD process are needed to meet these goals. Hence, an Internet-based PD system has to be built to support the PD process under a geographically distributed environment (Tu *et al.* 2000). This system can be used to support and facilitate communications and data flow at different stages throughout a PD lifecycle. To develop this PD system, new technologies have to be developed and integrated to improve the overall PD performance, *e.g.*, new CAD technology, collaborative design technology, Internet communication technology, information technology, artificial intelligence (AI), knowledge base technology, cost/lead time optimisation technology and decision support technology. These new technologies, which can continually improve product design and manufacturing ability, have become more and more essential for today's OKP manufacturing companies to compete effectively in the global market.

As an enabling technology, the Internet has been widely used in the PD field and can enable intelligent product design and agile manufacturing by updating traditional plant information systems and eliminating barriers to integration. It has launched a revolution in worldwide data transfer and communication. Its platform-independent architecture allows manufacturers to deploy applications across virtual organisations, which includes business partners/customers whose computing environments may be completely different. In the PD field, the application of the Internet has been used in numerous areas including marketing, management, design, manufacturing, production, planning, customer service, *etc.* (Su and Amin 2001). Internet-based RPD (IRPD) has been recognised as an efficient way to help manufacturing companies (especially OKP companies) stay competitive in the global market (Bullinger *et al.* 1996, Tu *et al.* 2000, Xie *et al.* 2001). However, there are problems that hamper the development of IRPD systems, *e.g.*, the integration of various technologies and tools with different enterprise data models (Huang *et al.* 2001b, Tu *et al.* 2001), worldwide knowledge and resource access among heterogeneous computer systems, the concurrent architecture for execution (Schaefer 1994), the 'seamless' integration of existing systems that are fragmented and nearly isolated which are employed at various stages of a whole PD process, information and knowledge sharing between various systems that are employed at different stages of a PD process (Xie *et al.* 2001) and remote access to algorithms or large application tools that run on different platforms (Wagner *et al.* 1997). Hence, to meet the needs of rapid development of OKP products, existing RPD technology and systems have to be further developed in terms of communication, collaboration, and system integration, *e.g.*, customer-oriented virtual manufacturing, distributed concurrent engineering, global optimisation and the integration of various functional systems at different stages of the PD process. These issues have been the focus of both scholars and practitioners.

Besides exploring and discussing the aforementioned questions, this chapter will also place emphasis on reviewing past research, systemically figuring out how current technologies have been used, and identifying the future trends for implementing the IRPD in today's industrial environment. A comprehensive summary and review of historical literature and projection of future trends are helpful for OKP companies to make decisions concerning the implementation of Internet-based systems or upgrading existing manufacturing systems. They are also very helpful in figuring out possible future research directions. For large companies that have their own research facilities, the work presented in this chapter can help facilitate their development work, underline current issues and point out the right direction. For medium or small companies that are looking for Internet-based systems, this work will help them in the selection of satisfactory Internet-based systems to meet their specific PD environment. In this chapter, the recent development of these systems are discussed. A table listing some of the existing Internet-based PD systems and research projects are presented in Appendix A1. The characteristics of the systems and recent research projects are listed. This chapter reviews recent approaches and research in the Internet-based PD area and presents a systematic approach for implementing IRPD systems, which can help resolve the issues mentioned above. The key research issue of how to build an Internet-based distributed integrated PD system to achieve the goal of "rapid development of OKP products" is also addressed.

The chapter is structured as follows. Section 2.2 presents an overview of the literature of OKP and Internet-based PD approaches. In Section 2.3, the emerging research topics that exist in the implementation of Internet-based systems are discussed. Sections 2.4 and 2.5 review the key technologies and approaches for IRPD systems. Research issues and future trends of IRPD are covered in Section 2.6. Lastly in Section 2.7, a reference system structure for developing Internet-based RPD systems is proposed.

2.2 Overview of OKP and Internet-based Product Development Systems

Literature and research reports regarding OKP and Internet-based PD systems have emerged in recent years. Here, from among a large amount of literature and reports, some typical ones are selected and discussed briefly in terms of the following aspects: research focuses and trend of OKP; requirements for the next generation OKP systems; and the role of IRPD systems.

2.2.1 Review of OKP

OKP was intensively researched under the ESPRIT basic research on the theory of the factory of the future (FOF) in the late 1980s and early 1990s (Rolstadas 1991). Most research efforts have focused on OKP production management because operating efficiency is believed to be one of the major bottlenecks in OKP. Rolstadas (1991) proposed a framework for managing OKP, in which an adaptive scheme to change the production resource structure (as presented by an R-graph) to meet the changes of product structure (as presented by a P-graph) was suggested. However, the limitations and static nature of this adaptive factory managing structure were also noted. Up till now, the research in OKP has been heavily focused on the managerial problems in OKP. Not much work has been carried out on how to rapidly and economically develop OKP products. William (1995) identified some interesting research topics in made-to-order products, which include cost estimation and design decision support through application of advanced computer technologies.

Customisation is commonly understood to be a company's ability to satisfy various needs of customers by economically changing its product or service design and manufacture. It has been recognised as one of the main focuses in an agile manufacturing system (Hasan *et al.* 2009, Kidd 1994, Hasan *et al.* 2007). Meredith and Francis (1998) believed that an ideal agile manufacturing company was able to make a product at the same unit cost of making it in a batch of 10,000 units. This has set a clear requirement for improving the production efficiency in OKP.

Mass customisation is another popular term that has been widely used to note the mass production of a customised product or OKP. Tseng and Du (1998) characterised mass customisation as recognising each customer as an individual, while extracting maximum commonality to achieve scale of economy. One of the main techniques to achieve mass customisation is to develop product families and link them with product functional trees, thus linking customer requirements with product families. Jiao and Tseng (1999) presented a product family data model to support mass customisation practices. Johan (1997) used functional trees to express design history information. However, both Jiao and Tseng and Johan did not take the downstream manufacturing applications into their considerations.

To support customisation, wider research has taken place on the methods of product design and manufacture. The typical achievements among these research efforts would be the so-called design for modularity (Exixon 1996, Marshall and Leaney 1999, Kamrani and Salhieh 2002), variant design (Fowler 1996, Mckay *et al.* 1996, Ding 2006), design for manufacture (Gupta *et al.* 1997, Chen *et al.* 1998), design for production management (Nielsen and Holmstrom 1995, Yang and Pei 1999), and computer aided process planning (CAPP) (Alting and Zhang 1989, Wu and Zhang 1998, Tu *et al.* 2000). Marshall and Leaney (1999) developed a systematic approach for product modularity. However, design for modularity solves only part of the customisation problems, the developed modules by no means meet all the customer needs, especially highly customised OKP products.

Variant design is a method to adapt existing design specifications to satisfy new design goals and constraints. This normally consists of two basic components, *i.e.*, an extensive database to record product families and the possible variations, and a sound reasoning method (*e.g.*, case-based reasoning) to find solutions from the historical data in the database. Mckay *et al.* (1996) provided an improved data model to support product variety and pointed out the limitations of using EXPRESS in STEP (Standard for Exchange of Product model data) to model product variety.

Design for manufacture (DFM) places its emphasis on the integration or links between product design and manufacture. Gupta *et al.* (1997) widely reviewed the research work on DFM and pointed out that the common features of various DFM methods were cross-functional teams, good/bad examples, feature-based evaluation and empirical parametric evaluation. In the DFM approach, the automatic feature recognition and feature-based design method have been recognised as interesting research topics and have been widely studied. The typical feature recognition methods include syntactic pattern recognition, state transition diagrams, decomposition method, set theoretic approach, graph-based approach, and external access directions (Maropuoalos 1995, Lin *et al.* 1997). Some successful applications of feature recognition and feature-based design have been found in the literature. Lee and Kim (1999), for instance, proposed a feature-based approach to generate alternative interpretations of machining features from a feature-based design model. Lau and Jiang *et al.* (1998) presented a method to recognise manufacturing features from a STEP 203 file. However, as pointed out by Shah and Mantyla (1997), overlapped feature recognition is still an unsolved problem. In practice, the definition, identification and automatic recognition of design and manufacture features are the main problems that impede the application of feature-based design methods.

A few research projects have addressed the problem of design for production management. Nielsen and Holmstrom (1995) presented an approach to separate product features and the feature interfaces to simplify the production control activities from a supply chain perspective. Yang and Lei (1999) gave an example of integrating a CAD system with an MRP system based on STEP protocol. Hence the design choices reduced the impact on the required materials. However, these works did not completely address the integration problems between product design and process planning, and between process planning and shop floor scheduling, which are critical problems for rapid and economic development of OKP products.

To respond rapidly to a market chance, efforts at integrating process planning and scheduling in OKP have drawn more and more interest and attention in the last few years. Mamalis *et al.* (1996) presented an on-line integrated process planning and production scheduling system. In Mamalis's work, simulation tools were used to support integrated process planning and production planning tasks. Saygin and Kilic (1999) also presented a method for integrated flexible process planning and scheduling. Howard *et al.* (1998) proposed a generic manufacturing planning and control system architecture for different manufacturing environments. However, this work was not concerned with the product design tasks involved. Aidakhilallah (1999) developed an architecture for concurrent product design, process planning, and production control. However, the approach oversimplified the problem domains.

Morgan *et al.* (1990) indicated that enterprise integration could lead to high agility in a manufacturing company. As suggested by Tu (1996), OKP companies need high agility to meet various customers' needs and high integration to shorten the PD cycle and to save PD costs. Lim *et al.* (1998) identified four integration domains: business integration, business data integration, software integration and computer system integration. Jordan and Michel *et al.* (1999) classified system integration into four staked layers, viz. manufacturing system layer, production and process layer, computer system layer, and communication system layer. Numerous research efforts have been made to develop integrated system architectures, *e.g.*, AMRF (automated manufacturing research facility) developed by the National Institute of Standards Technology in USA (Furlani *et al.* 1983), CIMOSA (open system architecture for CIM) (Kosanke 1991) in Europe, and Purdue architecture (Bernus *et al.* 1996). However, these integrated system architectures were developed according to a formal system hierarchy, which normally exists in a traditional product-focused or batch manufacturing company (Toh *et al.* 1998). Manufacturing systems that are built according to these architectures cannot handle the frequent system structure changes resulting from the customisation in OKP (Tu 1997). To solve the conflict between system flexibility and integration, distributed object technology of computer communication has been recognised as a promising solution. By using this technology, various production systems are treated as clients. These clients can be freely plugged into or taken off a heterogeneous computer communication network (Orfali *et al.* 1999).

More recently, it has been widely believed that the computer intra-/inter-net communication technology would lead to a communication revolution in manufacturing. Some research efforts have been made to develop WWW (world wide web)-based manufacturing systems. Indrusiak (1998) carried out a so-called CAVE project, which aimed to develop a WWW-based design automation framework. The implementation and integration of some CAD tools according to the CAVE standards were addressed in their report. CORBA (Mowbray and Ruh 1997, Shin *et al.* 2003) and DCOM (Zhou *et al.* 2006, Li and Williams 1997, Adamopoulos *et al.* 2008) have been used for the development and deployment of WWW-based systems integration in a distributed heterogeneous environment. Menzel and Geiger (1999) presented a hybrid system using distributed simulations, objects and fuzzy set theory to improve the system efficiency and flexibility. Dorador and Young (1999) explored the definition of the structure of information models for supporting the related processes of design for assembly and assembly process planning to achieve the intensive sharing of information required in concurrent engineering through the life cycle of a product.

From the above literature review it is obvious that various methods and techniques for RPD have been widely researched. Some methods and techniques are well established while others are still in the early stages of research. However, the research trend of OKP is clear, *i.e.*, to develop a new generation of production systems by using computer aided and integrated technologies and some management concepts to economically and rapidly develop a customised product.

2.2.2 Requirements for the Next Generation of OKP Systems

In recent years the manufacturing industry has experienced tremendous changes, *e.g.*, from mass production through batch production, flexible manufacturing and lean manufacturing towards agile manufacturing or OKP philosophy (Booth 1996, Tu *et al.* 2000, Cheng *et al.* 2001). These changes are directly driven by various requirements for low product cost, high quality, high performance, and customer's choice and requirements, *etc.* They result from the unexpected changes in the competitive market environment, *e.g.*, globalisation of the market, variety of customer demands, customer-designed products, and shortened product life cycles. These market factors have great impact on all of the PD-related activities, *e.g.*, product definition, design, manufacturing process planning, production, workshop floor control, quality control delivery, and marketing. To meet these changes and challenges, a new OKP system needs to be developed.

In today's manufacturing practice, products are rarely designed, manufactured and maintained by a single company. This is especially true of small or medium sized (SMS) companies. An SMS company does not normally have the breadth of knowledge and the capability to understand all aspects of a PD process (Geller *et al.* 1995). Due to this, the strategy of developing alliances and running the business or manufacturing with various "partners" who have the wanted knowledge/expertise and hardware/software resources has been widely adopted by SMS companies. This provides global competitive advantages for these SMS companies. This collaborative production strategy was well referred to as a virtual manufacturing concept. By adopting the virtual manufacturing concept, the manufacturing system environment becomes heterogeneous as partners may use different design/manufacture tools, quality inspection systems, control facilities and different production resources and technologies. Such a heterogeneous manufacturing system environment will inevitably result in problems in management, communication and production. It will also hamper the systematic upgrading of the design and manufacturing tools/facilities, as well as affect the competitiveness of a company (Kim *et al.* 1998, Feldmann and Gohringer 1999). Hence, a new generation of OKP systems is required to cope with this heterogeneous manufacturing system environment.

As mentioned previously in this chapter, the manufacturing market tends to be global and dynamic. New technologies are continually emerging, and the product life cycle is getting shorter. Manufacturing strategies should therefore shift to support global competitiveness, new product innovation, and rapid market responsiveness. Hence a new generation of OKP systems should contain the key features of agility and rapid responsiveness, to maintain competitiveness in the global marketplace and the ability to rapidly combine the strengths of "partners" to meet market needs. It will thus be more time-oriented, while still focusing on cost and product quality. The fundamental requirements for a new OKP system are identified in the following:

- Enterprise integration. In order to support global competitiveness and rapid market responsiveness, an individual OKP company will have to integrate its PD processes (with its partners via networks).
- Support for organisations that are globally distributed (Tu and Xie 2000). This includes distributed knowledge-base systems or product information systems that are needed to support PD processes.
- Being able to cope with the heterogeneous and distributed manufacturing system environments.
- Open and dynamic structure. The manufacturing systems should be able to dynamically integrate new subsystems into the environment for specific applications or remove existing systems from the systems without influencing the basic structure of the working environment. This requires open and dynamic system architecture.
- Support cooperation and collaboration. The manufacturing systems should support geographically distributed teamwork, which includes cooperation and collaboration among team members.
- Agility and high customisation. The manufacturing systems must be used to shorten the PD cycle time and to respond to customers' requirements quickly.
- Technical advancement. New OKP systems need to adopt new technologies that are developed for specific stages of the PD process to keep its advantages over existing manufacturing systems.
- Compatible with most existing PD software tools.
- Stable and easy to use and maintain. The system should be user-friendly and fault tolerant both at the system level and subsystem level so as to detect and recover from system failures at any level and minimise their impact on the working environment.

2.2.3 Role of IRPD

To develop production systems that can meet the requirements of the new generation of OKP systems as mentioned in previous sections, various attempts have been made from the two aspects of product design and manufacture (Reed and Afieh 1998, Chui and Wright 1999). The results have illustrated the great potential of using the Internet to build up a virtual OKP enterprise with the capability of RPD. Hence, using the Internet to achieve RPD has become the most promising solution to meet the requirements of the next generation of OKP systems.

The Internet and its relevant technologies (*e.g.*, world wide web, communication, software tools and hardware) have made great progress in the past few years. The outstanding features of the Internet make it the best supporting platform for RPD in the distributed heterogeneous manufacturing environment (Cheng *et al.* 2001). For example, the Internet platform has the capability to integrate diverse software tools to support applications. This facilitates the development of an IRPD system that is able to integrate current techniques for PD to support distributed team members working together to design, manufacture and support products cooperatively and concurrently. The IRPD systems that are built based on the Internet platform can help manufacturing companies achieve flexibility, rapid response to the dynamic global market and changing customer needs, and the ability to rapidly produce and deliver products to the market.

New technologies have been continually put forward and gradually updated to the Internet platform. They include Internet-based CAx technology, Internet communication technology, Internet-based design for X, Internet-based collaborative design, Internet-based decision support, Internet-based workflow management, Internet-based CE, CAD/CAPP/CAM technology, Internet-based virtual simulation technology, AI, knowledge bases, global optimisation technology and CSCW tools. An IRPD system that is developed through the integration of these technologies using the Internet platform provides a solution for developing the next generation of OKP systems. The IRPD system will have great technological advantages over the traditional standalone environment or Internet-based systems for supporting only a single stage of the PD process. This IRPD system can consider the development process of OKP products as a whole and can thus achieve better performance.

2.3 Review of Internet-based PD Approaches

Internet-based rapid PD (IRPD) requires the integration of people, business processes and information technology across the PD life cycle for the purpose of RPD. A typical PD life cycle includes understanding customer requirements, product definition, product design, analyses and test/simulation, process planning, manufacture and delivering the product. This section will review recent achievements in the areas of PD process and the emerging issues in terms of implementation to improve IRPD performance. This includes Internet-based collaborative design, Internet-based design for X (DFX), Internet-based decision support and concurrent engineering (CE).

2.3.1 Collaborative Product Design

The initial work on collaborative design appeared more than a decade ago (Sriram *et al.* 1991), but it was not until recently that industry, academia and government have demonstrated the benefits of collaborative work (with the help of the Internet) for product design and manufacture. Collaborative design is the process of designing a product through concurrent cooperation among engineers from different functional areas in a manufacturing company, *e.g.*, design, process planning, manufacture, assembly, testing, quality and purchasing as well as participants from suppliers and customers (Sprow 1992). Collaborative design with geographically dispersed participants based on Internet communication allows participants to exchange data and thereby reduces the disadvantages of geographical dispersion. The objectives of such a collaborative design team might include optimising the mechanical function of the product, minimising the production or assembly costs, and ensuring that the product can be easily and economically serviced and maintained (Hartley 1998). This research area has been of tremendous interest to many researchers. As reported by Hartley (1998), collaborative design can reap substantial benefits. This is demonstrated through the following data. Collaborative design led to a 75 % reduction in design changes at Northrop, a 60 % reduction in PD time at DEC, a 40 % increase in customer satisfaction ratings at Xerox and a 60 % reduction in scrap at General Motors.

Collaborative design has been used in various areas due to the substantial benefits it can offer. For example, collaborating via the Internet for design teams at Boeing-Rocketdyne (Carman 1998), research on using collaboration tools for manufacturers at NIST (Steves and Knutilla 1999), and efforts at understanding and supporting the design process at Stanford University (Cutkosky *et al.* 1996) all show encouraging results. With the development of other technologies (*e.g.*, virtual reality and communication (Smith 1998)), the team members can achieve better visualisation and communication, and thus better collaboration. These technologies enable collaborative design to take place and will also lead to more efficient collaborative design. In 1996, the NIST initiated a research and development programme to help US industry speed the transition to 21st century manufacturing capabilities. This programme is called National Advanced Manufacturing Testbed (NAMT). This testbed contains a facility in which scientists and engineers from industry, academia, NIST and other government agencies work together to solve measurement and standard issues in information-based manufacturing. To achieve better collaborative design, several researchers have studied how designers work together in teams. For example, Minneman (1991) discussed social interaction among designers, which includes how design members in a design team with different opinions negotiate to reach common understanding. He also discussed how the integration could be used to improve design communication. Frankenberger and Birkhofer (1995) described engineering and psychological influences on designers, which include the influences on communication, *e.g.*, leadership and group organisation and how good team interaction leads to a good decision. To achieve collaborative viewing of mechanical part models on the Web, Kim *et al.* (2001b) developed a collaborative system called 3D-Syn. This system can support synchronous communication and manipulation of 3D part models on the WWW and can be used as an open platform for 3D collaboration to inter-link part suppliers and buyers. An object-oriented database is used for the part library that stores part information. As a result of the extensive research and development, there are some software tools available in the commercial market for supporting collaborative design. For example, the Collaborative Virtual Product Development (CVPD) software suite was developed to link global design teams and partners in a PD chain. Oracle (www.oracle.com) has entered the collaborative PD and product life cycle management software market with the cooperation of several CAD vendors. PTC offers a version of its Windchill engineering management tool for the oracle PD exchange.

Despite the success of the research mentioned above, there are still some research issues unresolved in collaborative design and manufacturing. For example, it is not clear how collaborative design and manufacturing can be implemented in different manufacturing companies that vary in terms of company size, people, communication, various management styles and working environments. There are some cases reporting successful collaborative designs. These cases are generally concerned with engineers working at a single site and on a relatively uniform computer system platform. The question of how collaborative design can be best implemented in an integrated environment where there may be a wide variety of computing platforms still remains unanswered.

2.3.2 Design for X (DFX) via the Internet

Design for X (DFX) has been recognised as an effective approach for implementing concurrent engineering to achieve the goals of RPD (Tu and Xie 2001). Huang (1996) gave a clear definition of DFX. D in DFX is interpreted as product design in the context of DFA (design for assembly), which means to design a product for the ease of its assembly (Edwards 2002, Boothroyd *et al.* 1996). X in DFX stands for manufacturability, inspectability, recyclability, *etc.* Using the Internet to provide DFX services can support rapid and collaborative PD. DFX can be used to reduce the time and cost of redesign, assembly and manufacturing. As reported by Bralla (1986), DFX services are usually performed using DFX guidelines, which include simple guidelines, advisory guidelines and quantitative guidelines. The simple guidelines are simple instructions such as "avoid sloping surface", "cutting rules" and "minimise the number of tools used in manufacturing" (Matousek 1963). The majority of guidelines fall into these so-called advisory guidelines. An advisory guideline usually includes a verbal description as well as diagrams showing right and wrong design features. By following these guidelines, designers know what to avoid when designing products. Quantitative guidelines provide quantitative evaluation of how good or bad a design feature is in terms of certain criterion utilisation (Carlsson and Egan 1994, Boothroyd 1996). However, most guidelines developed in the past do not provide quantitative evaluation of the quality of a design in terms of criteria such as cost, time (Carlsson and Egan 1994, Boothroyd 1996) and/or other aspects (Ishii *et al.* 1994). The latest developments tend to provide quantitative guidelines to improve the quality of design. Recent approaches to DFX delivery using a guidelines application have been carried out in various DFX areas such as design for manufacturing (DFM) (Mottonen *et al.* 2009, Lazaro *et al.* 1992, Balasinski *et al.* 2007), design for retirement and recyclability (Ishii *et al.* 1994, Zhang and Kuo 1997), design for environment (Stanley 1996) and design for ergonomics (Allada *et al.* 1992).

Internet technology is becoming popular for DFX analysis (Wagner *et al.* 1997, Huang and Mak 1997). Recently, several Web-based systems have been set up, *e.g.*, a Web-based design for assembly (Wang and Tian 2007, Shi *et al.* 1995) and the Web-based failure mode and effects analysis (FMEA) (Huang and Mak 1997). The impact of Internet-based DFX is found throughout the overall product design and manufacturing process. For example, Internet-based DFA techniques can be used to reduce the cost and time of assembly in a distributed manner by simplifying the product and process through such means as reducing the number of parts, combining two or more parts into one, reducing or eliminating adjustments, simplifying assembly operations, designing for parts handling and presentation, selecting fasteners for ease of assembly, minimising parts tangling, and ensuring that products are easy to test. Use of Internet-based DFA to reduce the number of parts will reduce inventory, which will reduce inventory management effort. As a result, it will support activities such as "just in time" (JIT), aimed at improving shop floor performance.

Through Internet-based DFX systems, product developers are able to access information that will help them improve the design of the part they are working with. These Internet-based DFX systems can be called up to analyse the current state of their design, point out where the design is too complicated, and indicate possible areas of improvement. Companies using Internet-based DFX techniques (*e.g.*, DFA and DFM) have reported to be able to reduce the number of parts, the number of assembly tools, the number of assembly operations, the assembly space, the number of suppliers, and the assembly time by up to 85 %. DFM helps prevent unnecessarily smooth surfaces, radiuses that are unnecessarily small, and tolerances that are unnecessarily high. The DFA objective of reducing the number of parts may lead to highly integrated, complicated, and multi-functional parts. DFM aims to keep individual parts simple because overly complicated parts can result in hidden costs that are not initially apparent. At the early stages of the design, there may not be a lot of information to work with, but Internet-based systems that are based on DFX functionalities will make sure that whatever information exists can be made available to the product team. This is an important issue for RPD.

Although there are many advantages of developing Internet-based DFX (IDFX) systems for RPD, the implementation of IDFX systems is never an easy task. Several problems have been raised for implementing DFX systems. Among these problems, an important question is whether there is a basic pattern for the development of these DFX tools (Olesen 1992). Another important issue is how traditional DFX systems can be updated to meet the requirements of the global RPD environment. These problems have become the major hurdles for developing and implementing IDFX systems.

2.3.3 Internet-based Decision Support

Starting from the front-end of the PD process, the Internet-based approach is particularly suitable for some areas where decision support is required, such as in customer requirements management or market research, earlier design decision making and customer involvement in product design and manufacturing. The Internet-based systems that can help decision-making at various stages of the PD process are vital for IRPD. Internet technology enables us to develop virtually any type of decision support system, *e.g.*, PDM (product data management) systems, EDM (electronic document management) systems and visualisation and virtual tools (http://www.visualmining.com/). Owing to the multimedia capability (Ockerman *et al.* 1999, Weyrich and Drews 1999) of the WWW and the success of remotely executing large programs (SU *et al.* 2001), Internet-based decision support systems are functionally better than standalone counterparts and compatible with heterogeneous environment. Literature and reports on various decision support systems/tools for the different stages of the PD process can be found at the following Web sites (http://emt.doit.wisc.edu/decision.html and http://www.decision-support.net/).

An interesting approach at the Philips advanced development centre is the use of Internet as a communication infrastructure to involve users in the new PD process (Muller *et al.* 1996). Another related project is the use of WWW to analyse customer requirements for software development (Anton *et al.* 1996). As Internet technology is playing an increasingly important role in marketing and sales as well as after-sales customer services, an Internet-based product and component catalogue has been widely used. Active Catalogue (Will 1996) is a project aimed to develop WWW-based component catalogues including models to enable "try before you buy" simulation analysis during a product purchasing process, which can be regarded as a decision support system for rapid product sell. HKCAINS (Hong Kong Accessory Information Network System) is an industry-based project aimed at developing a WWW-based network for the Hong Kong apparel industry to mutually communicate and seek advice from each other as well as to assess the economic viability of establishing this network (HKTAIGA 1996). Wong *et al.* (1996) proposed methodologies for a rapid and accurate response to request-for-quotation and demonstrated the method on the Internet. The key issues that should be considered when developing Internet-based decision support systems are identified as follows: distributed network of computers; sharing data; sharing tools; tracking data (Cheng *et al.* 2001). The Internet can also deliver functionalities of decision support through developing Internet-based virtual systems. One potentially significant project is "Intelligent Manuals" (Pham 1998). The objective of this project is to supply the electronic information necessary to support the continuous use and maintenance of a product from its delivery to its disposal. Kim *et al.* (1998) proposed a WWW-based architecture for collaborative design in mixed platforms and dispersed geography environments. They used open data standards to allow users on a wide variety of platforms to access and visualise product information. Kalyanapasupathy *et al.* (1997) proposed a system using the Internet to support the generation of group technology codes for mechanical parts. IAMS (intelligent assembly modelling and simulation) aims to facilitate assimilability checking in a virtual, simulated environment in order to avoid expensive and time-consuming physical mock-ups. This is in fact a project within a broader effort to develop a collaborative open design system (IAMS 1998) and acts as a decision support system. Szykman *et al.* (2000) developed "design repositories" to support the integrated and concurrent design process. They also pointed out that "design repositories" was key to providing a comprehensive product knowledge representation during the design stage. It is obvious that developing products without sufficient support information in a broad set of disciplines would result in longer PD cycles, higher development cost and quality problems.

However, for these approaches, there existed some limitations to their applications. The first is the requirement for a formal WWW-based distributed database structure with suitable product data models for a manufacturing company, *i.e.*, a decision must be made on what information should be shared and how to represent and record the information from a PD and the relevant tool/mould making processes. In practice, the company production manager usually makes his/her decisions about what information is needed, when it is needed and how it will be used depending on the context of the current problem (Boynton 1993). Also, the data structure of the product information management system may vary with the structure and culture of the company. Considering these questions in advance, collecting all the critical pieces of information for the decision, and finally integrating the information and decisions into a product model is usually a hard job (Gruber and Russell 1996).

2.3.4 Internet-based Manufacturing Scheduling, Planning and Control

Manufacturing scheduling, planning and control is a difficult problem for developing OKP products, particularly when it takes place in an open and distributed environment. In the manufacturing process, things rarely go as expected and dynamic changes are often needed. To save product manufacturing time and cost, Internet-based manufacturing scheduling, planning and control has become an important research field in recent years. As the manufacturing environment in a company often involves a variety of machining systems, monitoring facilities, control equipment, and information resources, an efficient and easy-to-use client–server manufacturing scheduling, planning and control system is vital for global enterprises. Such a system will facilitate scheduling job tasks among different machining systems, and provide fast data or information exchange between subsystems and/or terminals and rapid changes from the network to a CNC machine. Thus, various manufacturing systems/facilities distributed at different companies can be organically organised and shared under their agreements.

In order to develop systems for distributed scheduling, process planning and control, tremendous efforts have been made by many researchers. As reviewed by Shen and Norrie (1999), there were over 30 projects that were being carried out all over the world. For example, Shaw proposed using agents in manufacturing scheduling and factory control. He suggested that a manufacturing cell could subcontract work to other cells through a bidding mechanism (Shaw and Whinston 1983, Shaw 1988). YAMS (yet another manufacturing system) (Parunak 1987) was also one of the earliest agent-based manufacturing systems where in each factory the factory component is represented as an agent. Each agent has a collection of plans, representing its capabilities. The "contact net" is used for inter-agent negotiation. Zhang and Kuo (1997) proposed an integrated approach for process planning and production scheduling by building up an integrated manufacturing environment. Euwe and Wortmann (1997) and Valckenaers *et al.* (1998) addressed the requirements of the new planning system. Collaborative information system application and integration is recognised as a key factor. Thus, Lin and Solberg (1992) presented an integrated shop floor control framework for information and process control. Eberts and Nof (1993) presented a distributed planning approach for collaborative production. Tsukada and Shin (1998) addressed the problem of distributed tool sharing in flexible manufacturing systems. Duffie and Prabhu (1996), Maturana and Norrie (1997) and Tharumarajah and Wells (1997) addressed the scheduling and control problem in distributed manufacturing systems. Sikora and Shaw (1997) presented multi-agent coordination mechanisms for the integration of manufacturing scheduling tasks. With the same preoccupations, Gyires and Muthuswamy (1996) and Pan and Tenenbaum (1991) proposed two multi-agent frameworks for the integration of widely dispersed enterprises.

There are also approaches to achieve production planning and control by using Internet-based supporting systems. Zhou and Besant (1999), for instance, proposed an information management system for production planning and control in a virtual enterprise under a distributed environment. Ockerman *et al.* (1999) used multimedia technology in their project titled factory automation support technology (FAST) to improve worker performance throughout the factory. The IPPI (integrated product processing initiative) project is another major effort aimed at developing and validating a prototype process planning system that will utilise form feature product models defined in STEP format and is capable of generating intermediate product models representing the state of the product prior to and subsequent to each manufacturing operation. The goal of the IP3S (integrated process planning/production scheduling) project is to dynamically convert standards-based product specifications into process plans and schedules that best accommodate current shop load, the status and allocation of machines, fixtures and tools, and raw material availability, while minimising production costs, lead times and inventories, and maximising due date performance.

Typical research issues raised by recent research in this area can be sorted into the following three categories: (1) no common model for developing Internet-based scheduling, production planning and manufacturing systems for various applications; (2) issues in agent technology for the dynamic manufacturing scheduling, planning and control, *e.g.*, integration of planning and scheduling (Maturana and Norrie 1997); (3) research issues in scheduling and planning algorithms, *e.g.*, dynamic executing of remote large programs and algorithms (Su *et al.* 2000).

2.3.5 Internet-based Concurrent Engineering

CE was defined by Cleetus (1992) as "a systematic approach to integrated PD that emphasises response to customer expectations and embodies team values of cooperation, trust and sharing in such a manner that decision making proceeds with a large number of parallel working by all life-cycle perspectives, synchronised by cooperatively brief exchanges to produce consensus". CE has been used in a wide variety of applications for the purpose of improving the performance of RPD, *e.g.*, reducing the PD cycle time and cutting down costs (Young and O'Grady 1992, Michel and Clermont 1997). It has been found from contemporary research in the fields of CE that significant benefits can be achieved if suppliers are involved in the new PD process as early as possible. However, recent investigation in manufacturing industries has also revealed that this approach is not widely practised in industries and its implementation has been a great challenge to researchers and practitioners.

In the distributed PD environment, CE teams would include various functions as disparate as design, planning, manufacturing, assembly, testing, quality and purchasing as well as the involvement of suppliers and customers (Sprow 1992). The overall goals of CE include shortening time to market, reducing the costs of the total product life cycle, and increasing quality. The objectives of CE might also include the optimisation of the mechanical function of the product, minimising the production costs, reducing the PD time and ensuring the product can be serviced and maintained both easily and economically.

To support CE in the distributed environment, the international research community has carried out extensive research in order to establish ways of supporting engineering activities. However, the term Internet-based distributed CE contains new concepts, as the team participants are globally distributed and use a wide variety of computer systems. A few research projects have been reported in the literature describing how Internet-based distributed CE can best be implemented, particularly for a globally distributed CE team that may be using a wide range of computer systems. The requirements for building an environment to support Internet-based distributed CE are identified as follows (Kang *et al.* 1997):

1. The environment should use open standards to allow participation by a wide variety of parties, including suppliers and customers.
2. The environment should allow natural communication between team members.
3. A product database management system (PDMS) should be used to manage the storage and retrieval of the large amount of information used by any reasonable sized company. Such a PDMS should be integrated with other functions within the company.
4. The environment should support interactive 3D graphical representations of parts and products; these parts and products should be available for users to manipulate.
5. The environment should have effective data conversion engines to support data transfer between data stored in the PDMS, the standards used for data transfer,

the visualisation standards and data standards of various CAD, CAPP and CAM systems.

These requirements pose some challenging issues for researchers in this area. As identified by Christiansen *et al.* (1996), CE can be realised through three different non-exclusive approaches:

1. organisational approach, where engineers from different departments are organised into PD teams;
2. non-IT approach, where specification activities are supported with, for instance, technical manuals describing a group of similar products (or components) and their manufacturing specifications based on a group technological analysis of products or components; and
3. IT approach, for example, using features and product modelling to structure knowledge and information for specifying products in the different phases of the product life cycle.

Among these approaches the IT approach is the most promising to achieve Internet-based CE. However, this research is still at a very early stage; further research in terms of the supporting technologies and the application of CE in the whole PD process are required.

2.4 Review of IRPD Approaches

Besides the research topics reviewed previously, great efforts have also been made in other fields of PD, *e.g.*, workflow management such as WebFlow (WebFlow Corporation, http://www.webflow.com/), PrISMS (NASA 1996), OzWeb (Kaiser *et al.* 1997), resource planning, marketing and supply chain management (Huang and Mak 2001), *etc.* These extensive research efforts made in different areas of PD raise the possibility that IRPD systems can be developed through the integration of these technologies. Some researchers have also made contributions towards improving the agility and responsiveness of manufacturing enterprises and enhancing the ability of rapidly combining the strengths of manufacturers and suppliers. However, the functionalities of Internet-based systems for such purposes are limited. Some of the research approaches are briefly reviewed in the following.

2.4.1 Collaborative Approach

An interdisciplinary team that works collaboratively is one of the features of RPD to shorten a PD cycle. Various attempts have been made to develop Internet-based collaborative systems to support rapid product design and manufacturing using collaborative technologies. They include the research and development of collaborative technologies (Steves and Knutilla 1999), CSCW tools (Jiang *et al.* 2008, Bullinger *et al.* 1996, Prante *et al.* 2002), collaborative engineering database, engineering information framework (Szykman *et al.* 2000), engineering database structure (Xue *et al.* 2006) and the development of Internet- and information-based collaborative systems (Cheng *et al.* 2000). These collaborative approaches aim to achieve better cooperation among distributed team members and are of great importance for RPD. There are also tools and methods proposed for collaboration over the Internet. Some of the typical works are reviewed as follows. Chui and Wright (1999) developed an Internet-based multi-media educational tool for the design of simple mechanical parts. Ho *et al.* (2000) proposed a multi-media communication framework for the selection of collaborative partners in global manufacturing. The proposed framework was generated from the CIM-OSA (open system architecture for CIM) approach and the model was developed with the aim of simplifying the enterprise's collaboration. Reed and Afjeh (1998) used a web-based interactive engineering tool for engineering simulation and teaching and learning purposes. One of the early web applications was to provide rapid prototyping services on the Internet (Bailey 1995, Wright and Burns 1997). Smith and Wright (1996) collected a number of WWW-based design and manufacturing services for their cyber cut experiment. Roy and Cargian (1997) presented experimental workbenches for WWW-based design to production, including activities such as conceptual and detail product design, process planning, design for manufacture, NC programming and rapid prototyping.

2.4.2 AI and Knowledge Approach

The increasing complexity of products and processes requires earlier decision-making and tools that can help decision-making at the early stages are essential for RPD. Due to this, knowledge and knowledge modelling of design, process planning, quality, *etc.* and AI for supporting decision-making have become an important research field of RPD. In recent years, work has been carried out to develop Internet-based intelligent systems to support RPD. For example, Pan *et al.* (1997) integrated AI and knowledge technologies to greatly improve the agility of product design and manufacturing. The integration of these technologies with Internet technologies enables distributed manufacturing companies to achieve short PD cycle times and to respond quickly to sudden market opportunities. Cheng *et al.* (1997) and Pan *et al.* (1999) proposed a Java and AI-based system for the implementation of design agility and manufacturing responsiveness. Joseph *et al.* (1997) proposed a WWW approach for capturing and deploying the preferential knowledge required to resolve design conflicts. Xie and Tu (2000) proposed a WWW-based integrated PD platform for intelligent and concurrent sheet metal design and manufacturing. The platform integrated knowledge base, AI and Internet technologies. Cheng *et al.* (2000) presented a novel approach to implement agile design and manufacturing concepts using Internet-based technology. The underlying philosophy of the approach is to use Web-based design and manufacturing support systems as smart tools from which design and manufacturing customers can rapidly and responsively access the system's built in design and manufacturing expertise.

2.4.3 Integrated and Concurrent Approach

As collaboration and partnership become increasingly important in modern manufacturing companies, CE (Young and O'Grady *et al.* 1992, Kim *et al.* 1998) and integration methodologies (Hsin 2000) are recognised as successful ways to support RPD. This area is also a strongly emerging research field. A few research projects have addressed the importance of the integration and concurrent approach in the PD process. For example, Xie and Xu (2008) proposed an integrated PD platform for intelligent and concurrent sheet metal design and manufacturing. This platform integrates various software tools for sheet metal design, unfolding, planning, cost optimisation, manufacturing and marketing. Zhang and Kuo (1997) proposed an integrated manufacturing environment for the integration of process planning and production scheduling. Tu *et al.* (2000) presented a virtual PD platform for RPD with the integration of specific models for the involvement of customers. Chui and Wright (1999) presented a WWW computer integrated manufacturing environment for rapid prototyping and education. Reed and Afjeh (1998) used a WWW-based interactive engineering tool for engineering simulation, teaching and learning purposes. Research has also focused on ways to support an integrated and concurrent approach, for example, the integration of available information is a key step for different partners in a PD cycle to share information effectively (Dong and Agogino 1998). Chen *et al.* (1998) developed an integrated graphical user interface for CE design of mechanical parts, and several typical models were also suggested in their paper, including a WWW-based on-line user's guide, a part library, a design guideline checklist, a part modeller linked to a CAD system (Pro/ENGINEER) and a knowledge-based design critique system. Su and Amin (2001) proposed an Internet-based system for geographically dispersed teams to collaborate over the Internet for the purpose of integration in design and manufacture. A CGI (common gateway interface)-based multi-user method was developed to remotely execute large-size software systems via the Internet.

Great research effort has also been made to develop an integrated information-sharing platform for the purpose of supporting integrated and concurrent PD in a computer network environment. Some prototypes have been developed although they are still far from being commercialised, *e.g.*, Boynton (1993) Cutkosky *et al.* (1993), Gruber and Russell (1996) Dong and Agogino (1998), Xue *et al.* (2006). Cheng (2000) presented an empirical study of the implementation and integration of information systems for production management in manufacturing. Shackleford and Proctor (1998) developed Java-based tools for the development and diagnosis of a real-time control system. New approaches have also been proposed to implement WWW-based integrated systems for product design and manufacturing. These have been used in different areas such as agile manufacturing (Cheng *et al.* 2000), sheet metal concurrent design and manufacturing (Tu and Xie 2000), and collaborative design of 3D mechanical parts (Kim *et al.* 2001b). With the support of the European Union, an international team has worked on a collaborative research project called "global engineering network", which aims to provide a global collaborative design platform across various EEC countries (Gausemeier 1996).

2.5 Review of Implementing Technologies for Developing IRPD Systems

Although the research into Internet-based PD systems have been around for several years, a united way to develop IRPD systems still has not been found. The implementation of IRPD systems covers a broad range of current technologies. Internet applications can be implemented in many programming languages, tools and environments. The most popular technologies that can be used to develop IRPD systems include, DCOM (distributed component object model), CGI, STEP, SDAI (STEP data access interface), CORBA (common object request broker architecture), ActiveX technology, JAVA and agent technology. This section gives a review of how these technologies can be used for developing IRPD systems.

2.5.1 ActiveX and DCOM Technology

2.5.1.1 ActiveX

The Microsoft ActiveX technology allows programmers to assemble reusable software components into sophisticated applications and services in an Intranet/Internet environment easily with minimum effort (Swank and Kittel 1997). This technology has been used in the development of various Internet-based research projects (Huang and Mak 2001b, Xie *et al.* 2001a). One of the main features is that the software components, called Active X components, can be reused by different software platforms. ActiveX components can be created using Microsoft Visual Studio toolkits such as Visual Basic, Visual InterDev, Visual Java++ and Visual C++. Several types of ActiveX components are available to develop in both applications at server and client sides. ActiveX components can be used to develop applications that integrate tightly with other elements of the Internet or intranet site. Application clients can be compiled into ActiveX controls which can then be embedded into HTML web pages, or into ActiveX documents that are attached to HTML web pages (usually executed in a separate container). The downloading, installation and execution processes of both ActiveX controls and documents are similar. When a user accesses a URL with an ActiveX component, it is downloaded from the server and then registered on the client machine along with the HTML page that uses script to invoke the ActiveX component.

Huang and Mak (2001b) detailed the functionalities of the ActiveX in documents and controls. ActiveX controls are usually embedded into HTML web pages. ActiveX documents are non-HTML documents that can be viewed and edited in a web browser. The HTML page is replaced by the ActiveX document, and the ActiveX document executes in the web browser as its application container. ActiveX documents also offer more complex client-side processing than HTML pages with ActiveX controls. For example, ActiveX documents may have all the elements of standalone systems, *e.g.*, pull-down menus. On the other hand, ActiveX controls do not have capabilities such as pull-down menus. An ActiveX code component is an object or objects exposed by an application that can be controlled programmatically by other applications. ActiveX code components can be used to add functionality to an HTML page on the client-side just as with ActiveX controls. However, it is more often used to deploy application servers as ActiveX code components. Server-side ActiveX code components can be used to customise the creation and return of an HTML page, just like CGI programs. In addition, they can also be used to manage a database connection, execute a standalone algorithm and marshal queries received and results returned.

2.5.1.2 DCOM

The DCOM architecture is widely used across a broader scale of multi-user applications. It can be used in most of the application cases when developing Internet-based PD systems. DCOM is an ideal technology for multi-tier applications because it enables ActiveX components to work across networks, enabling developers to easily build systems that span computer boundaries. The DCOM has thus three unique strengths that make it an important technology for IRPD:

1. DCOM is based on the most widely used component technology.
2. DCOM is simply "COM with a longer wire" – a low-level extension of the component object model, which is the core object technology within Microsoft ActiveX. Major development tools (*e.g.*, Microsoft, Borland, Powersoft/Sybase, Symantec, ORACLE, IBM, and Micro Focus) and the applications they produce automatically support DCOM, providing the broadest possible industry support. Additionally, over 1000 existing commercial software components that work with DCOM are already available for use by developers.
3. DCOM is an open technology that runs on multiple platforms. Microsoft is openly licensing DCOM technology to other software companies to run on all of the major operating systems, including multiple implementations of UNIX-based systems. As recognised by many developers, it is the best networking technology to extend component applications across the Internet.

The combination of these three advantages – the largest install based native support for Internet protocols, and open support for multiple platforms – means that businesses can gain the benefits of a modern component application architecture without having to replace investments in existing systems, staff, or infrastructure. Developers are able to add components together without having to worry about network programming, system compatibility, or integration of components built from different languages. This can lower the cost and complexity of building distributed applications from components. DCOM leverages the investments companies have already made in ActiveX by providing the following benefits:

1. Multi-platform support. DCOM is designed to run on Windows 95, Windows NT, Macintosh, UNIX, and legacy operating systems, providing companies with

the basis for a common application infrastructure across their entire IT environment, which can lower integration costs and reduce integration complexity.
2. Evolutionary technology. In addition to Java support, DCOM enables components written in other languages, including C, COBOL, Basic, and Pascal, to communicate over the Internet, providing a growth path for existing applications to support Web technology.
3. Common components for the browser and Web server. Since ActiveX components can be embedded into browser-based applications, DCOM enables a rich application infrastructure for distributed Internet applications using the latest browser technology.
4. Security. DCOM integrates Internet certificate-based security with rich Windows NT-based security, combining the best of both worlds.
5. Standards-based. Microsoft is working with Internet standards bodies, including the IETF and the W3C, to offer DCOM to the Internet community as an open technology. DCOM is based on the Open Group DCE RPC, an open and widely deployed communications technology. The DCOM wire protocol extensions have been submitted as an Internet draft and are available at http://www.dc.luth.se/doc/id/draft-brown-dcom-v1-spec-00.txt.

2.5.2 Java Technology

Java technology has a broad application in PD processes (Cheng *et al.* 1997, Shackleford and Proctor 1998, Pan *et al.* 1999, Lumpur 2010). Java was originally proposed by Sun Microsystems (java.sun.com). This software development platform can be used to develop Internet-based systems on the Internet platform. It is a portable, object-oriented, distributed and multi-threaded programming platform. All Java programs can run in a Java Platform that has two components: the Java virtual machine (Java VM) and the Java application-programming interface (Java API). The Java VM allows Java application programs to run in any operation systems. This is important for the integration of the different manufacturing systems that run in a heterogeneous environment. Java API provides a variety of functions for users to develop different applications, *e.g.*, Java SDAI provides tools for accessing STEP-based databases, client/server tools, *etc.* Figure 2.1 shows an integrated environment for Internet programming by using JAVA script in both server and client sides. This environment has connections with data resources from networked databases and provides efficient server-side programming. Java applets depend on web browsers for their installation (downloading) and execution in the client machine.

In this integrated environment, JavaBeans, which are components built in Java, can be used. This JavaBeans specification describes new component architecture for Java to facilitate component code development and reuse. As the JavaBeans component model is based on Java classes, this model can be extended and reused, for example adding rules and using these rules with the class. Beans can be deployed as servers and clients. A bean is a reusable component that can be used to create

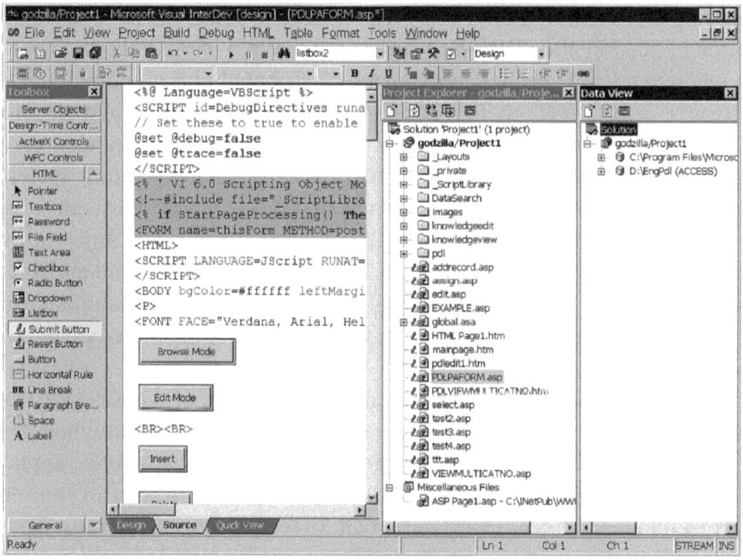

Figure 2.1 An integrated programming environment for developing IRPD systems

applets, applications or even HTML pages. The same bean should be able to play in any of these containers. The emergence of the JavaBeans component model is expected to further simplify web application development in the Java environment.

2.5.3 CORBA and Agent Technology

2.5.3.1 CORBA

CORBA, developed by the Object Management Group (OMG), has become a promising approach to the development and implementation of distributed systems (Harmon and Morrissey 1996). It has many distinguishing advantages over other approaches (*e.g.*, SQL, email-based groupware and transaction process monitoring approaches). It can avoid direct links between applications by defining a broker that acts as an object bus, and can also encapsulate existing legacy applications. These characteristics are extremely suitable for WWW-based systems development (*i.e.*, virtual enterprise). Another important feature of the CORBA (OMG 1998) technology is that it can be used for network communication between different computer system platforms. According to the CORBA standard, all software systems written in different programming languages on different platforms can exchange information via the interfaces defined by the CORBA IDL (interface definition language) and ORBs (object request brokers).

The OMG's CORBA technology can be used on the Internet as the communication mechanism. The ORB interoperability allows communication between independent implementations of the CORBA standards. The OMG also defines a specialisation of general inter-ORB protocol called Internet inter-ORB (IIOP). CORBA has its inherent advantages in heterogeneous distributed object computing environments (Steve *et al.* 1997) as the CORBA specification has defined six language mappings from OMG IDL to the most popular programming languages, and available CORBA products can support almost every popular operating system. The new generation of CORBA, as mentioned in http://www.cs.wustl.edu/~schmidt/corba-research-overview.html, will optimise its performance and improve its communication speed and safety. The integration of CORBA, STEP and agent technology promises to meet the various needs of developing advanced manufacturing systems. A more detailed technical introduction to CORBA technology can be found at http://www.omg.com.

As mentioned in Section 2.2, the characteristic feature of the RPD environment is heterogeneity in its computing and communication. This not only includes computing platforms, operating systems and network protocols, but also the applications utilised in different departments, *i.e.*, CAx, DFx, PDM, *etc.* In such an environment, CORBA technology becomes very important. In recent years, numerous approaches using CORBA technology in manufacturing have been found (Coulson 1998). For example, Whiteside *et al.* (1997) reported a CORBA-based manufacturing environment for Sandia's Agile Manufacturing Test-bed (SAMT). This environment uses CORBA technology to support information integration, sharing and cooperation among distributed manufacturing cells. Howard *et al.* (1998) used CORBA and STEP to develop the next generation standards and measurements needed for information-based manufacturing. Systems that were developed based on these standards can be used to support strategic partnerships among diverse and geographically dispersed companies. They are particularly effective at carrying out needed manufacturing operations and achieve integration by efficiently managing information flow between different manufacturing systems.

2.5.3.2 Agent Technology

In recent years, agent technology has become a popular method for improving PD process. Although the term "agent" has been used widely, it has no unified meaning, definition or structure (Lei *et al.* 1998). It can be a hardware or software component (Shoham 1993), or a combination medium between human users and software tools (Hao *et al.* 2006, Genesereth and Nilsson 1987, Khedro 1996). However, many researchers in different fields have implied or clearly suggested that an agent possesses certain fundamental characteristics and capabilities (Crowston and Malone 1988, Shoham 1993, Lei *et al.* 1998). Firstly, an agent possesses and maintains certain information. Secondly, it is able to interact with its environment and extract knowledge from it. Thirdly, the agent is able to communicate with other agents for information and knowledge exchange. Fourthly, it is able to process information, messages, and events and make decisions autonomously. Lastly, the structure and contents of an agent can be changed or defined by users through using a software platform. In the IRPD environment, an agent is defined as a software program that has fundamental characteristics and capabilities as defined above. These characteristics and capabilities of agent technology enable multi-task adjustment and control, which act as "coordinators" for the scheduling, planning and manufacturing in a distributed environment. As discussed in Section 2.3.4, agent technology has been widely used in diverse PD processes such as distributed manufacturing, scheduling, planning and control (Gyires and Muthuswamy 1996, Pan and Tenenbaum 1997, Sikora and Shaw 1997, 1998, Wang and Lin 2009).

2.5.4 STEP and SDAI

As a wide variety of software systems are applied in PD or production, *e.g.*, different kinds of CAD, CAE, CAPP, CAM, scheduling, and MRP systems, a PD computer system environment has been formulated. In this heterogeneous computer-based system environment, STEP is crucial for data communication and exchange between these systems. STEP (ISO 10303, 1994a, b) is an emerging international product data standard that uses a high level, feature-based and object-oriented approach to define products. It is able to provide a complete, unambiguous, computer-interpretable definition of the physical and functional characteristics of each unit of a product through its life cycle. STEP is organised as a series of parts according to description methods, application protocols, implementation methods, and conformance testing. STEP uses a formal specification language EXPRESS (Zhau and Liu 2008, Schenck and Wilson 1994) to specify the product definition information. The functionaries of STEP have been addressed by the ISO STEP handbook. Although STEP is still in its development stage (Mannisto *et al.* 1998), it has already been used in various industrial fields. Various STEP-based application protocol (AP) tools have been developed and used in PD processes, *e.g.*, modelling generic product structures (Mannisto *et al.* 1998), data migrating and translation (Mangesh *et al.* 2000), information modelling (Lee 1997), feature extraction from STEP geometry for agile manufacturing (Mangesh and Nagi 2000), building an integrated environment (Yang and Pei 1999) and Internet-based data exchange framework (Zhang *et al.* 2000).

SDAI is a tool provided by the STEP community interfaces for data sharing and exchange. Usually there are three kinds of data sharing methods: 1. exchange file; 2. SDAI; 3. database management system. Among these, implementation problems make the third option difficult, while sharing by means of files is just a static data sharing. A high level of standardisation is on the way for data sharing using SDAI, which is a dynamic way of sharing data. This feature is very important for dynamic information exchanges. SDAI standardises the runtime interface to STEP data, which enables application software to develop functions to dynamically operate STEP data. Commercial languages (*e.g.*, C, C++, FORTRAN and Java) provide application packages for SDAI. For example, SDAI/Java, as a binding language, has been used in various PDs. Portability of Java enables SDAI to be easily ported to any node in a heterogeneous enterprise network. Hence, SDAI has become an important tool for developing manufacturing systems that use STEP as a standard. Some case studies are available in http://mega.ist.utl.pt/~ic-sdai/docs/.

2.6 Research Issues and Future Trends of IRPD

Despite the contributions that have been made by numerous research and development projects to develop Internet-based systems for supporting various PD processes as reviewed in Section 2.5, there are still some major issues that hamper the development of IRPD systems, especially when it comes to the integration of technologies. These problems include the requirement for a formal Internet-based distributed data environment (Urban *et al.* 1999), knowledge and AI sharing and support, global optimisation algorithms (Tu *et al.* 2001) and Internet communication issues (Huang and Mak 2001b). These issues, which are briefly summarised in the following, need to be solved before IRPD systems can be developed successfully.

2.6.1 Implementation Issues

Internet technology is used by most global manufacturing companies. As reviewed in Section 2.5, some newly developed Internet communication technologies have been used to develop various PD systems. However, it is still not clear how these technologies and systems can be effectively implemented to support IRPD processes. It is also not clear what aspects of PD can be extended and added to the Internet-based systems. Principally, all the systems that are involved in the PD process can be implemented using Internet technology. In addition to these technical implementation problems, the integration of these technologies and their applications is a further problem for implementation of IRPD systems in manufacturing companies. People in the field of product design and manufacturing may have limited knowledge of Internet technologies, whereas IT people normally do not have enough knowledge of product design and manufacturing.

So far, to our knowledge, no literature or research report is available to review and summarise all the implementation issues, *e.g.*, how to develop Internet-based systems, identify what kind of systems are suitable for upgrading to the Internet, how to integrate design and manufacturing through the Internet and how to select Internet-based systems for different companies. Usually, the following questions should be asked before considering research or deploying Internet-based product design and manufacture systems. These questions are:

1. What product design and manufacture problems are most suitable for web applications?
2. How advantageous are the web applications over standalone and usual client-server applications?
3. How are sound web applications developed, implemented, deployed and applied in the field of product design and manufacture?

To develop IRPD systems, there are a number of issues that manufacturing companies have to face besides those given above. The first is the identification of the proper structure (*e.g.*, network structure, communication protocol) for IRPD systems under a heterogeneous working environment. The second is the selection of a programming environment that can work easily within the Internet environment and is supported by recent Internet technologies. The third is identifying how existing technologies described in Section 5 can be used to resolve IRPD issues. The fourth is deciding what kind of data environment the IRPD systems should be based on. The last is identifying how existing systems can be used (*e.g.*, how to upgrade existing standalone systems to the Internet, and connect existing product data management systems with the Internet). Other issues, *e.g.*, securities, ownership, *etc.* should also be considered before starting the development of the Internet-based approach.

2.6.2 Systems Integration Methodologies

As discussed in Section 2.2, IRPD requires the integration of people, business processes and information technology across the PD value cycle. This requires interfacing and integration across organisational functions as well as suppliers and customers. For the development of manufacturing technologies and the emergence of concepts like computer integrated manufacturing (CIM), there is a need for extensive cooperation between different engineering activities. More recently, the emergence of new manufacturing paradigms has taken the concept of cooperation to a higher degree. Product related information like design, manufacturing, utilisation, maintenance and disposal are not only shared among various departments within a company, but also among various "partner" companies around the world, for example, the sharing of new geometric algorithms (Wagner *et al.* 1997). This requires a new "generation" of systems integration, which not only includes information sharing, but also the integration of technologies for design, planning, simulation and manufacturing. As efficient and effective information exchange between different systems and different people in different PD teams becomes vital for IRPD, intensive investigation into how best to integrate various Internet-based systems that are used at different stages of the PD process through the Internet is required. Although the Internet has made a variety of types of information accessible to multiple users simultaneously, it still does not offer a simple way to use standalone applications (Tomarchio *et al.* 2001), particularly those computationally and graphically intensive applications often required at various levels in a manufacturing environment. There are many factors that affect the integration of Internet-based standalone support systems for design and manufacturing. The obstacles to their application are the heterogeneous manufacturing environments existing in a company or among enterprises. Furthermore, it is difficult for any single software supplier to have the necessary expertise to support the full suite of software modules that are used in an enterprise to facilitate its various functional activities. Hence, new system integration techniques are the key issue to address for developing IRPD systems.

In the early 1980s, manufacturers such as XeroX, Boeing, GE and the vendors Digital and Computervision realised that true integration at the workgroup level between multiple CAD and CAE applications was an extremely difficult problem to solve because of incompatible geometric representations and proprietary database and file structures. In recent years, despite great efforts having been made to enable system integration, *e.g.*, integration platform by Tu *et al.* (2001), smart drawing under networking environment by Dong *et al.* (1998) the PACT project by Cutkosky *et al.* (1993), *etc.*, further research and development are still needed since the issues in systems integration have gained more and more influence on the performance of RPD. The research issues for system integration include (Fowler and Luce 1995), technology development, standards development, and testbed and technology transfer. Technology development involves developing the necessary technologies for systems integration. The following research issues have been identified:

1. lifetime data model for systems integration (Tu *et al.* 2001);
2. determining attributes of a complete and consistent model for PD processes;
3. test platform for the integration mechanisms for systems integration.

For standards development, two research issues need to be addressed:

1. re-engineer critical standard development processes;
2. improve standards and system interoperability.

Finally for the testbed and technology transfer, three key problems need to be studied:

1. establish computer support systems for electronic networks, on-line databases, and user-friendly search techniques;
2. achieve transfer of technology over the Internet;
3. develop a supporting environment and communication testbed to perform tests.

These issues and problems will continue to be hot research topics.

2.6.3 CSCW and Interfacing Techniques

As IRPD requires the extensive cooperation of various PD systems over the network environment, computer supported cooperative work (CSCW) and interfacing techniques are becoming critical for developing IRPD systems. They directly affect the accuracy and efficiency of the data management through the whole PD cycle. However, the development of CSCW technologies so far has not been successful (Lubich 1995). The main problems include:

1. the technology is not available to all in the relevant community;
2. the current technology does not support multi-interaction among various systems;
3. the technology is not compatible with normal work; and

4. a set of rules to govern the CSCW work to meet with social and organisational norms is still missing.

A further problem with CWCS tools is the assumption of fixed models of users or organisational cultures, which do not meet the actual requirements of malleability and linkability (Simone and Divitini 1995).

The interfaces in the IRPD systems should have the following capabilities:

- workbench (common user interface to multiple PD applications);
- process modelling and workflow;
- application encapsulation and invocation; and
- application data translation and transfer and multi-vender networking and platform support.

The commonly used data exchange standards, *e.g.*, CAD-NT (CAD-Normteile), CGI (computer graphics interface), CGM (computer graphics metafile), SET (standard data exchange transport), PDES (Product Data Exchange Specification), STEP (Standard for Exchange of Product Data), and IGES (Initial Graphic Exchange Specification), do not have the functionalities to directly support interfacing via variant application software packages. In recent years, tremendous efforts have been made to solve this problem. For example, Lau and Jiang (1998) reviewed the state of the art data modelling and exchange in enterprise integration and found that companies mainly used neutral format file to exchange data between different systems. Tu and Xie (2001) presented an integrated product model that can be used to record all the necessary data through a whole PD cycle into a common object to achieve data sharing across diverse engineering applications. This product model or the common object can be manipulated directly by all kinds of computer aided engineering software systems. As the representation of non-geometric information using STEP is still at a preliminary stage (Wang and Bourne 1995) and industry's reliance on non-geometric information and knowledge-based design and manufacturing is increasing, Huaglory (2001) presented an advanced life-cycle model for complex PD. The life-cycle model called SICODEL was designed to organically incorporate the ideas from various existing paradigms, such as CE, life-cycle engineering, DFX, production planning and control, virtual prototype, intra-inter enterprise information, rapid prototype and evolutionary life-cycle models of software engineering into a whole integrated information system concept. Dorador and Young (1999) explored the structure definition of information models for supporting the related processes of design for assembly and assembly process planning to support the intensive sharing of information required in CE throughout the life cycle of a product. Lee *et al.* (1998) reported their experiences on developing information models for a variety of manufacturing domains, such as plant layout, process planning, discreteevent simulation (Ellis *et al.* 1998), and apparel pattern making for the SIMA (Support Initiative for Multimedia Applications) and NAMR (National Association of Manufacturers and Representatives) Programs. Three data-modelling methodologies were adopted in their research, these are the entity relationship (ER), the functional modelling and the object-oriented (OO). Three major information modelling languages, as IDEF1x (the Integrated Computer Aided Manufacturing (ICAM) Definition Language 1 Extended), EXPRESS and UML (the Unified Modelling Language) were used for building the information framework. They also suggested that effective information sharing and exchange between computer systems throughout a product life cycle is a critical issue. To achieve data communication and exchange between different systems under the Internet environment, Zhang *et al.* (2000) presented an Internet-based STEP data exchange framework for virtual enterprises. An Internet-based services translator for non-STEP and STEP data communication and exchange between heterogeneous computer systems was proposed. Urban *et al.* (1999) presented an integrated product data environment (IPDE). The IPDE goes beyond the functionality of current product data management tools and deals with product data at a semantic level using the STEP.

Despite the success of the above research, research into CSCW and interfacing techniques for the purpose of RPD is still in its developing stage. Some issues identified as main hurdles include: using STEP representation of non-geometric information is at a preliminary stage (Wang and Bourne 1995), the combination of the Internet modelling methodology (*e.g.*, HTML and CORBA) with STEP is not yet mature (Hardwick *et al.* 1996), and the Internet-based data translators between various data formats are still at an early research stage (Zhang *et al.* 2000). These are major research areas needed to achieve RPD.

2.6.4 Global Cost/Lead Time Optimisation

In order to make the PD process quicker and produce cheaper and better products, the optimisation of key parameters of the PD process is necessary and has become very important for PD. There are some optimisation algorithms proposed for improving the products' competitiveness, *e.g.*, multidisciplinary design optimisation (MDO) by Yang *et al.* (2001), collaborative optimisation by Braun (1996), multi-objective optimisation for engineering process by Suppapitnarm *et al.* (2000) and Shea (2001) and multi-criteria optimisation in product platform design by Nelson *et al.* (1999). Product cost and lead time are the most important parameters for a company to make a decision. They are indispensable for the success of a company. Hence, to optimise the PD process decisively, additional tools and methodologies for the optimisation of the PD process are required.

Instead of lead time optimisation, the literature and research reports are more concerned with the optimisation of cost. Traditional cost estimation systems require input data, process sequences, processing times, hourly rate for labour or machines, and other miscellaneous data. Usually, these data are stored in a cost database. With these data in place, the estimating process simply breaks down to a series of simple calculation and data storage. For example, Winbourne and Toolsie (1991) proposed a cost estimating system that can be used at the design stage. Utilising their costing system, the detailed specification of the process sequence is not necessary for the cost estimation. A major drawback in their system is the requirement for a user to estimate some initial parameters. No approach is given on how to estimate intelligently the values of these parameters. Wong *et al.* (1992) presented a cost estimating system, called the totally integrated manufacturing cost estimating system (TIMCES), which integrates CAD and process planning techniques into a unified system. Input for this system consists of the required processes, the process sequences, and cost databases. However, TIMCES still considers only predefined process sequences. Kamarthi *et al.* (1993) proposed a system to estimate the manufacturing cost of sheet metal parts in conjunction with their process plans. Based on the standard process sequence for the part family, the lowest production cost is found by selecting among the alternative machines in each process stage. Their system only considered 2D shapes or sheet metal parts. Examples of traditional cost estimating systems can be found in Casey (1987), Goldberg (1987) and Ostwald (1984).

These traditional cost estimation systems have the following drawbacks. Firstly, these traditional approaches do not have a global optimisation algorithm that can optimise the design, process and manufacturing process. Secondly, the traditional systems are never integrated with a CAD system for automatic data (cost and lead time) estimation. Lastly, the traditional systems are slow and time consuming. Under traditional cost estimating approaches, product costs may be misleading, performance measures inappropriate, cost control unfocused, and accounting systems irrelevant for cost management purposes. Yet accurate product costs are at the heart of pricing decisions for new product introductions, obsolete product withdrawal, and responding to competitive products.

From the foregoing discussion, it is clear that cost estimation is an area of great research interest. New technologies such as neural network (BODE 1998, 2000), activity based costing (Tamas *et al.* 2000), log-linear and nonlinear learning curve model (Timothy 1999), AI and mathematical models (Cheng *et al.* 1997, Xie and Xu 2008) have been widely used. Although several efforts have been made, there does not seem to exist a widely accepted system or a system with a wide applicability. In IRPD, not only the cost, but also other key parameters (lead time, quality, *etc.*) are very important for products to compete in the global market and help companies to make earlier decisions. The product developed with lowest cost, for example, may take longer and it may lead to losing the market. Thus, real-time global multi-objective optimisation is vital for achieving RPD goals as the market changes very fast. To meet the requirements, the global optimising system should have the following features:

- optimisation of the parameters (*e.g.*, cost and lead time) through the whole life cycle of a product;
- robustness;
- automatic generation of process alternatives;
- specification and representation of process alternatives;
- easy connection with commercial PD software packages;
- to gain wider acceptance;
- adaption to design, production, and manufacturing environment changes;

- manufacturability verification; and
- automatic selection of the process with optimal results.

Further research that leads to a multi-objective global optimisation system with these features is urgently required by industry.

2.6.5 Virtual Reality for Manufacturing

To be able to consistently handle the demand for increased efficiency, quality, and flexibility, and to reduce the development lead time and cost, it is becoming increasingly important to structure the PD process by following a well-developed methodology. One important aspect of this methodology is the use of models and architectures that provide an abstract, simplified representation of reality (Doumeingts *et al.* 1995). This leads to virtual reality (VR), which can be used to make better decisions (Mujber *et al.* 2004). Manufacturing processes can be defined, modelled, and verified (pre-run) before they are actually implemented. For example, virtual NC programming and tool path control (virtual NC), virtual robotics programming (IGRIP), virtual instrumentation and shop floor control are areas that have received considerable attention (Boman 1995). An improved decision-making process leads to increased efficiency and reduced costs, which are the two main goals of RPD. VR is an emerging technology with potential manufacturing applications in areas such as product design and modelling, process simulation, operating planning and real-time shop floor controls (Stampe *et al.* 1993, Pratt *et al.* 1994, Wang and Nee 2009).

In IRPD, VR plays an important role as it can be used to achieve prediction of system performance, evaluation of a certain feature in the system, design verifications and tests, comparison between alternatives, gaining knowledge of the system at different life-cycle phases, problem detection, and presentation of predicted results. This research area covers topics such as distributed VR through the Internet, the integration of VR with other PD tools or methodologies, and created VR manufacturing environments. Many researchers have made contributions in this research area. For example, Virtual Reality Modelling Language (VRML) has become an industrial standard (including the Internet version of VRML) and has been used in PD processes (http://3dgraphics.about.com/cs/vrmlbrowsers/index.htm). There are a number of VR tools available in the commercial market (reviewed by Klingstam *et al.* 1999). Some VR tools such as QUEST, IGRIP and virtual NC (Deneb Co.) have been developed and used in manufacturing. However, a VR manufacturing environment is a complex world which is made up of a large number of different 3D models. There are problems in creating or adopting such an environment for the RPD process. These problems call for further research. Firstly, it is a tedious and pains-taking process to create a virtual manufacturing world (Stampe *et al.* 1993, Evans 1994). Secondly, the integration of the VR world with other PD tools is not simple. Lastly, the speed of VR via the Internet has to be improved.

2.7 A Proposed IRPD System

The path to the IRPD paradigm can be paved by re-engineering the enterprise, re-design and an optimised PD process, and made more agile by using various enabling technologies. Through research, an Internet-based RPD system is suggested in this subsection for the rapid development of injection moulds. This system can be used in heterogeneous design and manufacturing environments. To minimise the cost of building IRPD systems, it is preferable to standardise the IT infrastructure as much as possible through the establishment of a software platform. This platform will be used to develop interoperability standards needed by the company to integrate the manufacturing processes. This proposed system does not intend to integrate all the Internet-based systems mentioned in this chapter. It focuses on the key technologies that can be developed and integrated to support rapid OKP development. The system has been developed and tested in a mould manufacturing company in Christchurch. Although this system is proposed for rapid injection mould development, the structure of the system extends well for the development of other OKP products. In fact, an Internet-based RPD system with a similar modular structure has also been developed (Xie *et al.* 2001) for the rapid development of sheet metal parts. Examples will be detailed in later chapters of the book. The effect of this research will enable the mould industry to gain the benefits from open, modular, and re-configurable integration of commercial software applications that support design, planning, and production within the environment.

Figure 2.2 shows the reference structure of the IRPD system. It consists of an IRPD software platform, a customer interface model that is developed based on QFD (quality function deployment) methods, an Internet-based integrated PD environment (IPDE), an Internet-based product design environment to support product design, an Internet-based virtual process planning/assembly environment, an Internet-based virtual simulation environment, an Internet-based virtual manufacturing environment, Internet-based design/manufacturing product data/knowledge bases, and a global cost optimisation model.

The IRPD system can be accessed from anywhere inside or outside of the company. This will first help the company's employees to work in an integrated information environment. An Internet-based integrated product data environment is built for data communication and exchange between various systems. The structure of the Integrated Product Development Environment (IPDE) was well addressed by Tu *et al.* (2001) and the information framework and data models have been well explained by Tu *et al.* (2000, 2001). They can share the PD software tools, data, and work concurrently to develop a product. This system brings together expert knowledge of various fields in the early phases of PD. The meetings and discussions through a PD cycle will automatically be recorded by the system when they communicate through the software platform. Secondly, the company's marketing staff can access this system via the Internet to get online engineering support from the company to address, discuss and meet customers' needs. In the system, the modules and technologies are integrated into a continual process chain. This process chain comprises

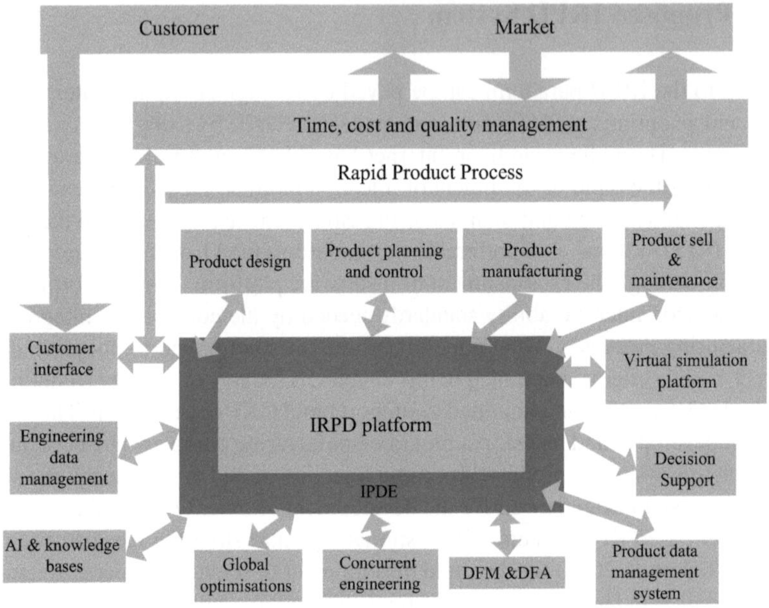

Figure 2.2 Reference structure of an IRPD system for rapid PD

all phases of the PD process, from customer requirements, first CAD draft and finally to the market. The system is equipped with all the necessary technology for fast and cost-efficient development of injection moulds. Tools like Internet communication, product data management and computer simulation are integrated into the continuous flow of data. The company's existing computer aided engineering and management systems such as CAD systems, knowledge management systems, and database management system can be integrated through the software platform of the proposed IRPD system.

The functions of the key modules in the system have been well addressed by Tu (1996) and Xie (1996, 1998, 2000) (as shown in Figure 2.3). In this system, a product can be designed, planned and virtually manufactured and tested with the applications of virtual technologies and methods using computer simulations. The customer interface model is defined to manage dynamic customer requirements so that the IRPD environment reacts more rapidly to the changing requirements of the market. Through the model, customer's voices are gathered through Internet-based customer interfaces and transferred to a shared QFD database (Tu and Xie 2001). A data analysis model is defined to match customer's requirements to engineering requirements. The data/knowledge are recorded in the design/manufacturing database after the product is developed. They can be further used as knowledge or references for similar PDs. Reuse of past experience and knowledge can greatly shorten the PD cycle time. An Internet-based DFM system is developed to improve product and manufacturing performance. An Internet-based DFA system is developed to cut down assembly time and cost. A cost optimisation algorithm

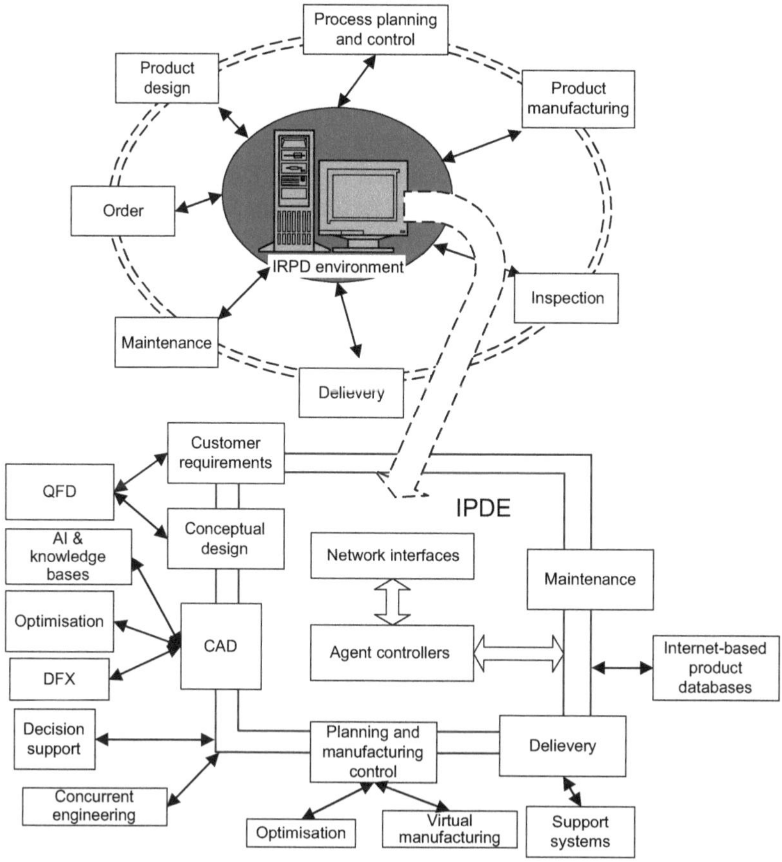

Figure 2.3 Structure of the IRPD environment

based on the shortest insert algorithm was proposed to reduce PD costs (Xie *et al.* 2001). As can be seen from Figure 2.3, the IRPD system integrates software tools for corresponding PD processes. Thus, with the integration of these technologies and models, the goals of the rapid mould development, *i.e.*, shorter lead time and cheaper cost, can be achieved through system integration and knowledge reuse.

2.8 Conclusion

This chapter provides an overview of the recent approaches of Internet-based PD methodologies for the purpose of rapidly and economically developing customised or OKP products. Although many research reports have been published in journals, and some Internet-based PD systems have been implemented in industry, a systemic overview and summary of these research and industrial implementations is not available. This chapter aims to fill this gap. As manufacturing companies, particularly OKP companies, rush to acquire Internet-based PD systems for the purpose of production integration, effective communication, resource sharing, increasing productivity and flexibility, and reducing costs and rework, some research and development directions have been identified in this chapter based on a review of existing research and on industrial needs. At the moment, the research and implementation of Internet-based systems are far behind industrial expectation. On the other hand, it is necessary for industries to understand the development conditions and currently available technology limits on the development of an IRPD system so they can properly select and adopt a suitable strategy.

References

Adamopoulos, D.X., Pavlou, G., Papandreou, C.A., 2008, DCOM and CORBA as the Software Infrastructure of New Telecommunications Services, http://citeseerx.ist.psu.edu/viewdoc/summary?doi=10.1.1.88.8026.

Aidakhilallah, K.A., 1999, Computer integrated process planning and scheduling: intelligent support for product design, process planning and control. International Journal of Production Research, 37, 481–500.

Allada, V., Kopardekar, P., Anand, S. and Mital, A., 1992, Implications of ergonomic and user considerations on manufacturing consumer. Proceedings of the Fourth International Conference on Design, Theory and Methodology, Scottsdale, AZ, 13–16 September, pp. 25–33.

Alting, L., Zhang, H.C., 1989, Computer aided process planning: the stateof-the-art survey. International Journal of Production Research, 27(4), 553–585.

Anton, A.I., Liang, E., Rodenstein, R.A., 1996, A Web-based requirements analysis tool. Proceedings of the 5th International Workshops on Enabling Technologies: Infrastructure for Collaborative Enterprises (WET ICE'96), 19–21 Jun 1996, pp. 238–243, Stanford, CA, USA.

Balasinski, A., Cetin, J., Karklin, L., 2007, Cost-performance tradeoff between design and manufacturing: DfM or MfD?. Design for Manufacturability through Design-Process Integration, Wednesday 28 February 2007, San Jose, CA, USA.

Bailey, M.J., 1995, Tele-manufacturing: rapid prototyping on the Internet. IEEE Computer Graphics and Applications, 5, 20–26.

Bernus, P., Nemes, L. and Williams, T.J., 1996, Architectures for Enterprise Integration (London: Chapman & Hall).

Bode, J., 1998, Neutral networks for cost estimation. Journal of Cost Engineering, 40, 25–30.

Bode, J., 2000, Neutral networks for cost estimation: simulations and pilot applications. International Journal of Production Research, 38, 1231–1254.

Boman, D.K., 1995, International survey: virtual-environment research. Computer, IEEE Computer Society, 28, 57–65.

Booth, R., 1996, Agile manufacturing. Engineering Management Journal, 6, 105–112.

Boothroyd, G., 1996, Book review on design for excellence by James G. Bralla. Journal of Manufacturing Systems, 15, 443.

Boynton, A., 1993, Achieving dynamic stability through information technology. California Management Review, 35, 58–77.

Bralla, J.G., 1986, Handbook of Product Design for Manufacturing (London: McGraw-Hill).

Braun, R.D., 1996, Collaborative optimization: an architecture for large-scale distributed design. PhD dissertation, Stanford University.

Bullinger, H.-J., Warschat, J. and Warner, K., 1996, Management of complex projects as cooperative task. In: R.J. Koubek, W. Karwowski (eds), Manufacturing Agility and Hybrid Automation: I. Proceedings of the 50th International Conference on Human Aspects of Advanced Manufacturing: Agility and Hybrid Automation, Maui, HI, August (Louisville: IEA Press), pp. 383–386.
Carlsson, M. and Egan, M., 1994, Design forManufacture and Assembly: AMissing Link between Quality and Product Development (Kungsbacka: Erlanders Electronic Print).
Carman, R.L., 1998, Boeing AIMS prospective on team collaboration. In: D.H. Brown (ed), Proceedings of the Implementation Roadmap: Getting IT Right the First Time, Dearborn, MI (Port Chester: D. H. Brown Associates) (http://www.dhbrown.com).
Casey, J., 1987, Digitizing speeds cost estimating. American Machinist and Automated Manufacturing, 131, 71–73.
Chen, K.-H., Chen, S.-J., Lin, L. and Changchien, S.W., 1998, An integrated graphical user interface (GUI) for concurrent engineering design of mechanical parts. Journal of Computer Integrated Manufacturing Systems, 11, 91–112.
Cheng, K., Pan, P.Y. and Harrison, D.K., 2000, The Internet as a tool with application to agile manufacturing: a web-based engineering approach and its implementation issues. International Journal of Production Research, 38, 2743–2759.
Cheng, K., Pan, P.Y. and Harrison, D.K., 2001, Web-based design and manufacturing support systems: implementation perspectives. International Journal of Computer Integrated Manufacturing, 14, 14–27.
Cheng, K., Harrison, D.K. and Pan, P.Y., 1997, Implementation of agile manufacturing – an AI and Internet-based approach. Proceedings of the 13th International Conference on Computer aided Production Engineering, Warsaw, June 1997, pp. 273–278.
Christensen, L.C., Christiansen, T.R., Jin, Y., Kunz, J.C. and Levit R.E., 1996, Modeling and simulation in enterprise integration – a framework and an application in the offshore oil industry. Journal of Concurrent Engineering: Research and Applications, 4(3), 229–246.
Chui, W.H., Wright, P.K., 1999, A WWW computer integrated manufacturing environment for rapid prototyping and education. International Journal of Computer Integrated Manufacturing, 12(1) 54–60.
Cleetus, K.J., 1992, Definition of Concurrent Engineering. Technical report CERC-TR-RN-92-003, CETRC, West Virginia University.
Coulson, G., 1998, A distributed object platform infrastructure for multimedia application. Journal of Computer Communication, 21, 802–818.
Crowston, K. and Malone, T., 1988, Intelligent software agents. Byte, 13, 267–271.
Cutkosky, M.R., Tenenbanum, J.M. and Glicksman, J., 1996, MADEFAST: collaborative engineering over the Internet. Communication of the ACM, 39, 78–87.
Cutkosky, M.R., Engelmore, R.S., Fikes, R.E., Genesereth, M.R., Gruber, T.R., Mark, W.S., Tenebaum, J.M. and Weber, J.C., 1993, PACT: an experiment in integrating concurrent engineering systems. Computer, 26, 28–37.
Ding, J.J., Tan, S.L., Zhang, H.H. and Song, X.F., 2006, Research on Module-Based Variant Design for Mass Customization. Knowledge Enterprise: Intelligent Strategies in Product Design, Manufacturing, and Management, 207, 1022–1029.
Dong, A. and Agogino, A.M., 1998, Managing design information in enterprise-wide CAD using "smart drawings". Journal of Computer Aided Design, 30, 425–435.
Dorador, J.M. and Young, R.L.M., 1999, Information models to support the interaction between design for assembly and assembly process planning. Paper presented at the 1999 IEEE International Symposium on Assembly and Task Planning, Porto, Portugal.
Doumeingts, G., Vallespir, B. and Chen, D., 1995, Methodologies for designing computer integrated manufacturing systems: a survey. Journal of Computers in Industry, 25, 263–280.
Duffie, N.A. and Prabhu, V.V., 1996, Heterarchical control of highly distributed manufacturing systems. International Journal of Computer Integrated Manufacturing, 9, 270–281.
Eberts, R.E. and Nof, S.Y., 1993, Distributed planning of collaborative production. International Journal of Advanced Manufacturing Technology, 8, 258–268.

Edwards, K.L., 2002, Towards more strategic product design for manufacture and assembly: priorities for concurrent engineering, Materials & Design, 23, 651–656.
Ellis, K., Jones, A. and Lee, T., 1998, Requirements Analysis: Process Plan Specification Workstation Level. Report of NISTIR 6172 (Gaithersburg: National Institute of Standards and Technology), June.
Euwe, M.J. and Wortmann, H., 1997, Planning systems in the next century (I). Journal of Computers in Industry, 34, 233–237.
Evans, D., 1994, AutoCAD in the Fast Lane. CADENCE, 72–84.
Exixon, G., 1996, Design for Modularity, Design for X-ability (London: Chapman & Hall), pp. 357–379.
Feldmann, K. and Gohringer, M.J., 1999, Multimedia system for remote diagnosis of complex placement machines. International Journal of Advanced Manufacturing Technology, 15, 722–729.
Fowler, J.E. and Luce, M.E., 1995, Systems Integration for Manufacturing Applications Program 1995 Annual Report. (http://www.mel.nist.gov/msidlibrary/doc/sima95rep/nu-annua.htm).
Fowler, J.E., 1996, Variant design for mechanical artefacts: a state-of-art survey. Engineering with Computers, 12, 1–15.
Frankenberger, E. and Birkhofer, H., 1995, Teamwork in engineering design practice. In: V. Hubka (ed), Proceedings of International Conference on Engineering Design (ICED), Zurich, Switzerland, 2, 828–833.
Furlani, C.M., Kent, E.W., Bloom, H.M. and Mclean, C.R., 1983, The automated manufacturing research facility of the national bureau of standards. Proceedings of Summer Computer Simulation Conference, Vancouver, BC, 11–13 July, pp. 557–572.
Gausemeier, J., 1996, Global engineering network-weltweiter Informationsverbund zur Staerkung der Innovationskraft in Produktentwicklungsprozessen. VDI-Berichte, 1302, 39–52.
Geller, T.L., Lammers, S.E. and Mackulak, G.T., 1995, Methodology for simulation application to manufacturing environments. Proceedings of the 1995 Winter Simulation Conference (WSC'95), VA, USA, 3–6 December, pp. 909–916.
Genesereth, M.R. and Nilson, N.J., 1987, Logical Foundations of Artificial Intelligence (Los Altos: Morgan Kaufmann).
Goldberg, J., 1987, Rating cost-estimating software. Manufacturing Engineering, 98, 37–41.
Gruber, T.R. and Russell, D.M., 1996, Generative design rationale: beyond the record and replay paradigm. Design Rationale: Uses, Techniques and Concepts. Lawrence Erlbaum Associates, pp. 323–349. ISBN:0-8058-1567-8.
Gupta, S.K., Regli, W.C., Das, D. and Nau, D.S., 1997, Automated manufacturability analysis: a survey. Research in Engineering Design, 9, 168–190.
Gyires, T. and Muthuswamy, B., 1996, Planning and coordination in distributed manufacturing. Journal of Computer Information Systems, 2, 16–29.
Hardwick, M., Spooner, D., Rando, T. and Morris, K., 1996, Sharing manufacturing information in virtual enterprises. Communications of the ACM, 39, 46–54.
Harmon, P. and Morrissey, W., 1996, The Object Technology Casebook (Chichester: Wiley).
Hartley, J.R., 1998, Concurrent Engineering: Shortening Lead Times, Raising Quality, and Lowering Costs (Portland: Productivity Press).
Hktaiga, 1996 (http://hktaiga.ust.hk/main.htm).
Ho, J.K.L., Fung, R., Chu, L. and Tam, W.M., 2000, A multimedia communication framework for the selection of collaborative partners in global manufacturing. International Journal of Computer Integrated Manufacturing, 13, 273–285.
Howard, A., Kochhar, A. and Dilworth, J., 1998, Application of a generic manufacturing plan and control system reference architecture to different manufacturing environment. Proceedings of the Institution of Mechanical Engineers, 212, Part B, 381–395.
Hsin, H.C., 2000, The implementation and integration of information systems for production management in manufacturing: an empirical study. International Journal of Computer Integrated Manufacturing, 13, 369–387.

Huaglory, T., 2001, Advanced life-cycle model for complex PD via stage aligned information substitutive concurrency and detour. International Journal of Computer Integrated Manufacturing, 14, 281–303.
Huang, G.Q. and Mak, K.L., 1997, DFX shell: a generic framework for developing Design for X tools. International Journal of Computer Integrated Manufacturing, 13, 271–280.
Huang, G.Q. and Mak, K.L., 2001a, Issues in the development and implementation of web applications for product design and manufacture. International Journal of Computer Integrated Manufacturing, 14, 125–135.
Huang, G.Q. and Mak, K.L., 2001b, Web-integrated manufacturing. International Journal of Computer Integrated Manufacturing, 14, 1–2.
Huang, G.Q., 1996, Design for X. Concurrent Engineering Imperatives (London: Chapman & Hall).
Iams, 1998, Intelligent Assembly Modeling and Simulation: IAMS (http://www.ndim.edrc.cmu.edu/CODES/Project/intelligent.html).
Indrusiak, L.S., 1998, Ambiente de apoio de circuitos integrados baseado no world wide web Master's thesis, Porto Alegre, CPGCC da UFRGS.
Ishii, K., Eubanks, C.F. and Di Marco, P., 1994, Design for product retirement and material lifecycle. Materials and Design, 15, 225–233.
ISO 10303-1:1994, 1994a, Industrial Automation Systems and Integration – Product Data Representation and Exchange – Part 1: Overview and Fundamental Principles.
ISO 10303-11:1994(E), 1994b, Industrial Automation Systems and Integration-Product Data Representation and Exchange – Part 11: The EXPRESS Language Reference Manual.
Jiang, X.Y., Yu, T.B., Zhang, Q.G., Wang, W.S., 2008, Product Quality Design Based on CSCW under Networked Manufacturing. Wireless Communications, Networking and Mobile Computing, 2008. WiCOM '08. 4th International Conference, 12–14 Oct. 2008, Dalian China.
Jiao, J. and Tseng, M.M., 1999, An information modeling framework for product families to support mass customization manufacturing. Annals of the CIRP, 48, 93–98.
Jordan, J., and Michel. F., 1999, Next Generation Manufacturing (NGM), Blue Book of SME, published by the Computer and Automated Systems Association of the Society of Manufacturing Engineers.
Kaiser, G.E., Dossick, S.E. and Yang, J.J., 1997, Process Enactment for the World Wide Web. Technical Report, Columbia University, (http://www.psl.cs.columbia.edu/ozweb.html).
Kalyanapasupathy, V., Lin, E. and Mins, I., 1997, Group technology code generation over the Internet (http://www.isr.umd.edu/Labs/CIM/profiles/lin/docs/gt/).
Kamarthi, S.V., Cohen, P.H. and Demetter, E.C., 1993, CMES: cost minimization and estimation system for sheet metal parts. International Journal of Flexible Automation and Integrated Manufacturing, 1, 69–79.
Kamrani and Salhieh, 2002, Product design for modularity.
Kang, S.-H., Kim, N., Kim, C.-Y., Kim, Y., and O'Grady, P., 1997, Collaborative Design Using the World Wide Web. Technical Report TR 97–02 (Ames: Iowa Internet Laboratory, University of Iowa).
Khedro, T., 1996, Distributed problem-solving approach to collaborative facility engineering. Advances in Engineering Software, 25, 243–252.
Kidd, P.L., 1994, Agile Manufacturing: Forging New Frontiers (Reading: Addison-Wesley).
Kim, C.-Y., Kim, N.K., Kim, Y. and Kang, S.-H., 1998b, Distributed concurrent engineering: Internet-based interactive 3-D dynamic browsing and markup of STEP Data. Journal of Concurrent Engineering, 6, 53–70.
Kim, Y., Choi, Y. and Yoo, S.B., 2001a, Brokering and 3D collaborative viewing of mechanical part models on the Web. International Journal of Computer Integrated Manufacturing, 14, 28–40.
Kim, Y., Kang, S.H., Lee, S.H. and Yoo, S.B., 2001b, A distributed, open, intelligent product data management system. International Journal of Computer Integrated Manufacturing, 14, 224–235.
Klingstam, P., 1999, Overview of simulation tools for computer-aided production engineering. Journal of Computers in Industry, 38, 173–186.

Kosanke, K., 1991, Open system architecture for CIM (CIM-OSA) standards for manufacturing. Proceedings of the International Conference on Computer Integrated Manufacturing, Singapore, 2–4 October, pp. 77–80.

Lumpur, K, 2001, A Java-EE based Product Design Management System. 2010 International Conference on Intelligent Computing and Cognitive Informatics, June 22–23, Malaysia.

Lau, H. and Jiang, B., 1998, A generic integrated system from CAD to CAPP: a neural file-cum-GT approach. International Journal of Computer Integrated Manufacturing Systems, 11, 67–75.

Lazaro, A., De Sam, D. and Engquist, T., 1992, Knowledge-based advisor for the design of sheet metal parts. In: Design for Manufacture, Winter Annual Meeting of the ASME, Anaheim, CA, 8–13 November (New York: ASME), pp. 35–42.

Lee, J.Y. and Kim, K., 1999, Generating alternative interpretations of machining features. International Journal of Advanced Manufacturing Technology, 15, 34–48.

Lee, R.S., Chen, Y., Cheng, H.Y. and Kau, M., 1998, A framework of a concurrent process planning system for mould manufacturing. International Journal of Computer Integrated Manufacturing, 11, 171–190.

Lee, Y.T., 1997, Data Sharing Implementation based on the Information Model for Apparel Pattern Making. Report of NISTIR 6139 (Gaithersburg: National Institute of Standards and Technology), January.

Lei, M., Yang, X., Tseng, M.M. and Yang, S., 1998, Design an intelligent machine center strategy and practice. Mechatronics, 8, 271–285.

Li, H., and Williams, T.J., 1997, Some extensions to the Purde enterprise reference architecture: explaining the Purde architecture and the Purdue methodology using the axioms of engineering design. International journal of computers in industry, 34, 247–259.

Lim, S.H., Juster, N. and Pennington, A.D., 1998, A role interaction approach to support manufacturing enterprise integration. Proceedings of the Institution of Mechanical Engineers, 212 Part B, 439–451.

Lin, A.C., Lin, S.Y. and Cheng, S.B., 1997, Extraction of manufacturing features from a feature based design model. International Journal of Production Research, 35, 3249–3287.

Lin, G.Y.J. and Solberg, J.J., 1992, Integrated shop floor control using autonomous agents. IIE Transactions, 24, 57–71.

Lubich, H.P., 1995, Towards a CSCW Framework for Scientific Cooperation in Europe (Berlin: Springer).

Mamalis, A.G., Malagardis, I. and Kambouris, K., 1996, On-line integration of a process planning module with production scheduling. International of Advanced Manufacturing technology, 12, 330–338.

Mangesh, P.B., Downie, B., Hardwick, M. and Nagi, R., 2000a, Migrating from IGES to STEP: one to one translation of IGES drawing to STEP drafting data. Journal of Computers in Industry, 41, 261–277.

Mangesh, P.B. and Nagi, R., 2000b, STEP based feature extraction from STEP geometry for Agile Manufacturing. Journal of Computers in Industry, 41, 3–24.

Mannisto, T., Peltonen, H., Martio, A. and Sulonen, R., 1998, Modeling generic product structures in STEP. Computer-Aided Design, 30, 1111–1118.

Maropuoalos, P.G., 1995, Review of research in tooling technology: process planning. International Journal of Computer Integrated Manufacturing Systems, 8, 13–20.

Marshall, R. and Leaney, P.G., 1999, A system engineering approach to product modularity. Proceedings of the Institution of Mechanical Engineers, 213, Part B, 847–851.

Matousek, R., 1963, Engineering Design: A Systematic Approach. (transl. from the German by Burton, A.H. and edited for British readers by Zjohnson, D.C.). Blackie, London.

Maturana, F.P. and Norrie, D.H., 1997, Distributed decision-making using the contract net within a mediator architecture. Decision Support Systems, 20, 53–64.

McKay, A., Erens, F. and Bloor, M.S., 1996, Relation product definition and product variety. Research in Engineering Design, 2, 63–80.

Menzel, T. and Geiger, M., 1999, Hybrid simulation objects using fuzzy set theory for simulation of innovative process chains. Proceedings of 2nd International Conference on Intelligent Processing and Manufacturing of Materials (IPMM'99), 1, pp. 641–647.

Meredith, S. and Francis, D., 1998, Journey towards manufacturing agility-exploring the essential sixteen elements of the agile wheel. World Innovation and Strategy Conference, Sydney, Australia, pp. 23–37.

Michel, A. and Clermont, P., 1997, Cost and Cycle Time Reduction in C.E. Local Reacitivity: Concepts and Case Study, Oakland University, Rochester, Michigan, U.S.A. 245–251.

Minneman, S.L., 1991, The social construction of a technical reality: empirical studies of group engineering design practice. PhD thesis, Stanford University.

Morgan, D., Lach, M. and Bushnell, R., 1990, ISDN as enabler for enterprise integration. IEEE Communication Magazine, April, 23–27.

Mottonem, M., Harkonem, J., Belt, P. and Haapasalo, H., 2009, Managerial view on design for manufacturing. Industrial Management & Data, 109, pp. 859–872.

Mujber, T.S., Szecsi, T. and Hashimi, M.S.J., 2004, Virtual reality applications in manufacturing process simulation. Journal of Materials Processing Technology, 155–156, pp. 1834–1838.

Muller, P.C., De Poorter, R., De Jong, J. and Van Engelen, J.M.L., 1996, Using the Internet as a communication infrastructure for lead user involvement in the new product development process. Proceedings of WETICE'96, pp. 220–225.

NASA, 1996, NASA Program Information Systems Mission Services (PrISMS). Technical Report.

Nelson, S.A., Parkinson, M.B. and Papalambros, P.Y., 1999, Multicriteria optimization in product platform design. Proceedings of DETC99: 1999 ASME Design Engineering Technical Conferences, Las Vegas, NV, DETC99/DAC-8676, September, pp. 1–8.

Nielsen, N.P.H. and Holmstrom, J., 1995, Design for speed: a supply chain perspective on design for manufacturability. International Journal of Computer Integrated Manufacturing Systems, 8, 223–228.

Ockerman, J.J., Najjar, L.J., Thompson, J.C., 1999, FAST: future technology for today's industry. Journal of Computers in Industry, 3, 853–864.

Olesen, J., 1992, Concurrent development in manufacturing-based on dispositional mechanisms. PhD thesis, Technical University of Denmark.

OMG (Object Management Group), 1998, CORBA architecture and specification (http://www.omg.org).

Orfali, R., Harkey, D. and Edwards, J., 1999, Client/Server Survival Guide, 3rd edn (New York: Wiley).

Ostwald, P., 1984, Cost Estimating, 2nd edn (Englewood Cliffs: Prentice-Hall).

Pan, J.Y.C. and Tenenbaum, J.M., 1991, An intelligent agent framework for enterprise integration. IEEE Transactions on Systems, Man, and Cybernetics, 21, 1391–1407.

Pan, P.Y., Cheng, K. and Harrison, D.K., 1997, A neural-fuzzy approach to the selection of journal bearings. Proceeding of the 13th National Conference on Manufacturing research, Glasgow, UK, 9–11 September, pp. 427–431.

Pan, P.Y., Cheng, K. and Harrison, D., 1999, Java-based systems: an engineering approach to the implementation of design agility and manufacturing responsiveness. Proceedings of the 14th International Conference on Computer-Aided Production Engineering, Durham, UK, pp. 86–93.

Parunak, V.D., 1987, Distributed artificial intelligence. In: M.N. Huhns (ed), Manufacturing Experience with Contract Net (London: Pitman), pp. 285–310 (http://www.erim.org/_vparunak/YAMS87.pdf).

Pham, D.T., 1998, (http://intell-lab.engi.cf.ac.uk/manufacturing/ipm/ipm.html).

Prante, T., Magerkurth, C. and Streitz, N., 2002, Developing CSCW tools for idea finding – empirical results and implications for design. Computer Supported Cooperative Work, Proceedings of the 2002 ACM conference on Computer supported cooperative work, SESSION: From methods to design, 106–115.

Pratt, D.R., Barham, P.T., Locke, J., Zyda, M., Eastman, B., Moore, T., Biggers, K., Douglass, R., Jacobsen, S., Hollick, M., Granieri, J., Ko, H. and Badler, N., 1994, Insertion of an articulated human into a networked virtual environment. Proceedings of the 1994 AI, Simulation and Planning in High Autonomy Systems Conference, University of Florida, Gainesville, FL, 7–9 December.

Reed, J.A. and Afjeh, A.A., 1998, Developing interactive educational engineering software for the World Wide Web with JAVA. Computer in Education, 30, 183–194.

Rolstadas, A., 1991, ESPRIT Basic Research Action No. 3143 – FOF production theory. International Journal of Computers in Industry, 16, 129–139.

Roy, U. and Cargian, M., 1997, Tele-manufacturing: Object Slicing For Rapid Prototyping on the Internet, Oakland university, Rochester, Michigan, U.S.A. 520–524.

Saygin, C. and Kilic, S.E., 1999, Integrating flexible process plans with scheduling in manufacturing systems. International Journal of Advanced Manufacturing Technology, 15, 268–280.

Schaefer, O., 1994, Metadatabase integration in shop floor real-time environment. Diploma of Engineering thesis, Munich Technological University.

Schenck, D. and Wilson, P., 1994, Information Modeling the EXPRESS Way (New York: Oxford University Press).

Shackleford, W. and Proctor, F., 1998, JAVA-based tools for development and diagnosis of real-time control system. Proceedings of the 1998 ASME Design Engineering Technical Conference, Paper N0. DETC/CIE-5530.

Shah, J.J. and Mantyla, M., 1997, Parametric and Feature based CAD/CAM (London: Chapman & Hall).

Shaw, M.J., 1988, Dynamic scheduling in cellular manufacturing systems: a framework for networked decision making. Journal of Manufacturing Systems, 7, 83–94.

Shaw, M.J. and Whinston, A.B., 1983, Distributed Planning in Cellular Manufacturing Systems. Technical Report.

Shea, K., 2001, An approach to multi-objective optimization for parametric synthesis. International Conference on Engineering Design, ICED01 Glasgow, 21–23 August.

Shen, W.M. and Norrie, D.H., 1999, Agent-based systems for intelligent manufacturing: a state of-the-art survey. Knowledge and Information Systems, 1, 129–156.

Shi, J., Huang, G.Q., Mak, K.L., 1995, Web-based design for assembly, Proceedings of SPIE- The International Society for Optical Engineering Modeling, Simulation, and Control Technologies for Manufacturing. Philadelphia, PA, Oct 25–26, pp. 192–195.

Shin, J.H., Park, S.S., Ju, C.J. and Cho, H.B., 2003, CORBA-based integration framework for distributed shop floor control, Computers & Industrial Engineering, Volume 45, Issue 3, pp. 457–474.

Shoham, Y., 1993, Agent oriented programming. Artificial Intelligence, 60, 51–92.

Sikora, R. and Shaw, M.J., 1997, Coordination mechanisms for multi-agent manufacturing systems: application to integrated manufacturing scheduling. IEEE Transactions on Engineering Management, 44, 175–187.

Sikora, R. and Shaw, M.J., 1998, A multi-agent framework for the coordination and integration of information systems. Management Science, 44, S75–S78.

Simone, C. and Divitini, M., 1995, A notation for malleable and interoperable coordination mechanisms for CSCW systems. Proceedings of COOCS'95, Milpitas, CA, 13–16 August.

Smith, A., 1998, Metaworlds and virtual space-towards the collaborative virtual design studio. Proceedings of the Design Computing on the Net's 98, University of Sydney, Sydney, (http://www.arch.usyd.edu.au/kcdc/journal/vol1/dcnet/stream1/paper4/).

Smith, C. and Wright, P., 1996, CyberCut: a WWW based design to manufacturing tool (http://kingkong.me.berkeley.edu/_smythe/school/cybercut.html).

Smunt, T.L., 1999, Log-linear and non-log-linear learning curve models for production research and cost estimation. International Journal of Production Research, 37(17), 3901–3911.

Sprow, E., 1992, Chrysler's concurrent engineering challenge. Manufacturing Engineering, 108, 35–42.

Sriram, R.D., Logcher, R. and Fukuda, S., 1991, Computer Aided Cooperative Product Development (New York: Springer).

Stampe, D., Roehl, B. and Eagan, J., 1993, Virtual Reality Creations (Corte Madera: Waite).

Stanley, D.P., 1996, Application of product life cycle design guidelines at Navistar. Proceedings of the Air and Waste Management Association's 89th Annual Meeting and Exhibition, Navista, TN, 23–28 June.

Steve Vinoski., 1997, CORBA: Integrating Diverse Applications within Heterogeneous Environments. IEEE Communications Magazine, 35, 46–55.

Steves, M.P. and Knutilla, A., 1999, Collaboration technologies for global manufacturing. Proceedings of the ASME International Mechanical Engineering Congress and Exposition (IMECE): Symposium on Manufacturing Logistics in a Global Economy, Nashville, TN.

Su, D.Z. and Amin, N., 2001, A CGI-based approach for remotely executing a large program for integration of design and manufacture over the Internet. International Journal of Computer Integrated Manufacturing, 14, 55–65.

Suppapitnarm, A., Seffen, K.A., Parks, G.T. and Clarkson, P.J., 2000, A simulated annealing algorithm for multiobjective optimization. Engineering Optimization, 33, 59–85.

Swank, M. and Kittel, D., 1997, World Wide Web database developers guide on the Web (http://dev.mcp.com:8001/about/imprints/samsnet/).

Szykman, S., Sriram, R.D., Bochenek, C., Racz, J.W. and Senfaute, J., 2000, Design Repositories: NextGeneration Engineering Design Databases. Journal of National Institute of Standard and Technology. On the Web: http://www.mel.nist.gov/msidlibrary/summary/pubs00.htm

Tamas, K., Lozano, S., Guerrero, F. and Onieva, L., 2000, A flexible costing system for flexible manufacturing systems using activity based costing. International Journal of Production Research, 38, 1615–1630.

Tharumarajah, A. and Wells, A.J., 1997, A behavior-based approach to scheduling in distributed manufacturing systems. Integrated Computer Aided Engineering, 4, 235–249.

Toh, K.T.K., Newman, S.T. and Bell, R., 1998, An information system architecture for small metalworking companies. Proceedings of the Institution of Mechanical Engineers, 212 Part B, 87–103.

Tomarchio, O., Vita, L. and Puliafito, A., 2001, Mobile agents and legacy systems: how to integrate alternative communication paradigms. 3rd International Workshop on Mobile Agents for Telecommunication Applications (MATA2001), August, Montreal, Canada.

Tseng, M.M. and Du, X., 1998, Design by customer for mass customisation products. CIRP Manufacturing Technology, 17, 103–106.

Tsukada, T.K. and Shin, K.G., 1998, Distributed tool sharing in flexible manufacturing systems. IEEE Transactions on Robotics and Automation, 14, 379–389.

Tu, Y.L., 1996, A Framework for Production Planning and Control in a Virtual OKP Company. Technical Papers of the North American Manufacturing Research Institution of the SME, pp. 121–126.

Tu, Y.L., 1997, Production planning and control in a virtual OKP company. International Journal of Computers in Industry, 34, 271–283.

Tu, Y.L. and Xie, S.Q., 2001, An Information Modelling Framework for Sheet Metal Parts Intelligent and Concurrent Design and Manufacturing. International Journal of Advanced Manufacturing Technology, 18(12), 873–883.

Tu, Y.L. and Xie, S.Q., 2000, A WWW based integrated PD information management system. IFAC, MIMO2000, July, Greece.

Tu, Y.L., Xie, S.Q. and Liu, J., 2000, Virtual PD for one-of-a-kind. IFAC, 14–18 June, Aachen.

Tu, Y.L., Xie, S.Q. and Zhou, Z.D., 2001, An information modelling framework to concurrent product design and manufacturing. IAM'2001, 17–21 March (UAE: American University in Dubai).

Urban, S.D., Ayyaswamy, K., Fu, L., Shah, J.J. and Liang, J., 1999, Integrated PD data environment: data sharing across diverse engineering applications. International Journal of Computer Integrated Manufacturing, 12, 525–540.

Valckenaers, P., Bongearts, L. andWyns, J., 1998, Planning systems in the next century (II). Journal of Computers in Industry, 34, 239–245.

Wagner, R., Castanotto, G. and Goldberg, K., 1997, FixtureNet: interactive computer aided design via theWWW. International Journal of Human–Computer Studies, 46(6) (http://www.teamster.usc.edu/fixture/paper/wcg1-6.html).

Wang, C.H. and Bourne, D.A., 1995, Using features and their constraints to aid process planning of sheet metal parts. Proceedings of IEEE International Conference on Robotics and Automation, pp. 1020–1026.

Wang, L.C. and Lin, S.K., 2009, A multi-agent based agile manufacturing planning and control system. Computers & Industrial EngineerinG, 57, 620–640.

Wang, L.H. and Nee, A.Y.C., 2009, Collaborative Design and Planning for Digital Manufacturing. Springer, ISBN: 978-1-84882-286-3.

Wang, Q.Y. and Tian, L., 2007, A systematic approach for 3D VRML model-based assembly in Web-based product design. The International Journal of Advanced Manufacturing Technology, 33(7–8), 819–836, DOI: 10.1007/s00170-006-0494-6.

Weyrich, M. and Drews, P., 1999, An interactive environment for virtual manufacturing: the virtual workbench. Journal of Computers in Industry, 38, 5–15.

Whiteside, R.A., Pancerella, C.M., Klevgard, P.A., 1997, A CORBA-Based Manufacturing Environment. hicss, vol. 1, pp. 34, 30th Hawaii International Conference on System Sciences (HICSS) Volume 1: Software Technology and Architecture.

Will, P., 1996, Active catalogs project home page (http://www.isi.edu/active-catalog/index.html).

Winbourne, J.P. and Toolsie, G.M., 1991, Computer-aided tool cost estimating (CATE). Proceedings of the 1991 ASME International Computers in Engineering Conference and Exposition, Computers in Engineering (New York), vol. 1, pp. 617–621.

Wong, C.C., Veeramani, D. and Chalermdamrichai, V., 1996, QUESTER: a computer integrated system for virtual shopping through the Internet. 5th IE Research Conference Proceedings, Minneapolis, MN, 18–19 May, pp. 758–763.

Wong, J.P., Imam, I.N., Kamrani, A.K., Parsaei, H.R. and Tayyari, F., 1992, A Totally Integrated Manufacturing Cost Estimating System (TIMCES). In: H.R. Parsaei and A. Mital (eds), Economic Aspects of Advanced Production and Manufacturing Systems (London: Chapman & Hall).

Wright, D.T. and Burns, N.D., 1997, Rapid prototyping cellular systems for virtual global enterprises. Proceedings of the 5th International Conference on FACTORY 2000, 2–4 April, pp. 423–428.

Wu, R.R. and Zhang, Y.Y., 1998, A CAPP framework and its methodology. International Journal of Advanced Manufacturing Technology, 14, 155–260.

Xie, S.Q., Tu, Y.L., Liu, J.Q. and Zhou, Z.D., 2001, An integrated and concurrent approach for compound sheet metal cutting and punching. International Journal of Production Research 39(6), 1095–1112.

Xie, S.Q. and Xu, X., 2008, A STEP-compliant process planning system for compound sheet metal machining. International Journal of Production Research, 46(1), 25–50.

Xie, S.Q. and Tu, Y.L., 2000, An Integrated CAD/CAPP/CAM System for Compound Sheet Metal Cutting and Punching. 7th IFAC Symposium on Automated Systems Based on Human Skill, Joint Design of Technology and Organisation, June 15–17, 2000, Aachen, Germany.

Xue, D., Yang, H. and Tu, Y.L., 2006, Modeling of evolutionary design database. Journal of Computing and Information Science in Engineering, Transactions of the ASME, 6(1), 22–32.

Yang, Y.-S., Jang, B.-S., Cho, S., Ruy, W.-S. and Jung, H.-S., 2001, Managing approximation in multidisciplinary design optimization. International Conference on Engineering Design, ICED01 Glasgow, 21–23 August.

Yang, C.O. and Pei, H.N., 1999, Developing a STEP-based integration environment to evaluate the impact of an engineering change onMRP. International Journal of Advanced Manufacturing Technology, 15, 769–779.

Young, R.E., Greef, A. and O'Grady, P., 1992, An artificial intelligent-based constraint network system for concurrent engineering. International Journal of Production Research, 30(7), 1715–1735.

Zhang, H.C. and Kuo, T.C., 1997, Environmentally conscious design and manufacturing: concepts, applications, and prospectives. Proceedings of the 1997 ASME International Mechanical Engineering Congress and Exposition, Dallas, TX, 16–21 November, pp. 179–190.

Zhang, Y.P., Zhang, C.C. and Wang, H.P.B., 2000, An Internet-based STEP data exchange framework for virtual enterprises. Journal of Computers in Industry, 41, 51–63.

Zhao, W., and Liu, J.K., 2008, OWL/SWRL representation methodology for EXPRESS-driven product information model: Part I. Implementation methodology. Computers in Industry, 59, 580–589.

Zhou, Q. and Besant, C.B., 1999, Information management in production planning for a virtual enterprise. International Journal of Production Research, 37, 207–218.

Zhou, B.H., Xi, L.F. and Yu, C.M., 2006, DCOM and MMS–based control software architecture for automated manufacturing system. The International Journal of Advanced Manufacturing Technology, 27, 951–959.

Chapter 3
Product Modelling in Support of Rapid OKP Development: a Review

Abstract For OKP companies, how to better manage and record previous product development (PD) knowledge has become a core issue to address in order to improve the PD process, cut down development cost and reduce lead time. In recent years, considerable effort has been placed on developing new enabling technologies for OKP companies to achieve high quality and productivity, and to quickly respond to the changing market to meet customer requirements. Product modelling is a pivotal activity in the PD process. Well-defined product models organise product data, production information and knowledge to satisfy the requirements of rapid changes in the PD environment. In this chapter, a comprehensive review is carried out of recent developments in product modelling technology. Four types of product modelling methodologies are discussed in detail. Two object-oriented product modelling methods, including STEP-based product modelling and UML-based product modelling are reviewed and compared. The research gaps and issues are identified. A generic product modelling framework is proposed to implement product modelling into the current integrated manufacturing environment. Future research trends in product modelling are also discussed.

3.1 Introduction

Today, most manufacturing enterprises are facing intensive competition from the global market. How to better manage and enhance their PD processes has become a core issue for every manufacturing enterprise. It is essential to seek an appropriate way to improve PD processes, which include shortening product life cycle, cutting development cost, reducing lead time, achieving high quality and productivity, and quickly responding to market changes to meet customer requirements (Xie *et al.* 2003, Tu *et al.* 2006). Effective utilisation of product information is essential to achieve the above requirements in the new manufacturing environment. Product information sharing and exchange between various stages of PD processes can lead to optimal solutions for developing a product, and has become of paramount importance for a company to succeed.

In recent years, various techniques have been developed in support of the integration of product information (Lee 1997, Lee and Kim 1999, Xie *et al.* 2001, 2003, Lee *et al.* 2004). However, in manufacturing industry at present there are many different types and versions of software tools employed to support PD activities such as product design, process planning and manufacturing. Different systems may use different formats for information representation. This makes it very hard to integrate and share product information and knowledge even within a single company. Product modelling technology has been widely utilised in PD processes, and plays a fundamental role in supporting effective information and knowledge sharing.

The product model is an entire product information database, which describes the product completely and unambiguously. It contains two general types of information: physical product design information represented by the design model; and process information, represented by the process and data model (http://pdesinc.aticorp.org/glossary.html). Different manufacturing systems, such as computer aided (CAX) systems, and product data management (PDM) systems, can employ the product model as a proper information carrier to support various PD activities. The modelling processes are divided into different stages (Schenck 1994, Lee *et al.* 2004). Modelling methodologies and technologies as well as support facilities, software, and modelling technologies are defined and prepared. In the second stage, a product information model, or product data model, is built up. The product information model is an important concept in the modelling processes. It is defined as a mechanism to organise and structure product data for the various PD activities. The overall model contents and structure defined in the previous stage are detailed and represented by different modelling technologies. The product information model can be further utilised as a basic element to develop new manufacturing systems. In the last stage, the product information model constructed in stage two is implemented to generate the product model. The product model stores the actual data on a suitable media, such as neutral file or database. Product models are directly employed by manufacturing systems to support product information manipulation. Thus, the product model can be considered the core resource supporting PD processes.

Product modelling is an indispensable information technology in PD processes, and has been continuously developed and explored to adapt to changes in the current manufacturing environment. In the beginning, the product model can represent product data only in one aspect. Other systems cannot directly utilise it as the convention work must be done in advance. In the computer integrated manufacturing (CIM) environment, this type of product model may result in an "information island", and missed data and repeated data may occur. Data exchange barriers exist between different systems due to limitations of the conventional product model. Therefore, product modelling technologies must be optimised to adapt to the integrated manufacturing environment. Different systems can use the same model with the same representations to represent product data. In this way, the effectiveness and efficiency of product data utilisation throughout the entire PD process can be improved. Thus, it is necessary to analyse current research and implementation of product modelling technologies, and to incorporate new technologies.

The product modelling concept was first recognised at an early stage when trying to integrate systems in support of various stages of the PD process, such as product design, process planning and manufacturing (Brooks and Greenway 1995, Zhang *et al.* 1995, Liu *et al.* 2000). There were no standard modelling methods and the focus was placed on how to connect systems practically. Later, with the development of feature-based modelling methodologies and object-oriented programming methods, product modelling was identified as a research area, and gained more attention after "integration" became crucial to PD processes (Xue and Dong 1994, Chen and Wei 1997). The development of STEP has moved product modelling into a new area, which enables one to integrate various PD systems using STEP as a standard (Lin *et al.* 2005, Zha 2006). STEP also defines a formal modelling language, EXPRESS, to support represention of the product data, and is the basis of the entire STEP standard. Abundant standardised modelling resources are defined in STEP using EXPRESS. Recent developments in product modelling have seen a wide use of STEP as the basic modelling infrastructure. In this chapter, the focus is on a review of recent developments in product modelling after STEP was developed, with the intention of identifying research gaps and issues in the area. The main goal is to develop a generic product modelling framework that is able to model product information throughout the whole product life cycle in different manufacturing environments.

The chapter is organised as follows. Section 3.2 discusses the four basic product modelling methodologies. Section 3.3 reviews recent developments in the most popular information modelling technology, object-oriented product modelling. Section 3.4 introduces recent developments in product modelling methods. Section 3.5 proposes a generic product modelling framework, which can support product modelling in the current integrated PD environement. Conclusions and future developments in product modelling are presented in Section 3.6.

3.2 Product Modelling Methodologies

Many product modelling methodologies have been reported in the literature and they can be generically categorised as follows: solid product modelling, feature-based product modelling, knowledge-based product modelling, and integrated product modelling. In this chapter, the four modelling methodologies are discussed and their details compared.

3.2.1 Solid Product Modelling

Solid product modelling is a technique used to unambiguously represent the product information of 3D shaped objects. The most common solid product modelling methods are boundary representation (B-Rep) and constructive solid geometry (CSG).

The B-Rep modelling method considers the solid object as being bounded by a number of faces, which are bounded by a number of edges, which in turn are bounded by two vertices. The main advantage of B-Rep modelling is that product geometric information about the faces, edges and vertices is explicitly stored in the representation, which enables fast display. CSG modelling constructs the product based on a collection of primitive solids, such as cubes, cylinders and spheres. They are combined by applying set operations (union, difference and intersection) and represented by a binary tree called the CSG tree. The main advantage of CSG modelling is that it can define complex objects with a compact and simple data structure. It is possible to covert between the two solid modelling methods and there is literature discussing the integration of B-Rep and CSG (Pilz and Kamel 1989, Kodiyalam *et al.* 1990, Shapiro and Kossler 1993) (refer to Appendix A.2).

The solid modelling methods are very mature now and the solid product model is widely used in different PD phases. However, since solid modelling approaches such as CSG and B-Rep merely focus on building geometric models with low-level geometric details, they are quite different from how a human designer designs a product (Chen and Wei 1997). Even if some models contain parts of the technical product data, such as dimensions and tolerances, the solid model still lacks the mechanism to present other necessary information for the entire PD life cycle, especially the manufacturing data, product functionality information, *etc.* To solve this problem, the feature concept was generated for use in product modelling processes.

3.2.2 Feature-based Product Modelling

A "feature" can be viewed as information sets that refer to aspects of form or other attributes of a part, such that these sets can be used in reasoning about the design, performance or manufacture of the part or assemblies they constitute (Salomons *et al.* 1993). From the definition, it can be seen that the "feature" naturally associates product geometric data with other types of data, such as the machining feature containing manufacturing information, the assembly feature consisting of assembly process data, *etc.* Thus, it is believed that feature-based modelling has several advantages over conventional CSG and B-Rep approaches (Chen and Wei 1997), such as capturing design intents, associating functionality with product geometry, and working on high level shapes instead of geometric details. It was first introduced in computer aided process planning (CAPP) applications using shape patterns, such as pocket, step, *etc.*, which present rich information to the process planner because there are only a limited number of ways to machine them. Using features can provide sufficient information, both geometric and non-geometric, to speed up the PD process.

The feature, in general, can be distinguished as a design feature or a manufacturing feature. The design feature, represented as mechanical components and mechanisms, can support the designer to communicate with a design system easily according to the design intent, and to satisfy the design functions. The manufacturing feature, which is more important, is the interpretation and the combination of "form features from the viewpoint of manufacturing, assembling, and inspection" (Krause *et al.* 1993). An intelligent and comprehensive feature-based PD can be formed to support CAD/CAPP/CAM integration through cooperation with feature-related techniques, such as feature extraction, validation, mapping etc. Due to this, feature-based product modelling is now quite popular.

An abundance of research into feature-based product modelling has been reported in the literature. Primitive efforts used the solid modelling method, such as CSG and B-Rep, to present form features (Pratt 1988, Wingard 1991, Yang and Li 1992). Recently more effort has been concentrated on modelling non-geometric product data based on the feature concept. Xue and Dong (1994) introduced the design feature coding system and the manufacturing feature coding system in a feature-based concurrent design system. The design functions, product geometry and production operations are the modelling criteria used for the coding systems. Chep and Tricarico (1999) specified features as form feature, precision feature, technological feature and manufacturing feature. The first three features were called primary features and were used to generate the manufacturing features (with certain constraints). All feature data were extracted from CAD systems and manipulated in a generative CAPP (GCAPP) system. Van Holland and Bronsvoort (2000) and Chang and Ning (2001) both discussed the use of assembly features to support assembly product design and assembly processing. Gao *et al.* (2000) proposed an intelligent feature-based product modelling system in which the product model used an application feature and a system feature to present geometric information, non-geometric attributes and knowledge related to the product.

The above examples show a wide range of implementations of feature modelling. The feature objects can provide high density information sets for all PD activities. Ideally, features can logically integrate all kinds of product information at the product design stage. Feature-based modelling can be considered an extension of the solid modelling method and plays a more important role in the CIM environment.

3.2.3 Knowledge-based Product Modelling

Knowledge-based product modelling is characterised by the use of artificial intelligence technologies such as reasoning mechanisms, constraints, *etc.*, in the modelling processes. One important characteristic of knowledge integrated models is the ability to build abstract taxonomies of products or processes as objects and to store knowledge about former designs, possible alternative parts in an assembly or the abilities and validity of processes used for a specific class of products (Krause *et al.* 1993). The use of this knowledge can greatly reduce unnecessary re-analysis, re-design, and re-planning, simplify modelling tasks, and ensure modelling quality. The knowledge-based product model can be used in a knowledge-based system (KBS) for such applications as product design and process planning.

Salustri (1996) presented a formal theory for knowledge-based product model representation, namely Axiomatic Information Model for Design (AIM-D). Based on axiomatic set theory, the AIM-D provides a formal basis for the notions essential to design, such as features, parts and assemblies, systems, and subassemblies. It is not a product modelling system *per se*, but rather a logical product structure whose axioms define criteria for determining the logical validity of product models. Zhang *et al.* (2005) proposed the concept of knowledge component, which includes product configuration rules, constraints and knowledge reasoning mechanisms. The authors used this intelligent and integrated object to build up a knowledge representation model for product configuration purposes.

A key issue for knowledge-based product modelling is the product knowledge representation. Most of the knowledge involved in product process development can be represented in terms of production rules. The rule states the knowledge in two parts. The first part of the rule, called the antecedent or left-hand-side (LHS), expresses a situation, while the second part, called the consequent or right-hand-side (RHS), states the actions or conclusions that apply if the situation is true (Juri *et al.* 1990). A typical rule can be represented as "IF-THEN" actions. Other knowledge representation methods include frame-based knowledge representations and case-based knowledge representations. These three methods are often used in an integrated way to support complex manufacturing knowledge representation. Juri *et al.* (1990) integrated frame-based and rule-based knowledge representation in a feature-based modelling system. Lou *et al.* (2004) used a frame-rule structure to process the application knowledge in a KBS mould product design system. The frame represented the objects in project modelling and the rule represented field knowledge. The rules were integrated in a related frame. In addition, this mould-based KBS consisted of a cased library to support case-based reasoning. All these examples offer an efficient way to evolve a CAX integrated environment.

3.2.4 Integrated Product Modelling

The expression "integrated" denotes semantic integration, which means basic functionality offered by computer applications for design, planning and production extended with the semantics needed to support real enterprise integration (Gielingh *et al.* 1991). Thus, an integrated product modelling methodology can be a functional combination of different modelling methods. All kinds of product data, including geometric product data, feature information and product knowledge, can be modelled using the integrated modelling method and stored in a related integrated product model. In addition, the integrated modelling method can present product data at different stages of the product life cycle. Therefore, the integrated product model can provide sufficient product data to support the various computer application systems.

Much effort has been focused on integrating the feature concept and the knowledge base within the product model. This is because the feature concept can provide a mechanism to manipulate geometric information as well as non-geometric functional information; the knowledge base can support and make full use of the high level product data associated with feature concepts. Thus, PD can process in an informative and intelligent environment, and achieve high effectiveness and efficiency. Chen and Wei (1997) developed a feature-based design system for net shape product manufacturing. In the design evaluation module of the system, a feature-based product model (the form feature), a data model for representing the technique product data (the manufacturing feature) and a knowledge model were tightly merged together to ensure validation of the net shape product design. Song *et al.* (1999) used feature and manufacturing knowledge stored in an integrated product model to support the real-time concurrent product and its production process design system. Roy and Kodkani (1999) described an integrated collaborative product design system which operates under a single internet interface and consists of several servers geographically dispersed. The semantic and syntactic content of the product model can be accessed and manipulated through the single Internet interface to demonstrate the system's open architecture and interoperability. Brunetti and Golob (2000) presented a feature-based integrated product model that can handle conceptual design information. Through this model, the early design can link with downstream part and assembly modelling.

Integrated product models are widely used in PD activities. Baum and Ramakrishnan (1997) reported research on new 3D product modelling technology in a shipbuilding application. This modelling approach associates geometric information and non-geometric functionals based on the integrated database system. The new 3D product model can satisfy the information requirements for concurrent ship building. Chen and Jin (2005) analysed the characteristics of multidisciplinary collaborative design (MCD) of a product and proposed a new MCD-oriented product information model (MCDPM) that integrates physical structure, design semantics and collaboration management data. Oetter *et al.* (2004) discussed the merits of using an integrated product model that provided product manufacturing and life cycle information in the ship building industry. Horvath *et al.* (2000) discussed how to represent human design intent in the integrated product model. Lin *et al.* (2005) presented a framework for an integrated product model to combine both product static information and dynamic evolvement information to support data integration and process integration. The software prototype, SCC-ProModelling system was used for aerospace PD.

The above research used the integrated product modelling approach or the integrated product model to support different PD stages in different applications. The integrated product modelling method is therefore considered the most promising modelling method in today's CIMS environment.

These four methodologies take different views to structure the product data and represent them in proper formats. They appeared in the sequence listed above. The development history of these modelling methodologies indicates the changes in the kind of information required by the manufacturing environment. The primitive modelling methodologies focused on modelling the low level, detailed product design data, *e.g.*, the CAD 3D model. Because the current manufacturing environment requires comprehensive ranges of data including design data, manufacturing resources information, production knowledge, *etc.*, a product model with only geometric data cannot fully satisfy the requirements in the current manufacturing practice, and will lead to problems in the PD processes. Modelling of high level, semantic product data into a functionally integrated product model is now required. Appendix A.2 lists some of the papers reviewed in this chapter. The modelling methodologies, data format of the models, modelling purposes, tools used and the advantages of the product models are shown.

3.3 Recent Developments in Product Modelling

Object-oriented (OO) is an important concept in the engineering domain and has been widely used in many applications. The Object-oriented technology emerged in the late 1980s and was applied initially in the object-oriented programming language. The mainstream programming languages, such as C++, Java, *etc.*, are now all Object-oriented. The feature characteristic of the Object-oriented concept and technologies is the use of an "object" as the basic element to present a real-world thing or concept. The advantage of this is its ability to encapsulate the data, including its structure and its manipulations, into the object. Message transmissions among objects are the ways to invoke operators. No direct use of an operator within an object is applied, which can enhance information security. The object also has the character of polymorphism, which means the use of various names to point to the same object in order to reduce the complexity in system development and to increase the flexibility of the system. A set of objects sharing a common structure and common behaviour is defined as a "class". The class has the characteristic of inheritance, which means the common attributes and methods of a subclass are collected from its superclass. Inheritance implies that all subclasses can have data or methods transferred from the superclass, which can support the reuse of data and omit data repetition. The object-oriented methodology offers advantages such as process flow simplification, cost reduction and resources waste reduction (Sanchez and Choohineh 1997), design standardisation and reuse (Linthicam 1995, Amaitik and Kiliç 2007), improvement of efficiency and effectiveness, and repetitive sources reduction (Allen 1996). Due to this, the object-oriented paradigm has been widely extended to many engineering domains.

Object-oriented technology merged into the product modelling domain in the early 1990s, and was termed object-oriented modelling. Since then, it has become the most popular modelling method. Rumbaugh *et al.* (1991) described how the object-oriented method can be used for modelling. The authors used three models to describe a system (product): the *object* model, describing the objects in the system and their relationships; the *dynamic* model, describing the interactions of the system; and the *functional* model, describing the data transformations within the system. Since an "object" can consist of not only information but also functions, Object-oriented modelling techniques have many advantages over other methodologies. Thus in PD processes, the object-oriented method is the most popular modelling methodology, not merely for modelling product, but also for modelling production processes and PD systems.

3.3.1 Object-oriented Product Modelling

Object-oriented product modelling combines object-oriented concepts and technologies with the various modelling methodologies. All the advantages of object-oriented are naturally brought to this product modelling process. Hvam (1999) demonstrated an entire procedure for building up an object-oriented product model. Within the seven sequential phases, object-oriented technologies, such as object-oriented analysing, object-oriented design, and object-oriented programming are engaged as the key elements for constructing the product model. With the help of object-oriented product modelling technology, the modelling tasks become more precise, effective and efficient. Object-oriented modelling technologies are suitable tools for the implementation of product modelling methodologies, such as feature-based, knowledge-based and integrated modelling methodologies. These object-oriented technologies include object-oriented programming, object-oriented database, object-oriented database management system, and some specific object-oriented modelling languages, such as the Unified Modelling Language (UML), EXPRESS language (ISO 10303-11: 1994(E)), *etc.* Much implementation effort has been applied in this field.

Gu and Zhang (1994) presented an object-oriented product model, named OOPM, which was used in an object-oriented process planning system. The product model contained three parts: CSG tree object (part geometry), feature object (technical information), and solid primitive object. Chen and Wei (1997) merged product design and process knowledge into an object-oriented feature-based product model for concurrent net shape manufacturing. Chep and Tricarico (1999) discussed the object-oriented analysis and design approach for modelling product manufacturing features; they used an object-oriented system analysis method to specify eight models to represent the manufacturing features based on interactions between the primary features extracted from the CAD model. The manufacturing feature model could be employed by a GCAPP system to automatically generate the machining operation sequences. Sormaz and Khoshnevis (1997) recommended an object-oriented approach to model the process planning knowledge in a hierarchical structure. Chen (1999) presented an object-oriented infrastructure including a product model, a knowledge model, and a data model. These models were represented as a common schema object and developed by object-oriented modelling techniques. They were then integrated for implementation in a computer-aided concurrent net shape product and process development environment.

It can be seen from the above efforts that object-oriented modelling methods can form the product data into a well defined structure. All the data are more easily utilised and modified. An important research direction for object-oriented product modelling is the presentation of product data in a standardised format to make the product model more adaptable. Two major achievements have been obtained: the first is the development of STEP-based product modelling; the other is UML-based product modelling.

3.3.2 STEP-based Product Modelling

STEP is the acronym for Standard for the Exchange of Product model data, which are serials of International standards codified as ISO 10303. STEP is being developed to enable complete and correct interchange of product data between various CAD/CAM systems, other manufacturing related software, and vendors (Jordon 1994). It was formally published in 1994 and it is now widely accepted by academic researchers, system vendors, and industrial experts. The scopes of STEP are (ISO 1994-1 : 1994(E)):

- the representation of product information, including components and assemblies;
- the exchange of product data, including storing, transferring, accessing and archiving.

STEP defines a formal modelling language EXPRESS to represent product data. This is the basis of the entire STEP standard. EXPRESS is a data description language (DDL), and can precisely describe the data structure (including data type, data relationship, *etc.*) but not the data value. The entity-attribute structure is utilised for modelling. EXPRESS supports the utilisation of functions and constraints. EXPRESS is suitable for building up the product information model, especially for complex modelling tasks. It also has a graphical representation format, namely EXPRESS-G. These two are compatible and can be converted into each other.

Abundant standardised modelling resources are defined in STEP using EXPRESS. These resources can be classified into two categories, integrated resources and application protocols (APs). The integrated resources are a part of an International Standard that defines a group of schemas used as the basis for product data representation. The APs are a part of ISO 10303 which specifies an application interpreted model satisfying the scope and information requirements for a specific application. The two main approaches of implementing STEP-based product modelling are modelling based on APs, and modelling based on integrated resources.

3.3.2.1 Modelling Based on APs

APs provide well defined information models for a specific application. Thus, AP users can directly use this standardised model to organise the product data in this domain. For example, the AP203 (ISO 10303-203 : 1994(E)), which is for configuration controlled 3D designs of mechanical parts and assemblies, has been accepted by many CAD systems, such as Pro/Engineer, Solidworks, as one kind of formal product geometric data representation format. Shaharoun *et al.* (1998) employed

AP203 as the basis for geometrical data representation in a product data model for developing plastic products. Hur *et al.* (2002) used STEP AP203-based product models as the geometric product data input basis for process planning of a hybrid rapid prototyping system. This paper showed that manufacturing computer systems, which are compatible with the AP203, can naturally integrate with CAD product data in the AP203 format to form a CIM environment.

AP214 (ISO 10303-214: 2001(E)) and AP224 (ISO 10303-224: 2001(E)) are widely used as well. AP214 is for automotive mechanical product design and planning processes, but it can also be partially used in aviation manufacturing or other heavy industries. Meng *et al.* (1997) presented a STEP-based feature modelling system, based on AP214. Chin *et al.* (2002) also proposed an integrated product model (IPM) for concurrent engineering development. The four components of IPM (a geometric model, a feature model, a product definition model, and an integrated core model) were built on the basis of the AP214 framework. Hoehn *et al.* (2003) presented a product model based on AP214 for gear units. The model covered all data aspects of the gear product and its manufacturing processes. AP224 defines many machining features to support process planning. This part is naturally associated with the feature-based modelling methodology to support the presentation of product manufacturing data for process planning. Usher (1996) presented a STEP-based object-oriented product model based on AP224. This model supported the CAPP analysis process. Sharma and Gao (2002) developed a prototype STEP-enabled manufacturing planning system (SMPS). The system employed a feature-based product model based on AP224. The AP224-based product model can automatically support process plan generation in the SMPS without any user interaction. Xie *et al.* (2001) proposed a WWW-based PD platform for intelligent and concurrent sheet metal products design and manufacturing. A STEP-based object-oriented modelling method was used in their approach.

Much effort has been applied to use multiple APs to support modelling. These efforts discussed product models based on AP203 and AP224, which can support CAD/CAPP integration (Brooks and Greenway 1995, Zhang *et al.* 1995, Liu *et al.* 2000). AP203 provides a standardised representation method for the low level product geometric data, and AP 224 supports the representation of high level product feature data. Those two kinds of data are compatible and can be combined to support sufficient information for the whole PD processes. Liang *et al.* (1999) employed multiple STEP APs (AP203, 209, 214, and 224) to structure an integrated product data-sharing environment to support product information sharing and transferring. The use of multiple APs can easily model product data in different engineering domains or applications to support complex PD processes.

3.3.2.2 Modelling Based on Integrated Resources

Integrated resources are the generic information model about the basic product data. Due to this, a product model based on integrated resources has great adaptability. Gu and Chan (1995) developed a STEP-based generic product modelling (GPM) system. GPM is structured based on five libraries: product and version; product definition and relationship; geometric item; material; and tolerance. Integrated resources were used to model these five libraries. GPM supports a complete product information structure with standard data representation to achieve the requirements of the concurrent engineering environment. This model can adapt to different application uses. In some cases, the actual applications are beyond the scope of the currently available STEP APs. The STEP integrated resources can be used to construct new STEP-compatible modelling resources to satisfy the modelling requirement for a particular application. Tang *et al.* (2001) presented a STEP-based product modelling system for concurrent stamped part and die development, which consists of a three-layer architecture (application, logical and physical layers). They also proposed an integrated product model for concurrent stamped part and die development called DPIIM (Die&Product Integrated Information Model) whose key feature is the connection of integrated resources to provide an integrated information structure for creating a standard data model for concurrent stamped part and die development. Six EXPRESS defined schemas are constructed to describe DPIIM at the logical layer, which covers all the information pertaining to the stamped part, stamping die, actions, shape representation, manufacturing resource, tolerance and material. Thus, the DPIIM could provide a complete, comfortable and neutral definition of the modelled objects and the interrelationships among them, which can guarantee data consistency and completeness, and reduce data redundancy. Zha and Du (2002) selected STEP integrated resources and defined new STEP-compatible model entities to form a product information model for a mechanical product and assembly information. This STEP-based product model used five kinds of fundamental product data: nominal shape information; form feature information; tolerance information; mechanical part information, and assembly information, which provided sufficient information to support an integrated knowledge-based assembly planning system.

3.3.2.3 Development of STEP-based Product Modelling

STEP is still under development. New APs have been organised for other engineering applications, and some published parts have been updated to new versions. Some researchers have tried to integrate STEP with other standards to provide more comprehensive support for product modelling processes. Arnold *et al.* (1999) introduced a modelling method that used STEP and ISO Part Library to define an intermediate model to support product modelling in CAD/CAE applications. The STEP Tools Inc. launched a project named *Super Model* in October 1999. This project proposed to employ the STEP and STEP-NC (ISO 14649) to set up a comprehensive modelling environment and to provide standardised resources. The new model can support product data exchange and sharing within the whole life cycle of PD processes, including CNC machining processes. There have also been efforts to improve STEP-based product data exchange and sharing capability and efficiency. The HDF5 file has been researched and is considered a suitable mapping file format to represent a large EXPRESS product data set.

The STEP-based product modelling method provides a standardised, comprehensive, extendable and adaptable modelling environment for different engineering applications. Both product geometric data and manufacturing data can be modelled. The STEP-based product model is more informative than conventional product models and can provide the neutral data exchange facility to support data exchange and sharing between PD systems, thereby removing the existing barrier between current systems. Therefore, STEP and its enabling product model are the most important basis for the development of CIMS.

3.3.3 UML-based Product Modelling

UML is a formal modelling language defined by Object Management Group (OMG). It is a graphical modelling language rather than a programming language. UML is used for specifying, visualising, constructing, and documenting the artefacts of software systems, as well as for business modelling and other non-software systems. Within a UML-based model, both static and dynamic data can be modelled using UML class notation. The inner structure of the model can be explicitly represented by UML diagrams, even in the case of modelling very large and complex products like aircraft and shipbuildings. Many commercial software toolkits support UML model development. UML notation can be mapped into other programming languages, and the UML model can be implemented as the product data model for various manufacturing computerised systems. There are many reports in the literature using UML as the modelling tool to model products.

Canciglieri and Young (2003) used UML diagrams to represent three product data models to support injection moulding product development. The entire model was structured based on the viewpoints of mouldability, mould design, and mould manufacturing. These submodels are naturally linked and form a comprehensive multi-viewpoint product model to support team-based design, and adopt various manufacturing systems. Lee *et al.* (2004) developed a UML-based feature-based semantic product model for the hull design of a very large crude oil carrier. The model consists of four submodels; the generic model, the feature model, the part model, and the assembly model. This model can provide necessary information for the CAE system to analyse the ship hull structure model. The National Institute of Standards and Technology (NIST) of USA developed a UML-represented product model known as the core product model (CPM) (Fenves 2002). It is a generic product model, and models the most basic product data needed for any engineering application. The CPM model can be used as the core and extended for particular applications, *e.g.*, the product family evolution model (Wang *et al.* 2001) and the open assembly model (Rachuri *et al.* 2003) extended the UML-based CPM to model product family information and product assembly information, respectively.

Both object-oriented modelling languages UML and STEP EXPRESS/EXPREDSS-G, are computerisable, and are able to be converted to executable programming languages. The two can also map to each other. Both UML and STEP have the ability to model complex products as they have sufficiently defined notations. However, unlike UML, which only provides a standardised graphical data representation method, STEP offers more functions. STEP has lexical and graphical data representation methods, which means that STEP-based product models can contain and represent more detailed and comprehensive information. In addition, STEP defines standardised resources for different PD activities and different engineering applications. All engineers can use these standardised resources to carry out PD processes. Thus, the information exchangeability of the STEP-based model is much higher than that of the UML-based model in the CIM environment. The STEP-based modelling approach is the more promising.

3.4 Research Issues and Future Trends in Product Modelling

Section 3.3 reviews the contributions that have been made by numerous research and development projects to develop product modelling standards and methodologies to support the integration of various PD systems. However, there are major issues which hamper the development of a possible product modelling framework that can be used to model products of different types, product related information at different stages, and information exchange among various PD systems, especially when it comes to the integration of PD systems. These problems include the integration of STEP, STEP-based modelling methodologies and resources, modelling products of different types, product modelling and implementation methods to support the integration of PD systems. Also, there is still work to be done on the integration of product modelling, database technology and internet technology in support of product data sharing in distributed environments.

The STEP standard has been implemented in product modelling since 1994. As discussed, research has been very active in the last two decades, using the standardised product data representation of STEP to construct product models. However, most of the research work is still very limited in terms of how the STEP standard, STEP-based modelling methodologies and resources can be integrated to form a complete STEP-based modelling framework. The focus has been on how products at various stages can be modelled, the integration of CAD, CAPP and CAM, and the further development of STEP for various types of products. With the development of the STEP standard, STEP-based modelling methodologies and resources, we believe it is now possible to develop a generic STEP-based modelling framework which can provide an effective infrastructure for modelling products of various types in different manufacturing environments.

In today's cooperative manufacturing environment, different manufacturers need to exchange and share data on different types of products. However, due to the variability of product models, preliminary work needs to be done to convert the product data from different models. In many cases, this data conversion cannot ensure that all product information are properly converted into the required format. Missing data and conversion faults greatly impact on cooperative manufacturing activities. Thus, research to develop a generic product model capable of adapting requirements to model various types of products is essential.

Currently, product models have been developed to support an individual manufacturing process but do not have the capability to support a range of manufacturing processes. This limitation of current product modelling affects the implementation of product models to integrate different manufacturing systems and to realise concurrent engineering. Therefore, new product modelling and implementation methods for integrating various PD systems at different stages of the product life cycle are required.

Some preliminary research work has attempted to address some of these product modelling issues and to indicate future development trends in product modelling.

3.4.1 Modelling Product Family to Support Customised Manufacturing

The development of product families has received increasing interest in recent years because by sharing components across products, a family of products can be derived to provide variety while maintaining economy of scale (Du *et al.* 2002). It is also a cost-effective and time-efficient way to implement customised manufacturing. Nayak *et al.* (2002) presented a variation-based platform design method to identify the common platform parameters and the non-platform variables (scale factors) in a tradeoff between the standardisation requirement and the need to satisfy a range of performance requirements during product family design, which aims to satisfy a range of performance requirements using the smallest variation in product designs in the family. Du *et al.* (2002) introduced graph grammar formalisms for product family modelling, which defines a product family as a structured system to create a variety of products with shared core product technologies. It not only involves the shared base product, but also encompasses customisation modules, standard designs, and primary patterns of variety to generate custom designs. Jiao and Tseng (1999) developed an information modelling framework for product families through a generic product family architecture representation to minimise data redundancy and to support mass customisation manufacturing. The framework is based on a combination of elements of semantic relationships with the object-oriented data model. Zhang *et al.* (2006) proposed a functional modelling approach to support identification of both shared and individual modules across a family of products for module-based product family design. After separate functional modelling of each product variant, the individual functional models are merged into a single, coherent, and unified family functional model to determine all the shared and individual behavioural modules. This approach provides a fast method for generating new concept variants during conceptual design of a family of products.

3.4.2 Product Modelling in Conceptual Design

As many companies have demonstrated over time, a conceptual design for a new product contributes greatly to an improvement in competitiveness, because it permits a reduction of costs, an increase in quality, and often, a shortening of the time necessary to get the product on the market.

In Tay and Gu (2002) a function-based product model for conceptual design support is proposed. This enables modelling of a product in both the functional domain and physical domain at different levels of abstraction corresponding to the information requirements in different stages of the product design process. In each domain, the information is organised hierarchically according to the product decomposition structure. The two domains are interrelated and linked through function–form mapping relations, which allow designers from different backgrounds with various interests to access the design information and to identify the possible design solutions through function reasoning. A prototype system was implemented to demonstrate the proposed product model. Ayag (2005) proposed an integrated approach to evaluating conceptual design alternatives in a new PD environment. The analytic hierarchy process (AHP) technique is used to evaluate the conceptual design alternatives by ranking them using scores obtained from the process. A simulation generator integrated with AHP is the use to perform economic analyses for the AHP's high-score alternatives. Finally, the results of both simulation and AHP are used in a benefit/cost analysis to reach a final decision on the conceptual design alternatives. Brunetti and Golob (2000) introducd an approach towards a feature-based integrated product model that incorporates a feature-based representation scheme for capturing product semantics handled in the conceptual design phase and links early design with downstream applications of the PD process, such as part and assembly modelling.

3.4.3 Modelling Assembly Product

Assembly product is a different type of product from a mechanical product. The manufacturing process of assembly product requires both product geometric information (the entire product and its components) and assembly-related information (*i.e.*, assembly methods, assembly hierarchical relationship, *etc.*). Smith *et al.* (2001) used a three-matrix (connection matrix, support matrix and interference-freeness matrix) based product model to represent the required assembly information. With the support of a genetic algorithm, the stability problems that occur in mechanical product assembly sequence planning can be resolved. Yang *et al.* (2002) discussed an assembly-oriented product model for assembly planning and simulation of stamping die tools. The entire model consists of configuration information, all the solid models of component parts, and assembly relations represented by an assembly graph. The assembly of stamping die tools is divided into several subassemblies and individual parts. A hierarchical assembly tree is generated based on the assembly graph and assembly sequence planning can be generated by reasoning on the assembly product model and the assembly tree.

More recently, the STEP standard has been merged into assembly product modelling. It brings standardised data representation and a strong data exchange capability to the assembly model. Baysal *et al.* (2004) discussed the NIST Open Assembly Model (OAM). The OAM emphasises the nature and information requirements for part features and assembly relationships. It uses the model data structures of STEP to provide a standard representation and exchange protocol for the assembly product. The OAM can realise seamless integration of assembly product information throughout all phases of a product design. Zha (2006) developed a STEP-based product model for electro-mechanical assemblies. This model uses EXPRESS/XML schemas to represent the hierarchical structure of an assembly product (product–subassemblies-part) and the assembly information (using entity connector). It is used to evaluate the assimilability and assembling sequence by applying a fuzzy AHP approach.

3.4.4 Product Model in PDM/PLM Systems

Product data management (PDM) systems and product life cycle management (PLM) systems are two important types of management systems used in manufacturing enterprises. They employ product models to structure product data throughout the entire life cycle of a product and support various manufacturing activities in the PD processes. Thus, the corresponding product model needs the capability to model all product data throughout the entire product life cycle.

Brissaud and Tichkiewitch (2001) discussed how product models can be enabled to support the view of life cycle engineers by considering design actors in a whole PD cycle. The capitalisation and analysis of quality discrepancies occurring during the product life cycle is key to realising the objective of life cycle product model. In Li and Liu (2006), a product integrated model for all phases of the life cycle is proposed. Based on the integrated models and J2EE, a PLM prototype system is implemented. Weber *et al.* (2003) used a property-driven development/design approach to model products and PD processes. The core of this approach is a clear distinction between characteristics and properties of a product. It is claimed that a formalised representation along these lines, which also includes formalised interdependencies and external conditions, may significantly improve product and process models in computers.

3.5 Generic Product Modelling Framework

In the current manufacturing environment, product models require greater integration of information. This should include at least two levels of integration. The first level aims to support the information flow between the different stages of the product life cycle within a company. The data conversion between conventional product models employed by different manufacturing systems must be eliminated. The design, process planning, manufacturing, inspection, *etc.*, can use the same product model and form a CIM environment. To reach this level of integration, a product model should have the ability to present product data and information within the entire product life cycle (Brissaud and Tichkiewitch 2001). The second level aims to support the information exchange between different companies to cooperatively develop a product. The recent organisation of manufacturing activities has changed from centralised manufacturing to distributed manufacturing. It is more economical and efficient for medium and small size manufacturing companies to use outside resources to support their own PD. In this case, product modelling must adapt to this distributed PD environment. Distributed PD requires much higher levels of information exchange and sharing. In addition, these companies may employ different computerised manufacturing systems or system platforms and may be in completely different engineering domains. This results in an information exchange barrier. To solve this problem, the product model must support integrated PD in a heterogeneous manufacturing environment.

3.6 Conclusions and Future Work

In this chapter, a comprehensive review of recent developments in product modelling is presented. Four kinds of product modelling methodologies (solid product modelling, feature-based product modelling, knowledge-based product modelling and integrated product modelling) are discussed and compared in detail. The different modelling methodologies model product data from different viewpoints. The development of modelling methodology has changed from modelling low level detailed product geometric data to modelling high level semantic product data and knowledge. This change calls for more advanced product modelling technologies to be used in today's manufacturing environment. The object-oriented modelling method was discussed as an important implementation approach for the above modelling methodologies. Two popular object-oriented modelling implementation methods (STEP-based product modelling methods and UML-based product modelling methods) are reviewed. These two modelling methods provide standard data representation and are widely used in practical modelling processes.

From the above discussions about product modelling methodologies, technologies and their implementations, it can be concluded that product modelling is the key activity required to support the integration of PD processes. Modelling technologies are developed following the changes in the PD environment. In addition, it can be seen from the current literature that the integrated product modelling methodology (the STEP-based product modelling method) can be integrated to form a standardised modelling environment to support rapid PD in a distributed environment. However, three main research issues have not yet been resolved by current product modelling methods. The authors proposed a generic product modelling framework (GPMF) which can fit various products and integrate many types of information to support more manufacturing functionalities, such as PDM/PLC, machining processes, assembling, *etc.* The GPMF aims to support more information modelled in the product model throughout the entire PD processes. The GPMF can be used in different applications; it is open architecture and built up based on STEP standards to form a complete, standardised data exchange and sharing environment. Thus, the GPMF has strong extensibility and adaptability.

With the current rate of changes in the manufacturing environment, conventional product modelling technologies cannot satisfy all the requirements needed for future manufacturing, such as distributed manufacturing, concurrent engineering, *etc.* Manufacturing systems now require higher information integration throughout the whole product life cycle, and more flexibility and adaptability to the distributed manufacturing environment. Therefore, as the information supporter of various manufacturing systems, the product model must merge new techniques to be able to adapt to these new changes. Future research efforts should concentrate on the following areas:

- research into how high level product data can be modelled, especially modelling production functional data and knowledge to support intelligent PD;
- integration of different product data throughout the entire product life cycle of PD to support better information flow within the CIM environment;
- utilisation of web-based technologies to generate a distributed product modelling environment; implementation of STEP-based product modelling technologies to support collaborative manufacturing.

References

Allen, L. E., 1996, OO management, *Application Development Trends*, **3**, 50–55.

Amaitik, S. M. and Kiliç, S. E., 2007, An intelligent process planning system for prismatic parts using STEP features, *The International Journal of Advanced Manufacturing Technology*, **31**(9–10), 978–993.

Arnold, J. A., Teicholz, P. and Kunz, J., 1999, Approach for the interoperation of web-distributed applications with a design model, *Automation in Construction*, **8**, 291–303.

Ayag, Z., 2005, An integrated approach to evaluating conceptual design alternatives in a new PD environment, *International Journal of Production Research*, **43**, 687–713.

Baum, S. J. and Ramakrishnan, R., 1997, Applying 3D product modelling technology to shipbuilding, *Marine Technology*, **34**, 56–65.

Baysal, M. M., Roy, U., Sudarsan, R., Sriram, R. D. and Lyons, K. W., 2004, The open assembly model for the exchange of assembly and tolerance information: overview and example, *Proceedings of the ASME Design Engineering Technical Conference and Computers and Information in Engineering Conference 2004: Volume 4: 24th Computers and Information in Engineering Conference*, 7, 59–770.

Brissaud, D. and Tichkiewitch, S., 2001, Product models for life-cycle, *CIRP Annals – Manufacturing Technology*, **50**, 105–108.

Brooks, S. L. and Greenway, R. B., 1995, Using STEP to integrate design features with manufacturing features, *ASME Database Symposium Computers in Engineering*, Boston, MA, USA, Sept. 17–20, 579–586.

Brunetti, G. and Golob, B., 2000, Feature-based approach towards an integrated product modelling conceptual design information, *Computer-Aided Design*, **32**, 877–887.

Canciglieri, O. J. and Young, R. I. M., 2003, Information sharing in multi view point injection moulding design and manufacturing, *International Journal of Production Research*, **41**, 1565–1586.

Chang, X. Q. and Ning, R. X., 2001, Assemlby planning and evaluation oriented assembly feature modelling, *Jisuanji Jicheng Zhizao Xitong/Computer Integrated Manufacturing Systems*, **7**, 35–38 (in Chinese).

Chen, L. and Jin, G. D., 2005, Product modelling for multidisciplinary collaborative design, *International Journal of Advanced Manufacturing Technology*, **30**, 589–600.

Chen, Y. M. and Wei, C. L., 1997, Computer-aided feature-based design for net shape manufacturing, *Computer Integrated Manufacturing Systems*, **10**, 147–164.

Chen, Y. M., 1999, Integrating knowledge, data and geometry: a computer-aided concurrent engineering infrastructure, *Integrated Computer-Aided Engineering*, **6**, 189–212.

Chep, A. and Tricarico, L., 1999, Object-oriented analysis and design of a manufacturing feature representation, *International Journal of Production Research*, **37**, 2349–2376.

Chin, K. S., Zhao, Y. and Mok, C. K., 2002, STEP-based multiview integrated product modelling for concurrent engineering, *International Journal of Advanced Manufacturing Technology*, **20**, 896–906.

Du, X. H., Jiao, J. X. and Tseng, M. M., 2002, Graph grammar based product family modelling, *Concurrent Engineering Research and Applications*, **10**, 113–128.

Fenves, S. J., 2002, A core product model for representing design information, National Institute of Standards and Technology, NISTIR 6736, Gaithersburg, MD 20899.

Gao, Z. F., Qin, X. J. and Ou, Z. Y., 2000, Intelligent feature-based product model framework and its implementation, *Dalian Ligong Daxue Xuebao/Journal of Dalian University of Technology*, **40**, 193–197 (in Chinese).

Gielingh, W. F., de Bruijn, W. J., Bohms, H. M., Suhm, A., Cremer, R. and Bassan, J., 1991, Open architecture for information Integration of CIMModules, *Proceedings of the International IFIPTC5 Conference on Computer Applications in Production and Engineering: Integration Aspects*.

Gu, P. H. and Chan, K., 1995, Product modelling using STEP, *Computer-Aided Design*, **27**, 163–179.

Gu, P. H. and Zhang, Y., 1994, OOPPS: an object-oriented process planning system, *Computers & Industrial Engineering*, **26**, 709–731.

Hoehn, B. R., Steingroever, K. and Jaros, M., 2003, Product model for gear units, *Proceedings of the ASMR Design Engineering Technical Conference*, **4**(A), 385–395.

Horvath, L., Rudas, I. J. and Couto, C., 2000, Human intent models in integrated product modelling, *26th Annual Conference of the IEEE Electronics Society IECON 2000, Industrial Electronic Conference (IECON) Proceedings*, Nagoya, Oct. 22–28, **2**, 1274–1279.

Hur, J. H., Lee, K. W., Zhu, H. and Kim, J. W., 2002, Hybrid rapid prototyping system using machining and deposition, *Computer-Aided Design*, **34**, 741–754.

Hvam, L., 1999, A procedure for building product models, *Robotics and Computer-Integrated Manufacturing*, **15**, 77–87.

ISO, 1994a, Industrial Automation Systems and Integration: Product Data Representation and Exchange: Part 1: Overview and Fundamental Principles, First Edition, Reference Number: ISO 10303-1:1994(E), ISO, Switzerland.

ISO, 1994b, Industrial Automation Systems and Integration: Product Data Representation and Exchange: Part 11: Description Methods: The EXPRESS Language Reference Manual, First Edition, Reference Number: ISO 10303-11:1994(E), ISO, Switzerland.

ISO, 1994c, Industrial Automation Systems and Integration: Product Data Representation and Exchange: Part 203: Application Protocol: Configuration Controlled 3D Designs of Mechanical Parts and Assemblies, First Edition, Reference Number: ISO 10303-203:1994(E), ISO, Switzerland.

ISO, 2001a, Industrial Automation Systems and Integration: Product Data Representation and Exchange: Part 214: Application Protocol: Core Data for Automotive Mechanical Design Processes, First Edition, Reference Number: ISO 10303-214:2001(E), ISO, Switzerland.

ISO, 2001b, Industrial Automation Systems and Integration: Product Data Representation and Exchange: Part 224: Application Protocol: Mechanical Product Definition for Process Planning Using Machining Features, Reference Number: ISO 10303-224:2001(E), ISO, Switzerland.

Jiao, J. X. and Tseng, M. M., 1999, An information modelling framework for product families to support mass customization manufacturing. *CIRP Annals – Manufacturing Technology*, **48**, 93–98.

Jordon, L. A., 1994, A study in the exchange of product data using STEP and EXPRESS, Master thesis, Queen's University of Belfast.

Juri, A. H., Saia, A. and Penington, A. D., 1990, Reasoning about machining operations using feature-based models, *International Journal of Production Research*, **28**, 153–171.

Kodiyalam, S., Finnigan, P. M. and Kumar, V., 1990, Hybrid CSG/B-rep approach to three-dimensional shape optimization, American Society of Mechanical Engineers Paper.

Krause, F. L., Kimura, F., Kjellberg, T., Lu, S. C. Y., Van derWolf, A. C. H., Ating, L., ElMaraghy, H. A., Eversheim, W., Iwata, K., Suh, N. P., Tipnis, V. A. and Weck, M., 1993, Product modelling, *CIRP Annals: Manufacturing Technology*, **42**, 695–706.

Lee, J. Y. and Kim, K., 1999, Generating alternative interpretations of machining features. *International Journal of Advanced Manufacturing Technology*, **15**, 34–48.

Lee, K. Y., Lee, W. J. and Roh, M. I., 2004, Development of a semantic product modelling system for initial hull structure in shipbuilding, *Robotics and Computer-Integrated Manufacturing*, **20**, 211–223.

Lee, Y. T., 1997, Data sharing implementation based on the information model for apparel pattern making. Report of NISTIR 6139, Gaithersburg: National Institute of Standards and Technology, January.

Li, H. Y. and Liu X., 2006, Product modelling during its whole lifecycle for collaborative design in PLM, *Proceedings of the 6th World Congress on Intelligent Control and Automation*, J Dalian, China, 6890–6894.

Liang, J., Shah, J. J., D'Souza, R., Urban, S. D., Ayyaswamy, K., Harter, E. and Bluhm, T., 1999, Synthesis of consolidated data schema for engineering analysis from multiple STEP application protocols, *Computer-Aided Design*, **31**, 429–447.

Lin, L. F., Gao, P., Cai, M., Lou, Y. C., Wu, J. and Dong, J. X., 2005, Product modelling for integrated PD, *Zhejiang Daxue Xuebao (Gongxue Ban)/Journal of Zhejiang University (Engineering Science)*, **39**, 1168–1173 (in Chinese).

Linthicam, D. S., 1995, The object revolution, *Database Management Systems*, **8**, 46–52.

Liu, N. R., Li, S. P. and Dong, J. X., 2000, Research of a STEP-based tool for the integration of CAD/CAPP, *Jisuanji Fuzhu Sheji Yu Tuxingxue Xuebao/Journal of Computer Aided Design and Computer Graphics*, **12**, 286–290 (in Chinese).

Lou, Z., Jiang, H. and Ruan, X., 2004, Development of an integrated knowledge-based system for mold-base design, *Journal of Material Processing Technology*, **150**, 194–199.

Meng, M. C., Yang, L. and Bai, L. K., 1997, Feature modelling system based on STEP, *Jisuanji Jicheng Zhizao Xitong/Computer Integrated Manufaturing System, CIMS*, **3**, 34–38.

Nayak, R. U., Chen, W. and Simpson, T. W. 2002, A variation-based method for product family design, *Engineering Optimization*, **34**, 65–81.

Oetter, R., Barry, C. D., Decan, L. A. and Sorensen, P. F., 2004, Integrating manufacturing and life cycle information into the product model, *Journal of Ship Production*, **20**, 221–231.

Pilz, M. and Kamel, H. A., 1989, Creation and boundary evaluation of CSG-models, *Engineering with Computers (New York)*, **5**, 105–118.

Pratt, M. J., 1988, Synthesis of an optimal approach to form feature modelling, *Proceedings of ASME Conference Computers in Engineering (CIE)*, San Francisco, CA, August, 263–274.

Rachuri, S., Han, Y. H., Feng, S. C., Roy, U., Wang, F. J., Sriram, R. D. and Lyons, K. W., 2003, Object-oriented representation of electromechanical assemblies using UML, National Institute of Standards and Technology, NISTIR 7057, Gaithersburg, MD 20899.

Roy, U. and Kodkani, S. S., 1999, Product modelling within the framework of the World Wide Web, *IIE Transactions (Institute of Industrial Engineers)*, **31**, 667–677.

Rumbaugh, J., Blaha, M., Premerlani, W., Eddy, F. and Lorensen, W., 1991, Object-oreinted modelling and design, Prentice Hall Inc., Englewood Cliffs, NJ.

Salomons, O. W., Van Houten, F. J. A. M. and Kals, H. J. J., 1993, Review or research in feature-based design, *Journal of Manufacturing System*, **12**, 113–132.

Salustri, F. A., 1996, A formal theory for knowledge-based product model representation, *Knowledge-Intensive CAD II: Proc. IFIP WG5.2 Workshop*, eds Finger, S., Tomiyama, T., and Matyla, M., Chapman & Hall, London, 59–78.

Sanchez, N. G. and Choobineh, J. 1997, Achieving reuse with OO technology, *Information Systems Management*, 14, 48–55.

Shapiro, V. and Vossler, D. L., 1991, Construction and optimization of CSG representations, *Computer Aided Design*, **23**, 4–20.

Schench, D. A., 1994, Information modelling: the EXPRESS way, Oxford University Press, Oxford and New York.

Shaharoun, A. M., Razak, J. Ab. and Alam, M. R., 1998, A STEP-based geometrical representation as part of product data model of a plastics part, *Journal of Materials Processing Technology*, **76**, 115–119.

Sharma, R. and Gao, J.X., 2002, Implementation of STEP 224 in an automatic manufacturing planning system, *Proceedings of the Institute of Mechanical Engineers, Part B: Journal of Engineering Manufacturing* **216**, 1277–1289.

Shapiro, V. and Vossler, D. L., 1993, Separation for boundary to CSG conversion, *ACM Transactions on Graphics*, **12**, 35–55.

Smith, S. S. F., Smith, G. C. and Liao, X. Y., 2001, Automatic stable assembly sequence generation and evaluation, *Journal of Manufacturing Systems*, **20**, 225–235.

Song, Y. Y., Chu, X. P. and Cai, F. Z., 1999, Real-time concurrent product and design system for mechanical parts, *High Technology Letters*, **5**, 74–80.

Sormaz, D. N. and Khoshnevis, B., 1997, Process planning knowledge representation using an object-oriented data model, *International Journal of Computer Integrated Manufacturing*, **10**, 92–104.

Tang, D., Zheng, L., Li, Z. and Chin, K. S., 2001, STEP-based product modelling for concurrent stamped part and die development, *Computers in Industry*, **46**, 75–94.

Tay, F. E. H. and Gu, J. X., 2002, Product modelling for conceptual design support. *Computers in Industry*, **48**, 143–155.

Tu, Y. L., Xie, S. Q. and Fung, R. Y. K., 2006, PD cost estimation in mass customization, *IEEE Trans on Engineering Management*, **53**, 256–263.

Usher, J. M., 1996, A STEP-based object-oriented product model for process planning, *Computers and Industrial Engineering*, **31**, 185–188.

Van Holland, W. and Bronsvoort, W. F., 2000, Assembly feature in modelling and planning, *Robotics and Computer-Integrated Manufacturing*, **16**, 277–294.

Wang, L. H., Wong, B., Shen, W. M. and Lang, S., 2001, A web-based collaborative workspace using Java 3D, *Proceedings of the International Conference on Computer Supported Cooperative Work in Design*, London, Ontario, Canada, July 12–14, 77–82.

Weber, C., Werner, H. and Deubel, T., 2003, A different view on product data management/product life-cycle management and its future potentials, *Journal of Engineering Design*, **14**, 447–464.

Wingard, L., 1991, Introducing form features in product models: a step towards CADCAM with engineering terminology, Licentiate thesis, Dept. of Manufacturing Systems, Royal Institute of Technology, Stockholm.

Xie, S. Q., Tu, Y. L., Aitchison, D., Dunlop, R. and Zhou, Z. D., 2001, A WWW based PD platform for intelligent and concurrent sheet metal products design and manufacturing, *International Journal of Production Research*, **39**, 3829–3852.

Xie, S. Q., Tu, Y. L., Fung, R.Y. K. and Zhou, Z. D. 2003, Rapid one-of-a-kind PD via the Internet: A literature review of the state-of-the-art and a proposed platform, *International Journal of Production Research*, **41**, 4257–4298.

Xue, D. and Dong, Z., 1994, Coding and clustering of design and manufacturing features for concurrent design, *Computers in Industry*, **34**, 139–153.

Yang, J. G. and Li, B. Z., 1992, Research on transformation of workpiece information in CAD/CAPP, *Zhongguo Fangzhi Daxue Xuebao/Journal of China Textile University*, **18**, 1–7 (in Chinese).

Yang, X. H., Ou, Z. Y., Lu, P. D. and Han, F., 2002, Assembly sequence planning and assembling simulation of stamping die tools, *Proceedings of SPIE-The International Society for Optical Engineering*, **4756**, 274–279.

Zha, X. F. and Du, H., 2002, A PDES/STEP-based model and system for concurrent integrated design and assembly planning, *Computer-Aided Design*, **34**, 1087–1110.

Zha, X. F., 2006, Integration of the STEP-based assembly model and XML schema with the fuzzy analytic hierarchy process (FAHP) for muti-agent based assembly evaluation, *Journal of Intelligent Manufacturing*, **17**, 527–544.

Zhang, J. S., Wnag, Q. F, Wan, L. and Zhong, Y. F., 2005, Configuration-oriented product modelling and knowledge management for made-to-order manufacturing enterprises, *International Journal of Advanced Manufacturing Technology*, **25**, 41–52.

Zhang, S., Hou, X. and Wang, Z., 1995, STEP application protocol interoperability for CAD/CAPP integration, *ASME Database Symposium, Computers in Engineering*, Boston, MA, USA., Sept. 17–20, 687–690.

Zhang, W. Y., Tor, S. Y. and Britton, G. A., 2006, Managing modularity in product family design with functional modelling, *International Journal of Advanced Manufacturing Technology*, **30**, 579–588.

Part II
System and Framework

Part II of this book focuses on providing solutions for those OKP companies producing products that are highly customised. The solutions are at system and framework level, and target issues such as how to best communicate with customers, how to integrate and optimise the OKP product development process, and how to manage product manufacturing processes more efficiently.

Chapter 4 introduces an Internet-based integrated product development system for rapid development of sheet metal products, a typical example of an OKP product. The structure of the system is discussed and case studies are carried out to test the idea of the Internet-based integrated system. Several major modules of the system are described. These modules include the structure of the integrated product development system, an integrated data environment, a real-time computer aided process planning (RTCAPP) module, QFD-based global customer interfaces and design/manufacturing knowledge bases to support product design and manufacturing. The aim of the chapter is to introduce an example of how to establish or develop a system for developing OKP products. Sheet metal parts are highly customer oriented and the technologies involved can be expanded to the development of other products.

In the sheet metal processing and manufacturing industry there are many small or medium sized enterprises (SMEs). These enterprises have to make every effort to shorten product development lead time, improve production efficiency, approach high quality standards, while cutting down the costs at the same time. Chapter 5 presents a new compound cutting and punching production method supported by an integrated CAD/CAPP/CAM system. The solutions include an integrated data integration platform based on Pro/INTRALINK and STEP, and a knowledge-based RTCAPP system for compound sheet metal cutting and punching.

Chapter 6 presents an agent-based process planning system for developing OKP products. The focus is on the structure of the system. A number of agents are defined, including an unfolding agent, a feature recognition agent, a task agent, a nesting agent, a path planning agent, a bending agent, a machining method selection agent, a machine selection agent and a fixture/jig selection agent. Intermodules of typical agents (nesting agent, path planning agent, bending agent) are built for the system. In addition, the way agents cooperate with others to implement tasks is investigated. This chapter provides an example of how modern technology can be used to develop an Internet-based system for rapid product development.

Chapter 4
Integrated OKP Product Development System

Abstract To further develop the reference system structure of the Internet-based integrated PD system, as outlined in Chapter 2 into a working PD system, this chapter proposes an integrated PD system for rapid development of OKP products. The structure of the system is discussed and case studies are carried out to test the idea of the Internet-based integrated system. Several major modules of the system are discussed in this chapter. These modules include the structure of the integrated PD system, an integrated data environment, a real-time computer aided process planning (RTCAPP) module, quality function deployment (QFD)-based global customer interfaces and design/manufacturing knowledge bases to support product design and manufacturing. This chapter explores the definition and the structure of a system for the rapid development of sheet metal parts.

4.1 Introduction

Today companies in the sheet metal industry are under great pressure due to globalisation of the market, the short life cycle time of products, increasing product diversity, high demand for quality and short delivery times. To be competitive in the market, sheet metal companies must gain enough production flexibility to rapidly produce sheet metal products with acceptable quality. Multi-national sheet metal corporations especially must use new technologies for their PD, re-engineer their organisational structures and enhance their ability to make decisions correctly throughout the whole PD cycle, particularly at the early product design stages. These new technologies include Internet technology, CAD/CAPP/CAM integration technology, computer simulation technology, rule based reasoning (Dixon 1986, Wu *et al.* 2010), constraint networks (Young *et al.* 1992), knowledge bases and optimisation theory (Dowlatshahi 1992).

Usually, sheet metal product design and planning of manufacturing processes are carried out in a series of stages using different software packages, especially for bent sheet metal products. Firstly, the product is designed using a 3D CAD system, usually using wire-frame representations. Then, an unfolding software package is used to extend the model into 2D drawings of the flat parts. A computer aided process planning (CAPP) system is employed to plan manufacturing processes and to select the necessary tools. To minimise production waste and to improve production efficiency, optimisation software systems (*e.g.*, optimal nesting software systems, path-planning systems) are used. CAM software systems are also often used in a sheet metal manufacturing company to generate NC programs for CNC sheet metal processing machines, *e.g.*, a punching, shearing, or folding machine. Before delivering the NC programs to a CNC sheet metal processing machine, a simulation software package is used to simulate the processes and consequently to modify or refine the NC programs and to adjust the tool paths. Usually, in a sheet metal PD cycle, these processes are carried out separately and sequentially in different branches of the company. With this traditional design and manufacture method, a sheet metal product is frequently designed without systematic consideration of downstream PD requirements, such as process planning, manufacturability, production scheduling and manufacturing optimisation. Also, the feedback from these downstream processes to the product designer is made after the product is designed or even after it is manufactured. This often results in expensive and time consuming rework. Consequently, it affects the quality, cost and delivery time of the product.

To fill the gap between design and other downstream PD processes, quite a few research projects have started to focus on concurrent product design and manufacture and intelligent decision support in early product design stages (Kusiak 1993, Prasad 1996, Bartolotta *et al.* 1998, Dong and Agogino 1998, Tu *et al.* 2000). Concurrent product design and manufacture means that a designer should consider all possible requirements from the downstream PD processes and incorporate them into the product design. However, there is no platform or system available that can be directly used to support concurrent sheet metal design and manufacturing process planning by different branches or partners that may be globally distributed.

An integrated PD system that can support the intelligent decision process in early product design stages can be realised by applying advanced computer technologies, such as computer aided design and engineering technology (*e.g.*, Pro/ENGINEER), manufacturing optimisation software packages (Dowlatshahi 1992), computer simulation technology, and knowledge base technology (Bartolotta 1998). By using these computer software packages or systems, a designer can simulate and virtually observe the downstream processes for a possible product design on a desktop computer. This integrated PD system can help the designer to understand the downstream process requirements and to design a product right at the first instance.

However, simply putting together several computer aided sheet metal design and manufacture software systems does not provide a sound overall solution for achieving concurrent engineering in a modern sheet metal manufacturing company as its "partners" are globally distributed. A main problem to realising concurrent engineering in such a company is providing intensive real-time information interchanges (Devarajan *et al.* 1997) and optimisation (Tu *et al.* 2000) in the company. Also, the complexity of a sheet metal product produced by the company makes it impossible for a single designer to manage the complete PD process. Successful development of a complicated sheet metal product needs contributions and expertise from almost all the people in the company. On the other hand, the various computer software packages or systems are often developed by different vendors, and some software packages are not suitable for special requirements. Hence to achieve concurrent design and manufacturing of sheet metal parts, a system that contains several application modules to meet special requirements and can support efficient and effective information interchange among different software systems, different departments and different people in the company, is very important. Correct and effective information exchange throughout the PD life cycle can shorten the PD lead time, improve the production efficiency, achieve high quality, and at the same time, cut down the production cost.

The purpose of this research is to develop an integrated PD system for intelligent concurrent design and manufacturing of sheet metal parts. The structure of the integrated PD system is addressed in Section 4.3. To support the development of this system, several major modules will be discussed in the following sections, including an Internet-based integrated data environment to support information (data) sharing, and QFD-based global customer interfaces and a WWW-based knowledge bases module to support intelligent concurrent sheet metal PD.

4.2 Integrated Product Development System for Sheet Metal Design and Manufacturing

Because of the industry's increasing needs for other types of information or knowledge in the sheet metal design and manufacturing process, new classes of tools to support knowledge-based design, product data management and concurrent engineering are urgently needed in some companies. However, traditional CAD/CAM systems usually run independently and the information transfer among these systems is through particular file formats (STEP, DXF, *etc.*). Delays, errors and lengthy procedures for product definitions are also caused by a lack of supporting information. Also, traditional CAD/CAM tools focus mainly on database-related issues (Simon *et al.* 2000) and do not place particular emphasis on real-time information modelling for product representation in an integrated system.

Owing to the numerous and complicated interactions among the tasks of concurrent sheet metal PD processes, an Internet-based integrated sheet metal PD system is proposed as shown in Figure 4.1. The integrated PD system reported in this chapter aims to solve these problems and is based on a WWW platform that can be used by most operating systems and contains several modules, *i.e.*, an unfolding module, an Internet-based data integration platform, design/manufacturing knowledge bases, data communication tools among different modules, a RTCAPP module, a CAD module, a CAM module, a cost estimation module, computer simulation platforms, graphical user interfaces (GUIs), *etc.*

The unfolding module in this system adopts zero thickness and zero bend radii principles in order to facilitate the bent sheet metal part design process. The data in-

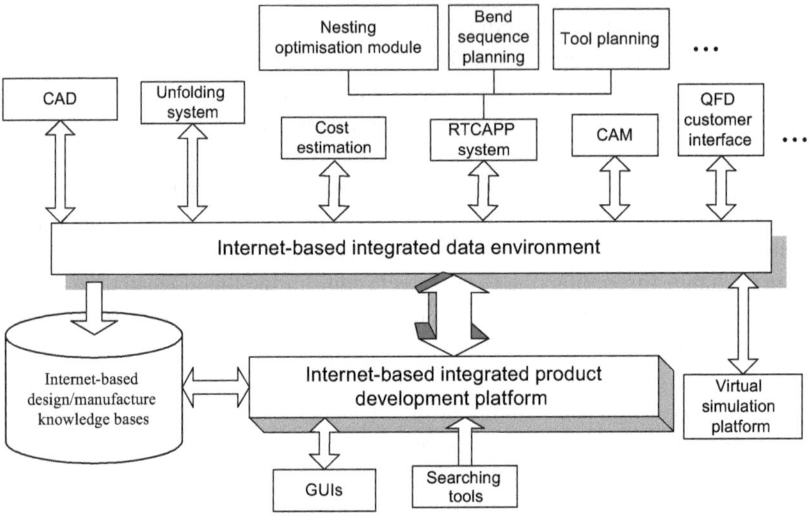

Figure 4.1 Integrated PD system for sheet metal concurrent design and manufacturing

tegration platform is a data integration environment for product design, unfolding, process planning and manufacturing process, which is developed based on the information integration framework, Pro/INTRALINK and the API of Pro/ENGINEER (Xie and Tu 2000). The data communication tools are developed for information exchange among different systems, *i.e.*, the data communication tool between the integration platform and CAPP software provides CAPP with suitable parametric data for the sheet metal parts and sends all this information back to the data integration platform in real time. This information, *e.g.*, as process plans, manufacturing parameters, tool sequences and cost, can be used by other modules. The data communication tool between the integration platform and CAD software captures related information from knowledge bases and provides CAD with suitable information such as cost, unfold ability, *etc.*, so that a decision can be made.

The customer interface module is provided to satisfy the demands of the customer. This is crucial these days to market competitiveness and business success, and this trend is set to become one of the prerequisites for business survival in this century. Organisations that manage to meet or improve on customer requirements will emerge as the leaders in the global market. Over the last decade, QFD has developed into a proven quality management concept and methodology that can facilitate the understanding and response to customer requirements at various stages of product design, development and manufacturing with effective allocation of enterprise resources.

The knowledge bases aim to provide the means to support human decision-making in both the early design process and the later design and manufacturing process by capturing and making available different types of design and manufacturing knowledge. The knowledge bases are built based on a "four-layer" structure of the information framework and are saved as part of the company's overall database. There are several knowledge bases involved in this project, *e.g.*, tool knowledge base, design knowledge base and manufacturing knowledge base.

The Internet-based cost estimation module aims to achieve PD cost estimates and optimal control under the complexities of a global manufacturing environment. The details of the cost estimation and optimisation algorithm are addressed in Chapter 10. This system is able to provide in-time cost estimates and optimisation results to product developers as important decision-making references in a concurrent engineering approach to PD. This concurrent approach enables product developers to design the product, plan the processes, schedule operations on the shop floor, manage logistics, and estimate and optimally control the cost feature by feature. This system can help product developers to clarify the PD life cycle data at early PD stages and hence avoid rework that often occur in the traditional sequential approach to PD. Avoiding rework in a PD cycle, particularly for customised products, will reduce the PD cycle time and cost significantly.

The RTCAPP module contains several submodules, including knowledge based auxiliary path automatic insertion, an optimisation module, a database, a tool library and a module library. This module will be explained in Chapter 5. The optimisation module is the most important module for achieving efficient and economic manufacturing. It includes the tool sequencing, tool path optimisation and nesting optimisation of sheet metal parts with complicated shapes.

Graphical user interfaces are also needed to serve as a bridge between the integrated PD system and human users. Creating GUIs for this integrated product system is very important and is largely a matter of describing where every interface component should appear on the screen, how it should appear and what should happen when the user interacts with it. The GUIs are designed using several platforms, and they can be viewed through a Web browser such as Internet Explorer. These interfaces are developed not only for information or knowledge sharing among different systems and users, but also for the "appointed" user to define suitable data structures for knowledge bases and information objects according to different applications.

The proposed integrated PD system can be accessed from anywhere either inside or outside a company. This will help company employees to work in an integrated information environment. They can share the PD software tools and data, and work concurrently to develop a product. The meetings and discussions through a PD cycle will automatically be recorded by the system when they communicate through this system. The company's marketing people can also access this system via the Internet to get online engineering support from the company to address, discuss and meet customers' needs.

In order to achieve this system, this chapter will discuss some key modules including the Internet-based integrated data environment, QFD-based customer user interfaces, simulation platforms and knowledge bases to support intelligent concurrent sheet metal PD. The zero thickness and zero bend radius principle and the information integration framework will be discussed in the next chapter.

4.3 Internet-based Integrated Data Environment

Data sharing in the RPD process in such a heterogeneous environment has always been a difficult task. For example, different design, analysis and planning tools, such as geometric modellers, analysis tools, planning tools, CAM tools, are all critical components of the overall PD process. A fundamental problem with data sharing across these tools is that the output of one tool is often required as the input of another tool. Dynamic information modelling and exchange is involved in the PD process. Also, as different tools have different data formats, it is difficult to move data from one step of the design process directly to another step without data translation. Hence, an efficient Internet data environment that can perform these functions will be key for supporting the development of the PD system.

An Internet-based integrated data environment is proposed to share product information across diverse software tools in the Internet environment. This Internet-based data environment aims to build an information bridge to fill the gap between sheet metal part design, process planning, simulation and manufacturing systems. This is an important issue in supporting integrated and concurrent PD. With development of the following technology and methodologies, the data environment that can be used to build information models for information sharing among different software packages can be created.

STEP plays an important role in supporting the functionality of the Internet-based integrated data environment. In order to maximise interoperability with most of the existing software packages used in industry, the framework makes use of standard representations when possible. For the representation of geometry, this project uses ISO 10303, which is more commonly known as STEP (the Standard for the Exchange of Product model data, officially ISO 10303) (ISO 10303-11:1994(E)). STEP is a family of standards specifying a description of product data throughout a product's life cycle in a computing-platform-independent manner; each individual standard in STEP can be used to define the information requirements for a particular area of design, manufacturing, engineering, or product support (ISO 10303-11:1994(E)). The conceptual schemas of different Application Protocols (APs) (ISO 1994a) within the STEP standard provide the semantic basis for the data managed by the integrated data environment through a standard schema definition for the data that is used within a particular design and/or analysis domain. We have also adopted and modified the STEP concept of "units of functionality" (UoFs) to define our notion of conceptual subcomponents of exchange files (Liang *et al.* 1997).

As the representation of non-geometric information is at a more preliminary stage (Wang and Bourne 1995, Bazargan and Falquet 2009) while industry's reliance on non-geometric information and knowledge-based design and manufacturing is increasing, new modelling methodologies for design and manufacturing knowledge such as function, behaviour, knowledge and other kinds of non-geometric data are needed. The use of UoFs and relationships will help to create data sharing information models for real-time data communication among the different software tools. This issue will be discussed in greater detail in Chapter 9.

An integrated global data structure is the key to building the Internet-based integrated data environment using STEP standards. This issue will be discussed in detail in Chapter 7. The emphasis of the global data structure is placed on the integrated data management and reuse of past PD experience to support a company's goal to shorten its PD cycle. The integrated data structure is modelled by EXPRESS from STEP. In terms of this data structure, a design/manufacturing knowledge base was developed as a major part of the WWW-based information management system. The basic principles and concepts of the knowledge base and the WWW-based information management system will be presented in later chapters.

4.4 QFD-based Global Customer User Interfaces

Satisfying the demands of the customer is crucial for market competitiveness and business success these days, and this trend is set to become one of the prerequisites for business survival in this century. Organisations that manage to meet or improve on customer requirements will emerge as the leaders in the global marketplace. Over the last decade, QFD has developed into a proven quality management concept and methodology that can facilitate the understanding and response to customer requirements at various stages of product design, development and manufacturing with effective allocation of enterprise resources.

The voice of customer (VoC) or customer requirements/attributes are often expressed in linguistic terms, which are usually non-technical in nature. Sometimes it is difficult for engineers to translate these requirements into definitive product and engineering specifications. QFD (Akao 1990, Cohen 1995, Lai *et al.* 2008) is a well-known quality management methodology that helps channel customer requirements through various stages of a manufacturing cycle, such as design, part planning, process planning and operation planning for a product. The process can be represented graphically by a series of interconnected structured matrices called the "houses of quality" (HoQ; Hauser and Clausing 1988). A typical HoQ will normally consists of eight parts, *i.e.*, VoC (or customer attributes), voice of engineers (VoE) (or product attributes), correlation between product attributes, competitive benchmarking, relationships between product attributes and customer attributes, priority of product attributes, target design values, and concept evaluation.

The semantics in the VoC may contain ambiguity and multiplicity of meaning (Khoo and Ho 1996), and makes the interpretation of VoC into precise product/engineering attributes complicated and difficult. Most of the recent research in QFD have assumed the customer and engineering/product attributes to be crisp, complete and independent of one another, although this is seldom the case in reality. These oversimplifications can lead to discrepancies in design priority setting and unbalanced resource allocation. Hence, a conventional HoQ can only give a picture of how the customer attributes could be met, and indeed many companies apply QFD and HoQ to do just that. However, in the proposed Customer Interface Model,

QFD is just the beginning of a series of inter-linking processes and planning stages. In this model, a focused HoQ is used.

The structural framework of the focused HoQ is similar to that of the basic HoQ, however, more detailed quantitative analyses would be conducted to establish the information it holds. First of all, a dictionary of key words abstracted from a company's VoC records has been developed to standardise the VoC and to clearly identify customer needs. An "affinity diagram" was applied to regroup the more essential customer attributes. A pairwise comparison exercise was carried out on the customer attributes, and the findings could be further analysed using the analytic hierarchy process (AHP) (Ho 2007, Saaty 1990, 1994). The resulting relative weights of importance for the customer attributes are normalised and expressed in percentages. A dictionary for VoE was also developed to standardise the technical languages used in the company.

One of the distinct merits in this focused HoQ lies in the fact that the attribute relationships are established quantitatively, instead of relying on qualitative interpretation. The attribute relationships are worked out based on the contributions and thus the importance of individual product attributes towards fulfilment of each customer attribute is considered. AHP plays an important role in the quantitative analysis in the focused HoQ and in the calculation of the relative weight of importance for each product attribute.

The target design values in QFD represent the design specifications for individual product attributes used to guide subsequent design and production activities. The output from this Customer Interface Model is a list of product attributes with relevant target design values, *i.e.*, features and the design specifications of a product required by the customer. These features will be classified into primary features and secondary features according to the design and manufacturing requirements or limits.

To design QFD-based customer interfaces for rapid development of sheet metal parts, case studies have been carried out in a sheet metal manufacturing company (KAM 2000). A complete customer database and an engineer database have been created as the input to the QFD process. A relationship matrix between VoC and VoE has been created using Microsoft Excel. In order to allow the link to takes place, an interface was designed. An interface is a program that enables the user to input data. The source code of the QFD-based customer interface is given in Appendix A.4.

4.5 Simulation Platforms

Some advanced computer simulation packages can be used as effective tools in the integrated PD system to approve some critical manufacturing processes and to derive the operational cost for manufacturing a product. A simulation platform for sheet metal cutting and punching has been developed by the authors (Xie and Tu 2000). The CNC machines that will be used to manufacture certain sheet metal parts can be built by using VNC simulation package from Deneb Co. Ltd. These machines are saved in a library and can be selected by users through the simulation platform. The

VNC is one of the four simulation products of Deneb Inc., USA. It has been developed to simulate all kinds of CNC machine tools and is a 3D simulation package. In the simulation platforms, a part can be machined just as on a real CNC machine. By using this simulation model to virtually machine a sheet metal part, some design mistakes and hidden production rework can be avoided, *e.g.*, to machine a mould with complicated sculptured surfaces (Lee *et al.* 1993) and some dead cutting points owing to cutting tool collisions can be detected. Through the simulation platform, the detailed times for manufacturing a product on the physical shop floor can be observed and recorded. These include machine times, set-up/tear-down times, transportation times, waiting times, machine break-down times, and idle times. Based on these times, the cost for manufacturing a product on the shop floor can be derived. Furthermore, a production schedule to produce a new product can be traced out through manufacturing this new product together with other products being produced in the company. Details of the simulation platform and other key application modules in the platform will be explained in the following chapters with case studies.

4.6 Knowledge Bases to Support Intelligent Design and Manufacturing

The design/manufacturing knowledge base for sheet metal parts can be built as an online data library or knowledge base to support all kinds of activities throughout a PD cycle. The data structure of the design/manufacturing knowledge base can be modelled according to the sheet metal part and is related to the information flow in its development process, such as resources, tool information, *etc.* After one product has been developed, all the related information can be stored in the design/manufacturing knowledge base and can be retrieved by data searching tools for use with similar PDs. The products in the design/manufacturing knowledge base represent the company's existing products. They are modelled and recorded in the data format according to the data standard provided by the integrated data environment. Other information objects or activities have been modelled from a database approach (Figure 4.2), for example the suppliers' information model includes all information about the company's suppliers, *i.e.*, contact information, product specification, reference prices, delivery lead times, reputation assessment and quality assessment. The inventory information object can also be modelled and used to record the company's present inventory level, which is used as a part of the company's manufacturing resource planning system. The tool information objects, including its relationship with certain features of the sheet metal product, are modelled and saved to the database as the company's tooling information. The resource objects are built and used to keep records of the company's equipment, such as cutting machine, punching machine, compound machine, employees and others. The present and past data are also recorded in design/ manufacturing knowledge bases. Based on the objects and relationships among objects, knowledge or useful information can be formed to support the development of new products.

Information models for design/manufacturing knowledge bases:

```
ENTITY Product;
Product_family: STRING;
Product_Title   : STRING;
Product_descrip: STRING;
Drawing_ID     : STRING;
DesignfileNo : INTEGER;
Productassemble_List : LIST[0:?] OF
Designverification;
Process_plan;
Salesinformation : SalesOrderForm
Toolmanufproduc: Tool;
 END ENTITY;

ENTITY Tool;
Tool_code       : STRING;
ToolTrailNumber: STRING;
staffcode : STRING;
Toolroommanager: Facility;
Setup_Time : INTEGER;
Operation Time: INTEGER;
Toolshape: INTEGER;
Toolmaterial: STRING;
END ENTITY;

ENTITY SalesOrderForm;
Salesordercode      : STRING;
Orderdate         : Date;
Deliverydate     : Date;
Customer         :Customer;
ProductCost       :STRING;
LeadTime        :STRING;
Placed By         : Employee;
Item List         :LIST[0:?]OFJob;
Order Quantity : INTEGER;
Order descrip    : STRING;
Status            : TRUE/FALSE;
END ENTITY;

ENTITY Supplier;
Suppliercode : INTEGER;
Sup_Name     : STRING;
Product        :STRING;
Contact        :STRING;
Tel_Number  : INTEGER;
Fax_Number : INTEGER;
Contactpeople: STRING;


END_ENTITY;

ENTITY ProcessPlanning;
PlanID           : STRING;
PlanDate         : Date;
Datemodified    : Date;
Employeename : STRING
Op_List          : LIST[0:?] OF
Machine Operation;
Plan_Desc       : STRING;
 Planningresult :INTEGER
END_ENTITY;

ENTITY Resource;
    ResourceID : STRING;
    ResourceName: STRING;
    ResourceDescrip : STRING;
    Supplier: Supplier;
    Resourcecost : INTEGER;
    Location: STRING;
    END ENTITY;
```

Figure 4.2 A selection of entities from the EXPRESS schema for the knowledge bases design

The relationships between features and knowledge units are established based on the object-oriented modelling concept by defining methods for knowledge activation. When the method or function is activated, a corresponding knowledge object is instanced to perform certain operations on the feature through running rules belonging to the knowledge object. Similarly, data are tightly associated with knowledge objects through defining methods or functions to activate relevant data objects from knowledge objects.

To illustrate the dynamic aspect of interactions among feature, knowledge and data, the punching tool sequencing model of sheet metal compound manufacturing (cutting and punching) is used as an example (Champati *et al.* 1996). Once the user enters this module, the tool process planning knowledge object is instanced to have the user select a tool from the tool library. Feature dimensions are instanced by feature objects; a tool object will be available after the user chooses a tool. The tool sequencing module will automatically determine if it can be used to punch the feature, and the tool-sequencing object will return certain information to show the status. The relationship between tool object and feature can be created, and can also be used as knowledge for the next tool selection of the same feature. The design and manufacturing knowledge bases can also be built in the same way.

The authors have developed a platform for storing knowledge using the object-oriented modelling language, as well as tools for examining, manipulating, searching and retrieving knowledge bases. The information access tool and collaborative communication tool have been developed as tools for efficient and effective knowledge access so that the knowledge bases can be used on a large scale in a distributed environment (*i.e.*, over the Internet). This WWW-based database/knowledge base system will be addressed in later chapters of the book.

4.7 Case Studies

This section discusses some case studies of the application of software packages developed by the authors, *e.g.*, a CAD/CAPP/CAM system (Xie and Tu 2000), a simulation software package, and a shortest non-machining path optimisation software package. Some application modules of the integrated PD system are also briefly discussed.

4.7.1 Programming Tools

The programming languages that have been used to achieve this integrated PD system include Visual C++, VBSCRIPT, JAVASCRIPT and EXPRESS. Visual C++ is an object-oriented language, which supports class and object operations. It is used for developing most of the application modules and tools. VBSCRIPT and JAVASCRIPT are used for design/manufacturing knowledge sharing and client–server communication. JAVASCRIPT is a cross-platform object-based scripting language for client and server application and can be used to create applications that run over the Internet. The use of these languages make it easy to achieve all the application modules of the integrated system.

4.7.2 Application Modules Test

In order to achieve an integrated PD system for sheet metal intelligent concurrent design and manufacturing, several modules have been developed. This chapter discusses the application results of some of the major modules using case studies, including the manufacturing cost estimation module, WWW-based design/manufacturing knowledge bases and a RTCAPP module consisting of several submodules, such as path planning, nesting, knowledge-based auxiliary path inserting. Some of the modules will be further addressed in later chapters.

The manufacturing cost estimation module estimates the manufacturing cost of the sheet metal part. This module takes input from the NC programming and the knowledge based CAPP modules. The cost object includes several elements. These elements are used to calculate the total cost of the product, and the machining time is calculated from the virtual simulation platform. The set-up times are estimated times associated with a particular operation. The cost of using different manufacturing methods can be calculated from machining times and expenses, *e.g.*, for a flame cutting processes, the manufacturing cost can be calculated if the manufacturing time and expenses (cutting gas, auxiliary gas) are known. The cost of sheet metal utilisation ratio can be calculated according to the data input from the nesting optimisation algorithms submodule. The total manufacturing cost of the part can be calculated by adding the estimated material cost to the sum of the costs of all processes. As one can see, all the optimisation algorithms have an influence on cost. The function of cost estimation has been defined in the data integration platform, which can be sent to CAD or other modules.

The RTCAPP module in this chapter presents a knowledge-based RTCAPP system based on the integrated system, which includes several optimisation modules that aim to improve the manufacturing efficiency and cut down costs, and knowledge-based auxiliary paths and manufacturing parameter selection for sheet metal cutting processes. Chapter 5 gives a detailed definition and explains how the RTCAPP module can be achieved. The functions in the RTCAPP system are designed to achieve different application tasks. These functions can access all required information through the data integration platform. Tool database, cutting parameter database and process rule database are all connected with the data integrated platform through open database connectivity. Functions that have been defined include automatic tool selection, tool path optimisation, nesting optimisation, and manufacturing parameters. For the sheet metal manufacturing RTCAPP system, real-time optimisation modules are very important for manufacturing efficiency and cost, while manufacturing parameters and the auxiliary path will influence the manufacturing quality of sheet metal parts.

All the key modules and tools have been tested in this system, *i.e.*, a communication tool with CAD (Pro/ENGINEER), CAM, knowledge capturing tool, shortest manufacturing path optimisation, nesting optimisation, *etc.* Figure 4.3 shows some case studies of the integrated PD system. Figure 4.3a shows the result of a simulation platform for shortest path optimisation, process planning and nesting algorithm. This system takes the geometric parameters as input and simulates the manufacturing process while choosing different CNC punching machines, bending machines and cutting machines. In order to improve the utilisation of sheet metal parts and manufacturing efficiency, a RTCAPP module has been developed. This module contains a nesting optimisation, auxiliary path inserting and a shortest non-machining path-planning module. Figure 4.3b shows the output from the RTCAPP software package that has been developed by the authors. The test results in several sheet metal companies show that the system can significantly shorten PD time, cut down cost and improve machining efficiency. Figure 4.3c shows a WWW-based QFD-based customer interface which can be used by product designers to make decisions at an early design stage. This product knowledge view platform is supported by knowledge bases that are built based on the information data environment.

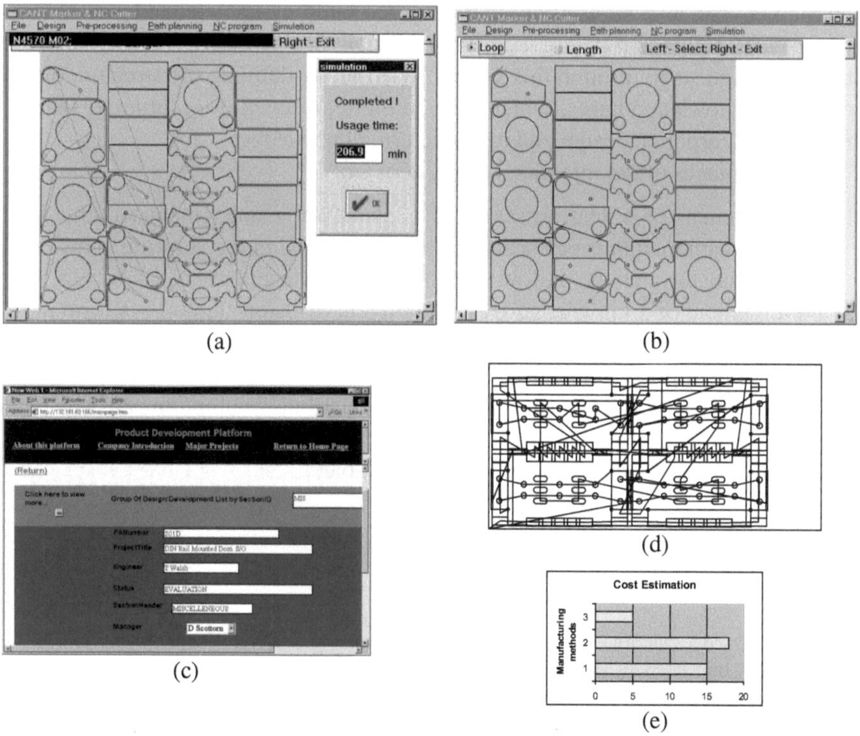

Figure 4.3 Case studies for the integrated PD system: (a) simulation platform and NC code generation; (b) RTCAPP module; (c) WWW QFD customer interface; (d) path optimisation; (e) cost estimation

Knowledge mining tools have been developed in order to search and locate related knowledge. With the support of product knowledge, it becomes easier for product designers to make decisions at early design stages. Figure 4.3d shows the simulation results of the shortest path planning optimisation module. Different optimisation algorithms have been used and tested in this system. Figure 4.3e shows the cost estimates for a flame cutting method using three different optimisation algorithms when given information such as cost of using a cutting method, tool set-up and exchange time and cost, *etc.* The utilisation ratio of a sheet metal part can be calculated by the nesting algorithm in real time, so costs can be optimised and sent back to the designer through the communication tool and economical design can be achieved.

4.8 Summary

In order to rapidly develop sheet metal parts, an Internet-based integrated PD system has been proposed in this chapter. An integrated data environment that aims to provide an information methodology to support sheet metal parts intelligent concurrent design and manufacturing has been described and explained. This integrated data environment can also be used in other areas for building knowledge bases to support product design and manufacturing. Intelligent decision support is achieved at an early design stage through this system by correctly defining the structure of the company's knowledge bases, which include the representation, capturing, sharing and reuse of corporate design and manufacturing knowledge.

The advantages of the integrated PD system are:

1. Establishment of a platform-crossed client/server management pattern. Distributed team members can share product design and manufacturing knowledge in real-time through the uniform graphic user interface.
2. Development of a knowledge-based intelligent decision support platform at an early stage of product design.
3. Development of an integrated PD system that can support concurrent design of sheet metal parts and manufacturing in a distributed computing environment through information integration.
4. The system is multi-vendor software compatible.

References

Akao, Y., 1990 (ed.), Quality Function Deployment: Integrating Customer Requirements into Product Design, translated by Mazur G. H. and Japan Business Consultants, Ltd., Productivity Press, Cambridge, MA.

Bartolotta, A., McLean, C., Lee, Y. T. and Jones, A., 1998, Production systems engineering: requirements analysis for discrete-event simulation. NISTIR 6154. National Institute of Standards and Technology, Gaithersburg, MD, April.

Bazargan, K. and Falquet, G., 2009, Specifying the Representation of Non-geometric Information in 3D Virtual Environments. Human-Computer Interaction. Novel Interaction Methods and Techniques Lecture Notes in Computer Science, Volume 5611/2009, 773–782, DOI: 10.1007/978-3-642-02577-8_85.

Champati, S., Lu, W. F. and Lin, A. C., 1996, Automated operation sequencing in intelligent process planning: a case-based reasoning approach. *International Journal of Advanced Manufacturing Technology*, **12**, 21–36.

Cohen, L., 1995, Quality function deployments: how to make QFD work for you, Addison Wesley Publishing Co., Reading, MA.

Devarajan, M., Kamran, M. and Nnaji, B. O., 1997, Profile of setting for feature extraction and feature tool mapping in sheet metal. *International Journal of Production Research*, **35**, 1593–1607.

Dixon, J. R., 1986, Artificial intelligence and design: a mechanical engineering view. *Proceedings of Fifth National Conference on Artificial Intelligence*,. AAAI-86/2.

Dong, A. and Agogino, A. M., 1998, Managing design information in enterprise-wide CAD using 'smart drawings'. *Computer Aided Design*, **30**, 425–435.

Dowlatshahi, S., 1992, Product design in a concurrent engineering environment: an optimisation approach. *International Journal of Production Research*, **30**, 1803–1818.

Hauser J. R. and Clausing D., 1988, The House of Quality. *Harvard Business Review*, May–June 1988, pp. 63–73.

Ho, W., 2008, Integrated analytic hierarchy process and its applications – A literature review. *European Journal of Operational Research*, **186**(1), 211–228.

ISO, 1994a, ISO 10303-1, Industrial Automation Systems and Integration-Product Data Representation and Exchange, Part 1: Overview and Fundamental Principles (EUSTEP Limited).

ISO, 1994, ISO 10303-11:1994(E), Industrial Automation Systems and Integration-Product Data Representation and Exchange, Part 11: The EXPRESS Language Reference Manual (EUSTEP Limited).

Kam, J. J., 2000, Project Report, Department of Mechanical Engineering, University of Canterbury, Christchurch, New Zealand. Project No. 31.

Kusiak, A. (ed.), 1993, Concurrent Engineering: Automation Tools, and Techniques, Wiley, New York.

Lee I. B. H., Lim B. S. and Nee A. Y. C., 1993, Knowledgebased process planning system for the manufacturing of progressive dies. *International Journal of Production Research*, **31**, 251–278.

Liang, J., 1997, STEP Piolet Project: Sonsolidated Schema for Airframe Engineering Analysis. *Proceedings of DETC97*: ASME Design for Manufacturing Conference, Sacramento, CA.

Prasad, B., 1996, Concurrent Engineering Fundamentals, 1, Prentice Hall, Englewood Cliffs, NJ.

Saaty, T. L.,1990, The Analytic Hierarchy Process, RWS Publications, Pittsburgh, United States.

Saaty, T. L., 1994, Fundamentals of Decision Making and Priority Theory, RWS Publications, 487 p., Pittsburgh, United States.

Szykman, S., Sriram, R. D., Bochenek, C., 2000, Design Repositories: Next-Generation Engineering Design Databases, *Journal of National Institute of Standard and Technology*: http://www.mel.nist.gov/msidlibrary/summary/pubs00.htm

Tu, Y. L., Xie, S. Q. and Liu, J., 2000, Virtual PD platform for one-of-a-kind production. International Federation of Automatic Control (IFAC), June, Aachen, Germany.

Wang, C. H. and Bourne, D. A., 1995, Using features and their constraints to aid process planning of sheet metal parts. *Proceedings of IEEE International Conference on Robotics and Automation*, 1, 1020–1026.

Wu B. Z., Gao, B., Song, R., 2010, Research on Rule-based Reasoning Intelligent Selection System of Injection Mould Standard Parts. *Advanced Materials Research*, **102–104**, 432–435.

Xie, S. Q. and Tu, Y. L., 2000, An Integrated CAD/CAPP/CAM System for Compound Sheet Metal Cutting and Punching. 7th IFAC Symposium on Automated Systems Based on Human Skill, Joint Design of Technology and Organisation, June 15–17, 2000, Aachen, Germany.

Young, R. E., Greef, A. and O'Grady, P., 1992, An artificial intelligent-based constraint network system for concurrent engineering. *International Journal of Production Research*, **30**, 1715–1735.

Chapter 5
Compound Machining Method for OKP Product Development

Abstract In sheet metal processing and manufacturing, there are many small or medium sized enterprises (SMEs). These OKP manufacturing companies have been facing keen competitive pressure in the market, which has forced these companies to make every effort to shorten PD lead time, improve production efficiency, approach high quality standards, and at the same time cut down the costs. To meet the needs of these companies, this chapter presents a compound cutting and punching production method supported by an integrated CAD/CAPP/CAM system for sheet metal manufacturing. Many existing commercial CAD/CAM systems are not suitable for this manufacturing method, especially under concurrent and global design and manufacturing environments. Some problems have to be solved before these CAD/CAM systems can be employed and integrated into this compound manufacturing method. This chapter deals with solutions to some of these problems. The solutions include an integrated data integration platform based on Pro/INTRALINK and STEP, and a knowledge-based real-time CAPP (RTCAPP) system for compound sheet metal cutting and punching. Within the proposed CAD/CAPP/CAM system, some key modules have been developed. They are the automatic tool selection and manufacturing sequencing module, a shortest tool path optimisation module, a cost estimation module and an automatic insertion of auxiliary path module based on knowledge bases. These modules will be described in this chapter.

5.1 Introduction

Sheet metal processing includes various manufacturing methods such as punching, blanking, bending, laser cutting, flame cutting, *etc.* Different manufacturing methods have different procedures, parameters and processes. Usually a sheet metal part is produced with different processes that may need to be carried out on different machines, this multi-machinery requirement influences not only the manufacturing efficiency but also the manufacturing quality and cost.

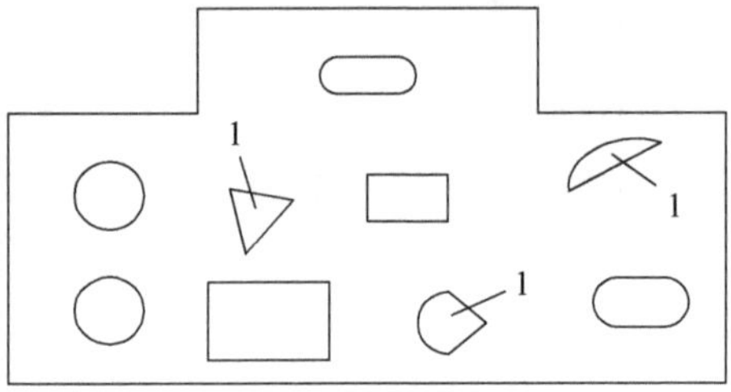

Figure 5.1 Sample sheet metal part for compound manufacturing

A compound manufacturing method means two or more manufacturing processes can be carried out on one machine. The compound sheet metal manufacturing method to be discussed in this chapter includes two common sheet metal manufacturing processes, *i.e.*, cutting and punching. By using this compound method, the cutting process and punching process can be carried out sequentially or concurrently in the same CNC sheet metal compound machine without altering the fixtures. This compound manufacturing method takes the high efficiency and low cost advantages of CNC punching and makes use of the high flexibility of the CNC cutting for complex contour cutting. Based on the authors' experience, this compound manufacturing method can be implemented economically. It is particularly suitable for those companies that produce sheet metal parts for a wide product range with short lead times. Figure 5.1 shows an example that needs to be processed through the compound manufacturing method, since the special-shaped parts 1 can only be cut, whereas all the other parts can be directly punched.

An integrated CAD/CAPP/CAM system is very important to these sheet metal processing companies to support implementation of the compound manufacturing method. Nowadays CAD/CAM systems that will enable a small- or middle-scale company to produce high quality products in a cost-efficient way are much needed. Many existing commercial CAD/CAM systems are either too expensive or too narrow in scope, in the sense that they provide only point solutions (Chaturvedi and Allada 1999). The design of parts and the planning of their manufacturing processes are highly specialised and knowledge-intensive in nature (Sakal and Chow 1994). The user is forced to use two or more CAD/CAM systems that must be compatible with each other. Many of these systems are standalone and expensive for small or medium sized companies. The integrated CAD/CAPP/CAM system should incorporate specific design and manufacturing methods in the companies.

Several integrated CAD/CAPP/CAM systems have been reported in the sheet metal manufacturing area. Wang and Bourne (1995) described automatic process planning for bent sheet metal parts. Ismail *et al.* (1993) developed a low-cost integrated computer aided press tool design (CAPTD) system which uses constraint-driven and object-oriented techniques to advise users on the design of press tools. This CAPTD system uses AutoCAD as the design platform, dBase IV for the selection of press tool components, and a C program-based routine for optimum layout of the workpiece. Lee *et al.* (1993) developed a knowledge-based process planning system (IKOOPP system) to automate and standardise the process planning function for the manufacture of progressive dies. The IKOOPP system receives part definition data from a commercial die design system. The relevant attributes of the manufacturing features within a progressive die model are extracted by a feature extractor and then used by the IKOOPP system to generate the process plan using knowledge bases. Lauwers and Kruth (1994) developed a computer aided process planning system for an electrical discharge machining process that can generate an optimal process plan based on the minimal cost criteria. An integrated intelligent process planning system for prismatic parts was reported by Sakal and Chow (1994); their system integrated AutoCAD and MasterCAM software packages using Autolisp programming language and artificial intelligence techniques. Park and Khoshnevis (1993) presented a real-time CAPP system to achieve concurrent design of prismatic parts and their manufacturing processes.

Despite the success of the systems described above, there exist some limitations to their application when extending them to compound manufacturing processes. These systems need to be further developed to overcome some key problems so that they can support compound manufacturing processes, especially for compound cutting and punching processes. These problems include automatic manufacturing process sequences generation, mathematical model and optimal algorithm for tool selection and manufacturing path optimisation, nesting heuristics and the automatic insertion of "suitable" auxiliary paths for cutting processes based on knowledge bases. Also, all these functions have to be considered in a timely manner in order to support concurrent design and manufacturing. To solve these problems, an integrated CAD/CAPP/CAM system has been developed and will be described in this chapter through discussion of its key components.

Data integration platform. This platform is developed based on Pro/INTRALINK database structure and its application programming interface (API). This platform is for the definition of application data and functions and aims to cover all data forms throughout the life cycle of sheet metal parts, from design to manufacture. The data integration in this platform is based on the STEP standard and object-oriented modelling methods. The platform involves multiple design and analysis applications that require data from several STEP application protocols (AP203, 209, 214, 224) and integration resources from different parts of STEP. Object-oriented modelling methods are used for data that are beyond the STEP standard, *i.e.*, aerodynamic analysis, dynamic data, parametric geometry and constraints. For information that is beyond the STEP standard, "extended" objects are created based on STEP and an object-oriented modelling method. The information modelling framework, as addressed in Chapter 4, is used to build information models for CAD/CAPP/CAM integration. Based on this framework, information being transferred among different modules in the CAD/CAPP/CAM integrated system can be represented using any object-oriented language. This approach has also been used in a computer aided system for concurrently designing and manufacturing a customised OKP mould or die that has complicated sculptured surfaces (Tu *et al.* 1998). The platform provides a data integration environment for the integrated CAD/CAPP/CAM system.

Knowledge-based RTCAPP module. This module includes several sub-modules such as automatic insertion of auxiliary paths based on knowledge bases, a nesting module and a submodule for tool sequencing and tool path planning.

Simulation platform. This has been developed for the integrated CAD/CAPP/CAM system. For this module, an example is provided to compare the compound manufacturing method based on this integrated CAD/CAPP/CAM system with traditional single manufacturing methods.

5.2 Integrated System Architecture for Punching and Cutting

Due to the increasing needs of industries for other types of information and knowledge in sheet metal design and manufacturing processes, new classes of tools to support knowledge-based design, product data management and concurrent engineering have begun to emerge in the engineering market. However, traditional CAD, CAM systems usually run independently and the information transfer among these systems is through specific file formats (STEP, DXF, *etc.*). Also, traditional CAD, CAM tools are mainly focused on database-related issues and do not place a primary emphasis on information models for product representation (Szykman *et al.* 2000). A favourable integration environment is critical to overcome these disadvantages. Most applications use commercial software packages and secondary development tools attached to them to establish the necessary integration environment.

The Pro/ENGINEER series from PTC is an integrated software package. One of its core technologies is parametric and feature-driven modelling and another is that design and manufacturing share a common related database using Pro/INTRALINK. The common space database (the shared database within the Pro/INTRALINK environment) and objected-oriented API can be used as a tool to develop an integrated data-sharing and communication environment where complete information integration overlapping the entire life cycle of a product can be achieved. The common space database (the shared database within the Pro/INTRALINK) is used as a collection point for design activities. The data in this environment can be shared by different commercial CAD and CAM platforms through file transfer, as well as accessed concurrently through Internet and Intranet communication. The CAD/CAPP/CAM integrated system for sheet metal compound manufacturing is developed based on the Pro/ENGINEER platform.

The overall structure of the sheet metal compound manufacturing CAD/CAPP/CAM integrated system is shown in Figure 5.2. There are several modules in this system, including an unfolding module, a data integration platform, a knowledge-based RTCAPP module, a CAM module, a cost estimation module and a simulation module. The unfolding module in this system adopts zero thickness and zero bending radii principles in order to facilitate sheet metal part design processes. The data integration platform is a data integration environment for data exchange and sharing between CAD/CAPP and CAPP/CAM that is developed based on Pro/INTRALINK

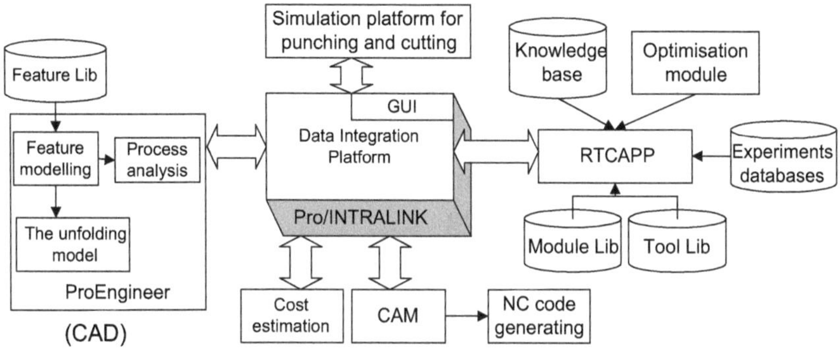

Figure 5.2 CAD/CAPP/CAM integrated system structure for sheet metal compound manufacturing

and API. The knowledge based RTCAPP module generates a process plan, manufacturing parameters, tool sequences, cost, *etc.* and sends all the information back to the data integration platform in real time so that other modules can use them. The simulation module is designed to test all the modules. This chapter discusses three major modules; the data integration platform based on STEP for sheet metal compound manufacturing, the knowledge based RTCAPP module, and the simulation module.

The knowledge-based RTCAPP module for sheet metal compound manufacturing contains several submodules, including knowledge-based auxiliary path automatic insertion, an optimisation module, an experimental database and a tool and module library. The optimisation module is the most important module required to achieve efficient and economic manufacturing and includes tool sequencing, tool path optimisation and nesting optimisation of sheet metal parts with complicated shapes.

There are also some other situations that need to be dealt with in this RTCAPP system, such as the selection of a suitable machining method (cutting or punching in this project). The production of the contour of a sheet metal part depends on many factors, including the degree of complexity of the contour, the batch size of the parts, machining precision, the thickness and material of the sheet metal part. To determine the sequence of the processes, the system must ensure that the outer contours are machined last so as to avoid machining parts that are not fixtured. Tool sequence planning is also important for the compound manufacturing process. Tools are often changed according to different features of a part or product, since different tools are usually needed to produce different features. The frequent tool changes lead to lower production efficiency. Hence, in order to save tool changing time, tool sequence planning is needed. Tool path planning is also very important to improve the efficiency through selecting the shortest machining paths and thereby shortening non-machining time. Knowledge bases are very important for the determination of cutting parameters, lengths and shapes of the auxiliary path and manufacturing processes. In the sheet metal laser cutting process, part materials, pressure of auxiliary gas and laser power are some factors which will influence cutting speed and acceleration. Some major modules will be further discussed in the following sections.

5.3 Data Integration Platform for Sheet Metal Compound Manufacturing

For CAD/CAPP/CAM integrated systems, data integration is very important from the standpoint of more effective utilisation of data in engineering enterprises based on computer aided systems (Chen *et al.* 1994, Xu and He 2004). Complete information integration overlapping the entire life cycle of a sheet metal product is the basis of data sharing and communication among CAD, CAPP and CAM subsystems. From Figure 5.2 we know that a convenient and common data integration platform is essential for sheet metal compound manufacturing, as there is more possible data communication and exchange among different modules.

The data integration platform is based on a step-structure information modelling framework. This framework is based on STEP and the object-oriented method. STEP is used to represent geometry information and information requirements for a particular area of design, manufacturing, engineering or product support (Duan *et al.* 1996). The framework contains four top-down information layers, which include the knowledge layer, the part layer, the feature layer, and the parametric layer. Information models can be built based on this information framework for data exchange and sharing between CAD, CAPP and CAM systems. The parametric layer contains the geometric data of the shape feature of the sheet metal parts and tool features. The feature layer contains all the feature information, which includes not only the feature information (*i.e.*, attributes, constraints) but also relationships with other feature-level information objects and objects defined by users. The part layer contains all the part information that includes feature information and relationships among different part-level information objects. The knowledge layer contains not only the part information, but also "knowledge-related" information objects and an inference engine. Application objects defined by users according to the requirements of the project can be entered into the feature and part layer. The detail of this framework has been well addressed by Tu *et al.* (2000) and has been used for information modelling of the concurrent manufacture of mould/die products with complicated sculptured surfaces (Tu *et al.* 1998).

Figure 5.3 shows an extended common object-oriented data class structure based on the information modelling framework; the relationships with other classes or objects have also been defined on this platform. Other features can also be defined based on the common class. For example, a new shape feature that contains information that is beyond STEP can be defined. Figure 5.4 shows an example using object-oriented modelling and STEP to build an information model for an inner feature and a related punching tool. Classes and relationships between the inner feature and the tool are clearly shown in Figure 5.3. For punch tool selection, each inner feature, outer feature and non-feature element of the outer contour has a related

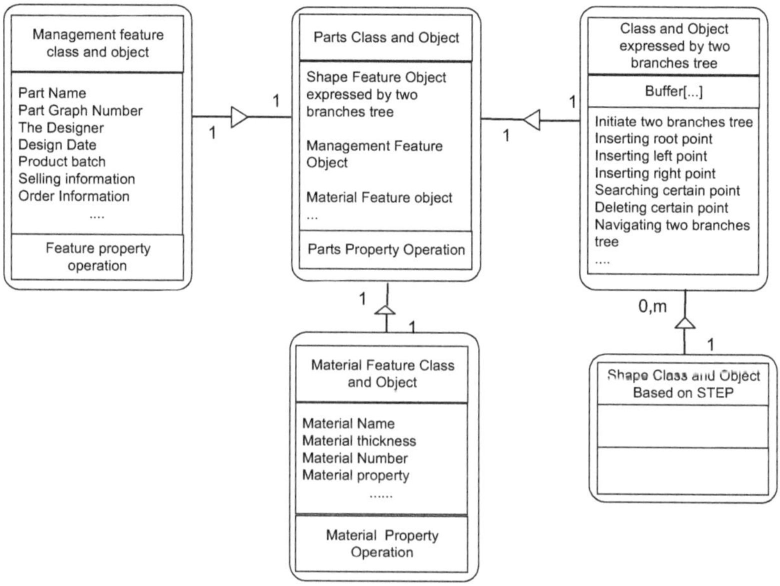

Figure 5.3 Object-oriented expression of sheet metal parts

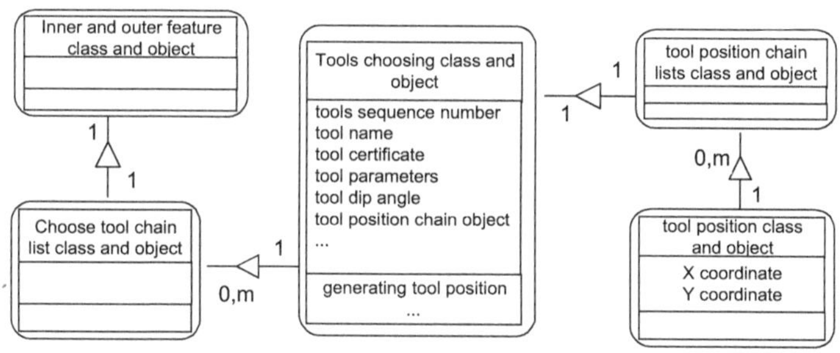

Figure 5.4 Object-oriented expression of inner holes feature – tool chain list

tool after a tool has been automatically selected. An information model that shows the feature–tool chain data structure and parametric element–tool data chain structure is also shown on Figure 5.4. Other information models that are needed for the CAD/CAPP/CAM system can be built based on this framework (Tu *et al.* 2000).

The data integration platform based on the framework for sheet metal compound manufacturing includes two integrated graphical user interfaces. One is a CAD/CAPP data interface, which includes feature data processing, feature data recognition/extraction and feature data transformation. The geometry information from CAD is saved in the parametric layer. The CAPP information (cost, attributes, constraints, machining sequences, collisions or machining time) from CAPP to CAD is defined through the data interface and saved in the feature layer (see Figure 5.3). The CAPP module can read the CAD information by accessing information in the parametric layer. The CAPP module can also add process information related CAD features, for example auxiliary paths for features to be cut automatically, and relationships to show the tool that would be used to punch the feature will be created. The other graphical user interface is the CAPP/CAM data interface. The main task to integrate CAPP and CAM under Pro/ENGINEER is to receive geometry and partial process information from CAD and machining parameters from CAPP. In this compound machine (laser cutter and punch), the information exchange between CAM and CAPP includes laser power, machine speed, laser focus position, punching tool information and other machining parameters. The information flow between CAPP and CAM is different in different systems and applications, but there is no standard protocol between CAPP and CAM that can be used (Papadimitriou and Steiglitz 1982). Therefore, these two interfaces have been designed to solve this problem. They can be used for non-standard protocol data and functions definition, such as cost information, specific manufacturing parameters and functions for specific manufacturing processes.

The data integration platform makes it possible for CAD, CAPP and CAM software to directly communicate with this platform concurrently. Other specific applications can also be carried out based on this platform. Real-time process planning (Joo *et al.* 1994) can be achieved based on this data integration platform if suitable information can be sent to the platform in real time.

5.4 Knowledge-based RTCAPP System for CNC Punching and Cutting

CAPP is one of the important and most complicated computer aided systems, especially for the sheet metal compound manufacturing industry. In recent years, advanced computer software techniques such as expert systems and artificial intelligence approaches have been used to develop generative process planning systems. The CMPP (Duda *et al.* 1997) and HZ-RCAP (Cai and Peng 1995) systems have a generative process planning functionality for cylindrical-shape parts, where knowledge of manufacturing processes and parameters are in the rule base. This chapter presents a knowledge-based RTCAPP system based on the data integration platform, which includes several optimisation modules that aim to improve the manufacturing efficiency and cut down costs, and provide knowledge-based auxiliary paths and manufacturing parameter selection for sheet metal cutting processes. Several typical functions of the CAPP system have been defined in Figure 5.5. The functions in the RTCAPP system are designed to achieve different application tasks. These functions can access all required information through the data integration platform. Tool database, cutting parameter database and process rule database are all connected with the data integrated platform through ODBC. In Figure 5.5, functions have been defined for automatic tool selection, tool path optimisation, nesting optimisation, manufacturing parameters, *etc.* For the sheet metal compound manufacturing RTCAPP system, real-time optimisation modules are very important for

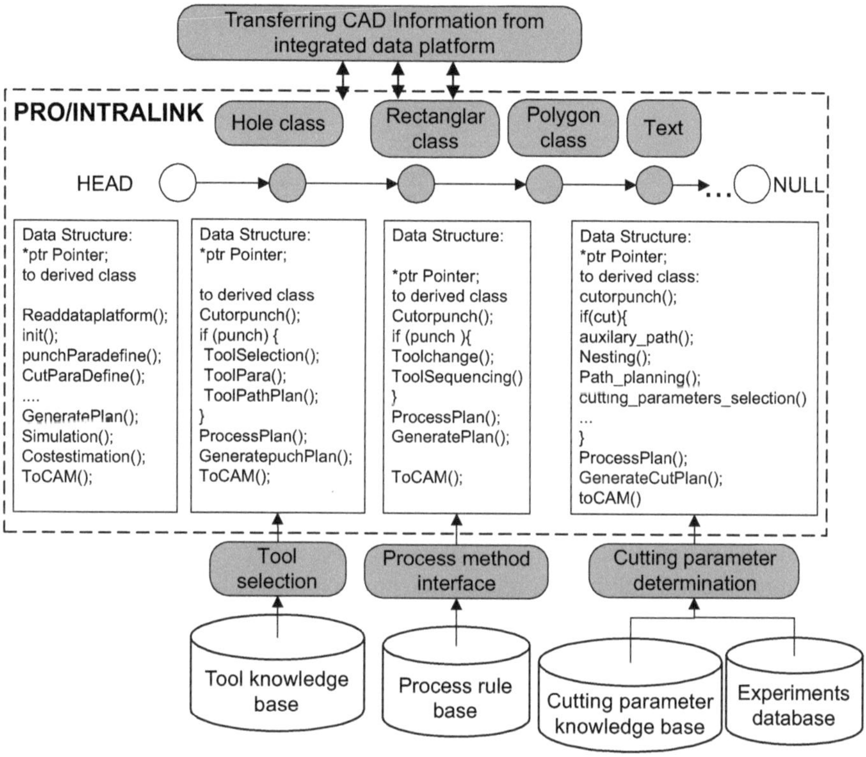

Figure 5.5 Data structure and functions of integrated CAPP system for sheet metal compound manufacturing

improving manufacturing efficiency and cutting down costs, while manufacturing parameters and the auxiliary path will influence the manufacturing quality of sheet metal parts.

5.4.1 Tools Sequencing Real-time Optimisation Algorithm for Punching

The main effort in generating a production plan for punching operations lies in the tool selection process. In the modern sheet metal industry the selection of suitable tools is a major economic issue for two reasons. Firstly, tooling is very expensive and should be minimised. Secondly, if the tool capacity of the tool rack is exceeded, expensive set-ups are needed, so tool sequencing and tool arrangement are very important. During the punching process of sheet metal parts, different tools are needed to punch different features, so tool change is needed frequently and will lead to lower efficiency. In order to save tool changing time, it is necessary to finish all

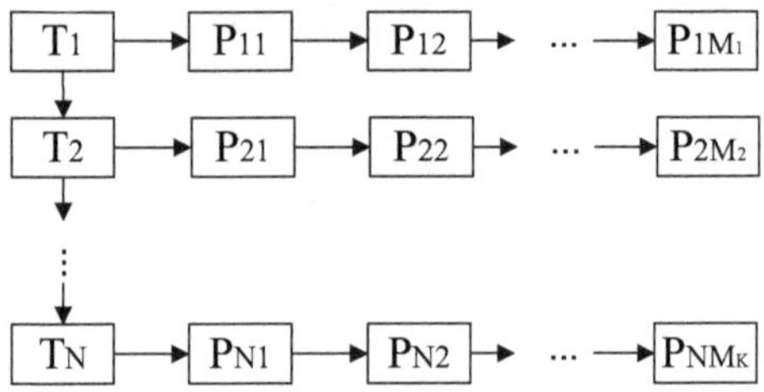

Figure 5.6 Tool–tool position chain-list

related feature machining procedures after choosing a tool before moving on to another tool. A tool sequencing algorithm is used to arrange tool sequences as follows:

1. Based on the feature–tool chain-list structure, determine all the tools that can manufacture all the inner and outer features for this part and construct a tool–tool position chain-list, as shown in Figure 5.6.
2. From tool–tool position chain-list structure, search for a tool position P nearest to the origin of a working coordinate.
3. Find related tool T_i according to tool position P_s.
4. Plan tool path for all tool T_i related tool position points; this algorithm will be given in a later section.
5. Decide if all tool machining sequences have been arranged; if finished, then algorithm ends, otherwise set the last tool position as P_E, then search the tool position P nearest to P_E from the tool positions on which the related tool sequence was not arranged, set $P = P_E$ and go to 3.

With regards to a single punching tool, the arrangement of the punching sequence must ensure that the total distance that the punching tool moves is the shortest, *i.e.*, a tool path optimisation problem. If a machining path can be arranged reasonably, then tool non-machining time can be reduced.

5.4.1.1 Mathematical Model

The punching tool path planning task can be described as: 1. a tool begins to punch from the current position; 2. it should finish all related punching operations, and the punching operations cannot repeat on any position; 3. an optimisation algorithm needs to be developed to automatically select a manufacturing sequence in which the distance moved by the tool is the shortest possible.

The tool path optimisation problem can be transferred to the mathematical problem of determining the shortest Hamilton return-loop in diagram theory. For convenience, the start position and all the related position descriptions for all punching operations of the tool can be described as a point set **P:**$\{p_1, p_2, \ldots, p_n\}$. Hence, the tool path optimisation problem can be described using diagram theory as follows:

Construct a whole diagram $K_n = (V, E)$, among them, V is a set of top points and E means the distance between each point. From all top points, the algorithm seeks a return-loop passing through all points only once, while ensuring the sum of distances the tool moves is the shortest. This is a Hamilton loop optimal problem.

In fact, this is an NP-completeness problem. Some algorithms could be used for this problem under different situations, such as dynamic programming and tree theory. There are no efficient algorithms that can find the exact solution within multinomial time (Summad and Appleton 1998, Papadimitriou and Steiglitz 1987, Hamel and Steel 1994). A reasonable method is to find an approximate algorithm to solve this problem. In our system, the "shortest insert" algorithm has been adopted, which can be implemented easily and meets the concurrent requirement of design and manufacturing.

5.4.1.2 Mathematical Description of the Shortest Insert Algorithm

The distance between a subreturn-loop T and a top point P which does not belong to the return-loop is defined as:

$$\mathrm{d}(T, p) = \min\{\mathrm{d}(x, p) : x \in T\}$$

Suppose the whole graph is expressed as $K_n = (V, E)$ and the number of points in this graph is n. Then the mathematical description of the shortest insert algorithm is given below:

1. Choose a top point v_1, and a subreturn-loop T_1 which is formed without side. Then find a top point v_2 which has shortest distance from v_1. Based on v_1 and v_2, another return-loop T_2 is formed, which includes sides (v_1, v_2) and (v_2, v_1).
2. If the subreturn-loop contains all n top points, then algorithm ends, otherwise, seek a top point $v_s \in V\ T_i$, this point meets the condition $\mathrm{d}(T_i, v_s) = \min\{\mathrm{d}(T_i, x), x \in V\ T_i\}$, $1 < i < n$.
3. Find side (v_x, v_y) from T_i, which ensures that $\mathrm{d}(v_x, v_s) + \mathrm{d}(v_s, v_y) - \mathrm{d}(v_x, v_y)$ is the shortest distance. Delete side (v_x, v_y) from T_i to give a new return-loop T_{i+1} formed by sides (v_x, v_s) and (v_s, v_y).
4. The top point number in this sub return-loop plus 1, then go to step 2.

5.4.1.3 Programming Implementation

The steps for implementing the shortest insert algorithm are:

1. Build a matrix $[\mathrm{d}_{ij}]$, which shows distances between different top points.
2. Create a linear list: "List1", and put all the top points in List1. Build a loop chain-list: "List2", put the first point **A** in List1 into List2, and delete this point from List1.
3. Calculate the distances between every point in List1 and the point **A** in List2. According to the calculation results, find the point **B,** which has the shortest distance. Insert **B** into List2 and delete point **B** from List1.

4. If List1 is NULL, then algorithm ends, otherwise compare the distances between all points in List1 and all points in List2, and find the two points with shortest distance (suppose the point in List1 is **x**).
5. For all neighbouring points M_i and N_i from List2, calculate the absolute value of $\mathrm{X}M_i + \mathrm{X}N_i - M_i N_i$. Select the two neighbouring points M and N with the smallest absolute value.
6. Insert x in the middle of M and N on List2, and delete **X** from List1. Go to (3).

After the algorithm finishes, linear chain List1 is empty, and all top points have been stored in return-loop list List2. The tool sequence and related position have been put in List2, and need to be put in related tool objects, a correct connection is automatically created between the tool object and related features to be punched by this tool. After the tool sequencing algorithm finishes, each feature to be punched should have been connected to a tool object.

5.4.1.4 Performance of the Algorithm

The performance of the shortest insert algorithm is very important in the CAPP system. Several tests have been carried out with regards to its speed, optimality and robustness. Table 5.1 shows three major characteristics of this algorithm compared with two other common algorithms. In this table, L_{a} is the length of the Hamilton return-loop and L_{b} is the optimum length of the Hamilton return-loop. As we can see, the nearest neighbour algorithm is easy to implement by programming, but its optimum performance is not good. The optimum weight spanning-trees algorithm can achieve better optimisation results, but is hard to achieve using programming and takes a long time to find the optimum result, especially when sheet metal parts are more complicated.

As discussed in the previous section, the shortest insert algorithm can be implemented easily by programming and performance is better than the nearest neighbour algorithm and is not time consuming (from Table 5.1), which is suitable for the real-time optimisation task in this project. In order to test the actual application results of this algorithm, a series of tests have been carried out in two sheet metal companies. A case study will be given in Section 5.5 (see Figure 5.10). These tests show that the algorithm can achieve good optimisation results and has high robustness and stability.

Table 5.1 Comparison of algorithms

	Time complex	Algorithm performance ($L_{\mathrm{a}}/L_{\mathrm{b}}$)	Programming implementation
Nearest neighbour algorithm	$\mathrm{O}(n^2)$	$\leq \frac{1}{2}(\log_2 n + 1)$	Very simple
Shortest insert algorithm	$\mathrm{O}(n^2)$	< 2	Simple
Optimum weight spanning-trees algorithm	$\mathrm{O}(n^3)$	$< 2/3$	Most complicated

5.4.2 *Knowledge-based CAPP for Sheet Metal Cutting*

Manufacturing parameters and auxiliary cutting paths are very important for the cutting quality of sheet metal parts if a cutting procedure is chosen. The automatic selection of manufacturing parameters using experiment results and a neural network reasoning method has been investigated by several authors (Raggenbass and Reissner 1991, Xie and Duan 1995). The auxiliary path is another difficult problem for sheet metal cutting, especially for flame or laser cutting. For sheet metal parts of different shapes, the auxiliary cutting path is quite different. Usually the auxiliary cutting path are of three types: "cut-in", "cut-out", and "closed".

The cutting process usually does not start from the contour directly, but from somewhere else. A hole is made (using laser or flame) first, and after cutting a segment of the auxiliary path, cutting of the contour begins. After contour cutting is finished, another exit auxiliary path is needed to be cut; these two segments of auxiliary path are the cut-in and cut-out auxiliary path. The hole is called "start hole", and should be set on waste material. When cutting inner contours, this hole should be set on the inner side of the contour, when cutting outer contours, the hole should be set on the outside of the contour. The closed auxiliary path is a special path that is added in order to avoid burning a sharp angle while cutting as outer contours. There are several different rules for different features, some examples are given below. All these rules have already been put into the knowledge base.

5.4.2.1 Cut-in and Cut-out Auxiliary Path

Choosing a cut-in point. The cut-in point is the first point in a contour after cutting the auxiliary path. This point is set automatically by the program. The CAPP platform also provides a user interface for setting this point. The automatic generation of the cut-in point is based on the principles as follows:

1. If the inner contour or the outer contour is circular, choose the point with least y coordinate as the cut-in point.
2. If the outer contour is organised by line segment and circular segment, a set of top points, U, will be formed by neighbouring parametric elements. Suppose U_X is a set of the points with least x coordinate from set U. From U_X the point P with least y coordinate can be chosen as the outer contour cut-in point.
3. If the inner contour is composed of line and circle segments, find top point **P** according to step **2**. From all the inner contours, find a related line or circle segment the end point of which is P, and then choose the middle point as the cut-in point.

Determination of cut-in and cut-out auxiliary paths. The cut-in or cut-out auxiliary path is actually a segment of a circle or line. We also call these cut-in or cut-out stretching arcs. Generally, for inner contours, the cut-in or cut-out stretching arc is a segment of a circle. For outer contours, the cut-in or cut-out stretching arc is generally a segment of a line.

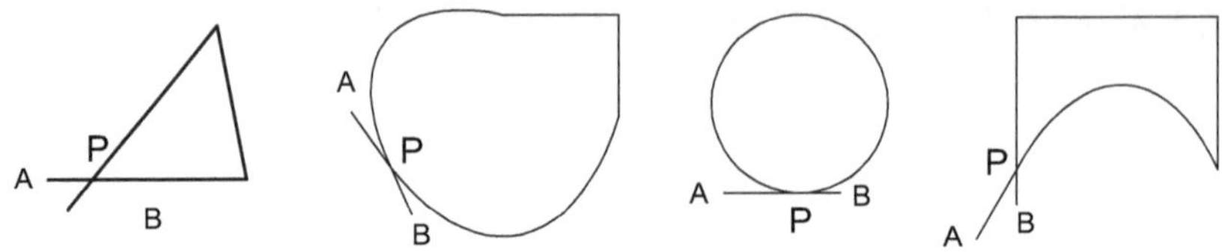

Figure 5.7 Cut-in and cut-out arcs for outer contour cutting

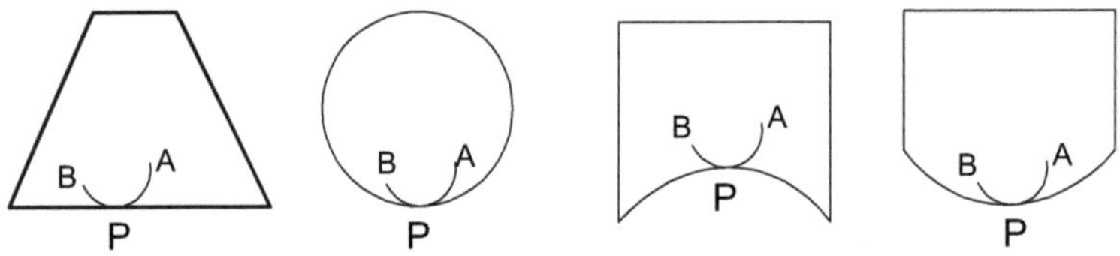

Figure 5.8 Cut-out and cut-in arcs for inner contour cutting

As shown on Figure 5.7, the cut-in and cut-out stretching arcs of the outer contour are extension lines of the tangent line to the outer contour at the cut-in point. On Figure 5.7, lines from point A to point P are cut-in stretching arcs, and the lines from point P to point B are cut-out stretching arcs.

As shown on Figure 5.8, cut-in and cut-out arcs of inner contours are circular segments, the directions being tangential to the original line or circle at point P; on Figure 5.8, lines from point A to point P are cut-in stretching arcs, and lines from point P to point B are cut-out stretching arcs.

In order to avoid burning the sharp angle while cutting sharp angles of outer contours, a circular auxiliary path is needed. A closed auxiliary path can be added on the convex point of outer contours. Adding an auxiliary path on a concave point is not allowed because it will "burn" the outer contours when the cutting tool enters from the auxiliary path to the contour. There are three kinds of sharp angle situations as shown on Figure 5.9, Line–line, line–circle and circle–circle. The added closed auxiliary paths as shown on Figure 5.9 are small triangles, the direction of the entry and exit paths of the auxiliary path are tangential to the original contour paths shifting at the corner.

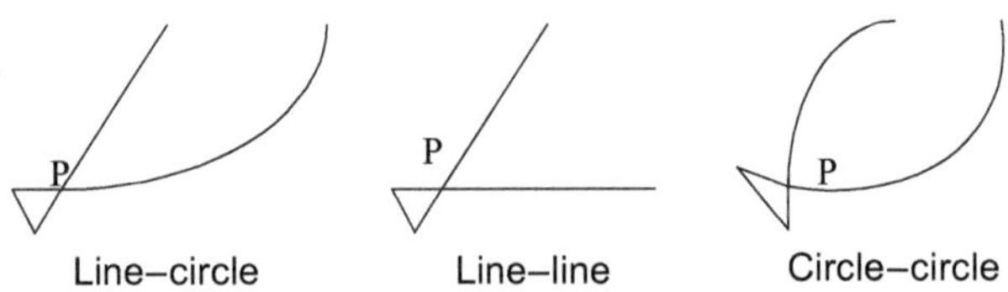

Figure 5.9 Circular auxiliary paths for outer contour manufacturing

5.4.2.2 Least Distance Between Start Hole and Contour

The "start hole" is a small hole made before cutting the outer contours. A specific amount of time is needed to make this small hole. There is an area called the heating influence area, and to ensure manufacturing quality, the heating influence area cannot extend into the contours of the part. This means a least distance between start hole and contour must exist. How this least distance is chosen influenced by the cutting pattern, part material and thickness, and so on. To automatically choose this least distance, a database was built based on experiments in the factory, and a neural network aggregated from these experiments was also developed in this CAD/CAPP/CAM system (Xie and Duan 1995).

The knowledge base containing the information on how to insert the cut-in and cut-out auxiliary path is very important for the quality of cutting processes. Once a knowledge base is built according to the aforementioned rules, a suitable auxiliary path can be inserted automatically according to this knowledge base and experiment databases. The experiment databases were built as a result of two years of cutting experiments in the factory; a reasoning module using a neural network with a self-learning function which can automate cutting process planning has been used in this system. This module is very important for compound manufacturing processes; without this module, the RTCAPP for compound sheet metal parts cannot be achieved.

5.4.3 Tool Path Optimisation for Cutting and Cost Estimation

Cutting tool path optimisation is important for cutting efficiency. Compared with the shortest tool path optimisation algorithm for the punching process discussed in Section 5.4.1, the cutting process for sheet metal parts is different from CNC punching and tool exchange is not needed. The principle of the shortest path optimisation algorithm for both cut and punch is almost the same, when some restrictions need to be considered when implementing the optimisation process for cutting. These restrictions are based on actual compound manufacturing requirements. The cutting and punching sequences need to follow some rules while the optimisation processes are considered, for example, both punching and cutting processes must obey the rule whereby the inner contour must be cut or punched first and outer contour cut or punched later. Hence, the tool path optimisation for CNC cutting includes not only path planning for the inner contours of each part, but also the machining sequence of parts after nesting. Generally, for compound cutting and punching, nesting optimisation of differently shaped parts should start first. The tool selection and punching path-planning processes have to wait until the nesting optimisation process finishes. The cutting process does not manufacture every part separately but depends on the shortest cutting distance optimisation algorithm. For the cutting sequence of each part, the entry point P of the part can be used for the arrangement of manufacturing sequences. For tool path planning of the inner contours of each part, the entry point P of the inner contours can be used for arrangements of machining sequence; the cutting path optimisation algorithm is the same as the punching tool sequencing algorithm discussed earlier.

The manufacturing cost estimation submodule estimates the manufacturing cost of the sheet metal part. This submodule takes its input from the NC programming and the knowledge-based RTCAPP submodule. The compound machining time is calculated from the NC part programs. The set-up times are estimated times associated with a particular punching operation, which include punching times and set-up times. All the cost information have been recorded in a database. The cost of using cutting processes such as the expense of cutting gas, auxiliary gas, cutting machine and laser machine for laser cutting can be calculated from the machining times. The cost of product design and process planning can also be calculated by related cost information in the database. The cost of sheet metal utilisation ratio can be calculated according to the data input from the nesting optimisation algorithms submodule. The total manufacturing cost of the part can be calculated by adding the estimated material cost to the sum of the costs of all processes. As we can see, all the optimisation algorithms strongly influence the cost. The function of cost estimation has been defined in Figure 9.5, the result of which can be sent to CAD or other modules.

5.5 Simulation Platform for Compound Manufacturing

Simulation is one of the most powerful analysis tools available for the design and operation of complex processes in a distributed environment (Tan and Hui 1998). For compound sheet metal cutting and punching, a simulation platform becomes more important because of the complex manufacturing processes resultant from both punching and cutting processes running concurrently. In this CAD/CAPP/CAM integrated system, a real-time simulation platform has been developed. This simulation platform is directly driven by geometric parameters, manufacturing parameters and process planning results. All key components such as punching tool selection, path optimisation and nesting optimisation have to be tested in this platform. Some modules (nesting and shortest insert algorithm) have been developed and tested in isolation in industry for more than two years before being integrated in this integrated CAD/CAPP/CAM system. A case study was undertaken with regards to tool path planning, and the performance of the optimum algorithms (nesting and shortest insert algorithm) and auxiliary cutting paths in two sheet metal companies. Single and compound manufacturing processes have been compared.

Figure 5.10 shows the different manufacturing processes of four frame parts after nesting optimisation using different manufacturing methods. As there was no nesting algorithm in these companies, the nesting program greatly improved the utilisation ratio of their raw materials, especially when the shape of the sheet metal part was more complicated. Auxiliary cutting paths were added automatically according to the knowledge bases discussed in Section 5.4.2, and can be clearly seen in Figure 5.10a and 5.10b. Figure 5.10a shows the performance of the laser cutting method with the shortest path optimisation algorithm and Figure 5.10b shows the performance of the laser cutting method using the nearest neighbour optimisation algorithm. For the example parts, the cutting distance difference between the two algorithms is shown in Figure 5.10h. It is obvious that the shortest path optimisation algorithm reduces the non-machining moving distance and its performance is better than the nearest neighbour optimisation algorithm. Figure 5.10d shows the punching machining method with shortest path optimisation, and Figure 5.10e shows punching machining without the optimisation

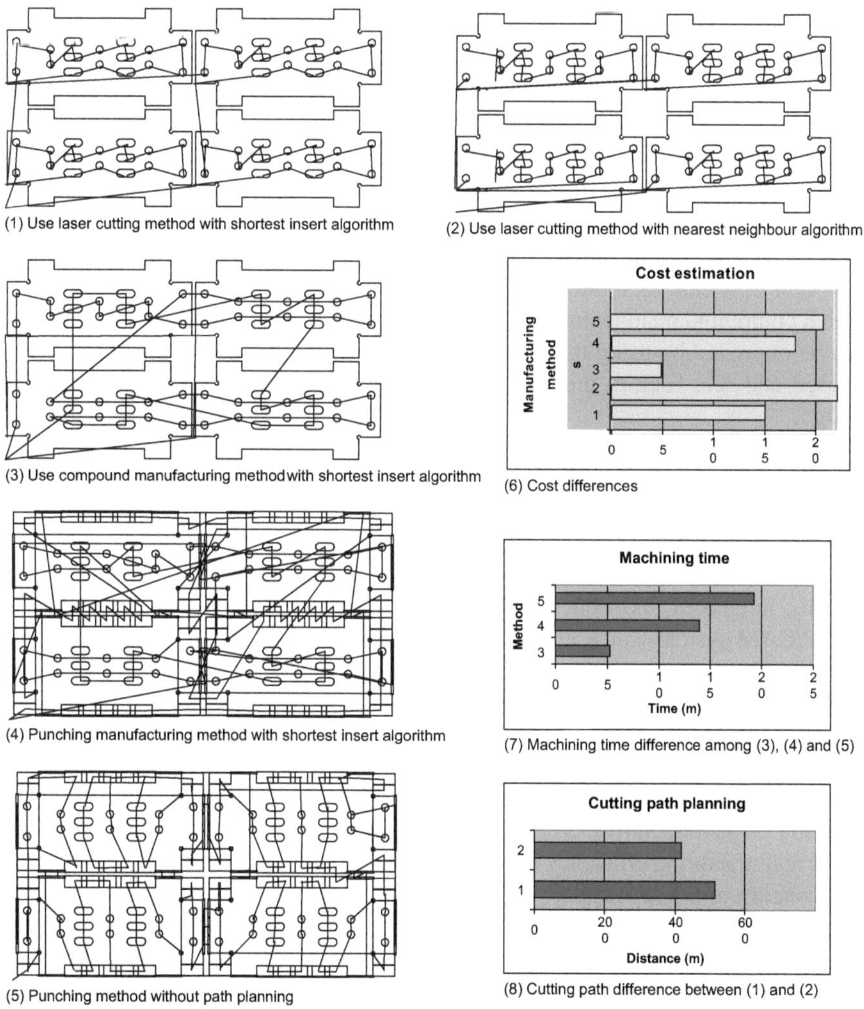

Figure 5.10 Simulation result for cutting, punching and compound manufacturing

algorithm. We can see the difference before and after the tool path-planning algorithm. Comparing Figure 5.10d with Figure 5.10e, the total path length after optimisation is longer than the total path length before optimisation. However, punching tools need to be changed more frequently according to Figure 5.10d, which results in a lot of waste of time. Figure 5.10g shows the machining time difference of the single punching machining method before and after optimisation when tool-changing time is set as 50 s and the moving speed of the machine is 5 m/s. The machining time difference is also shown in Figure 5.10g comparing the single punching method and the compound method. The better performance of the compound manufacturing method can be clearly seen when compared with single manufacturing methods. Figure 5.10c shows results using the compound manufacturing method containing all the functional modules. It is apparent that the total distances moved by the manufacturing tools shorter than when using other methods shown in Figure 5.10. This compound manufacturing method can greatly shorten the manufacturing time and distances that the tools move. Manufacturing efficiency is also greatly improved. After being given the related cost parameters detailed in Section 5.4.3, the overall cost can be automatically estimated. Figure 5.10f shows the cost estimation for the compound manufacturing method compared with the other two single manufacturing (cutting and punching) processes before and after optimisation. These simulation results show that manufacturing time and cost can be significantly reduced by using the CAD/CAPP/CAM integrated compound manufacturing method. The utilisation ratio of sheet metal parts can be improved by using the nesting algorithm in real time, and costs can be optimised and sent back to the designer in real time to achieve a more economical design.

5.6 Conclusions

A fully integrated and concurrent sheet metal compound manufacturing CAD/CAPP/CAM system which allows for the creation of a real-time process plan and definition of non-standard protocol information for sheet metal part design and compound manufacturing processes has been presented. The source code of the CAD/CAPP/CAM system is given in Appendix A.5. A data integration platform has been proposed for the definition of information models and functions for sheet metal punching and cutting. A knowledge-based real-time CAPP system with optimisation of manufacturing process and utilisation ratio has been developed to improve manufacturing efficiency and product quality, and to cut costs. This integrated and concurrent CAD/CAPP/CAM system eliminates the necessity for employing different software packages for simulation, NC generation and process planning and optimisation activities. It will enable "right first time" design, improve product quality, reduce product costs, and enable rapid response to market needs in terms of reaction time and product changes. This integrated CAD/CAPP/CAM system for compound sheet metal parts manufacturing has been developed using Visual C++

programming language. After the system was developed, it was implemented by two sheet metal manufacturing companies and used for over two years. The utilisation of their sheet metal parts has been improved by 15 %, costs and product design and manufacturing time have been greatly reduced, and the PD cycle time has been significantly shortened.

References

Cai, L., and Peng, L., 1995, CAPP system (HZ-RCAP) for rotatory parts under CAD/CAM integration environment. *Journal of Huazhong University of Science and Technology,* **23**, 83–87.

Chaturvedi, S., and Allada, V., 1999, Integrated manufacturing system for precision press tooling. *Journal Advance Manufacturing Technology,* **15**, 356–365.

Chen, C., Swift, F., Lee, S., Ege, R., and Shen, Q., 1994, Development of a feature-based and object-oriented concurrent engineering system. *Journal of Intelligent Manufacturing,* **5**, 23–31.

Duan, G., Wang, J., Liu, D., Lei, N., Bian, W., 1996, Research on an object-oriented CAD/CAPP/CAM integrated system based on STEP. *IEEE Conference on Industrial Technology,* pp. 29–33, 2–6 Dec 1996, Shanghai, China.

Duda, J., Habel, J., and Pobozniak, J., 1997, Control mechanism of expert system with representation of knowledge in the form of hierarchical decision nets. *Proceedings of Artificial Intelligence in Engineering XII,* Computational Mechanics Publications, pp.17–28.

Hamel, A.M., and Steel, M.A., 1994, *Finding a Maximum Compatible Tree is NP-Hard for Sequences and Trees* (Christchurch: Department of Mathematics and Statistics, University of Canterbury).

Ismail, H.S., Hon, K.K.B., and Huang, K., 1993, CAPTD: a low-cost integrated computer aided design system for press tool design. *Proceedings of the Institution of Mechanical Engineers, 207,* pp. 117–127.

Lauwers, B., and Kruth, J.P., 1994, Computer-aided process planning for EDM operations. *Journal of Manufacturing Systems,* **13**, 313–322.

Lee, I.B.H., Lim, B.S., and Nee, A.Y.C., 1993, Knowledge-based process planning system for the manufacture of progressive dies. *International Journal of Production Research,* **31**, 251–278.

Park, J.Y., and Khoshnevis, B., 1993, A Real-Time Computer-Aided Process Planning System as a Support Tool for Economic Product Design. *Journal of Manufacturing Systems*, **12**, 181–193.

Papadimitriou, C.H., and Steiglitz, K., 1982, *Combinatoral Optimisation: Algorithms and Complexity* (Englewood Cliffs, N.J.: Prentice Hall).

Raggenbass, A., and Reissner, J., 1991, Automatic generation of NC production plans in stamping and laser cutting. *Annals of the CIRP*, **40**, 247–250.

Sakal, R.L., and Chow, J.G., 1994, An integrated intelligent process planning system for prismatic parts using PC-based CAD and CAM software packages. *International Journal of Advanced Manufacturing Technology*, **9**, 166–174.

Summad, E., and Appleton, E., 1998, Applications of the Monte Carlo simulation method in tool selection for punching operations in the sheet metal industry. *International Conference on SIMULATION*, York, UK, 30 September–2 October.

Szykman, S., Sriram, R.D., Bochenek, C., Racz, J.W., and Sriram, R., 2000, Design repositories: next-generation engineering design databases. *Journal of National Institute of Standard and Technology*, on the Web: http://www.mel.nist.gov/msidlibrary/summary/pubs00.htm.

Tan, G.S.H., and Hui, K-L., 1998, Applying Intelligent Agent Technology as the Platform for Simulation. *The 31st Annual Simulation Symposium*, Boston, MA, April 5–9.

Tu, Y., Yang, W., and Xiong, Y., 1998, A concurrent manufacturing strategy for one-of-a-kind products with complicated sculptured surfaces. *International Journal of Advanced Manufacturing Technology*, **14**, 93–98.

Tu, Y., Xie, S.Q., and Liu, J., 2000, *Virtual Product Development for One-of-a-Kind.* IFAC, Aachen, Germany, 14–18 June.

Wang, C.H., and Bourne, D.A., 1995, Using features and their constraints to aid process planning of sheet metal parts. *Proceedings of IEEE International Conference on Robotics and Automation*, pp. 1020–1026.

Xie, S.Q., and Duan, Z.C., 1995, Automatic manufacturing parameters selection of sheet metal laser cutting by using experiments and neutral network method. *Journal of Mechanics*, **23**, 54–57 (in Chinese).

Xu, X., and He, Q., 2004, Striving for a total integration of CAD, CAPP, CAM and CNC. *Robotics and Computer-Integrated Manufacturing*, **20**(2), 101–109.

Chapter 6
An Agent-based Sheet Metal Process Planning System

Abstract The requirements for computer-aided process planning (CAPP) systems have changed in the current integrated manufacturing environment. More requirements such as openness, real time operation and distribution of process planning need to be satisfied by integration. New systems must be open and dynamic with the ability to adapt and accept radical unpredictable changes in structures and industrial practices. This chapter proposes an agent-based process planning system (ABPPS) for optimal sheet metal PD. The structure of the system is discussed in detail. A number of agents are defined, including an unfolding agent, a feature recognition agent, a task agent, a nesting agent, a path planning agent, a bending agent, a machining method selection agent, a machine selection agent and a fixture/jig selection agent. Intermodules of typical agents (nesting agent, path planning agent, bending agent) are built for the system. In addition, how agents cooperate with others to implement tasks is also investigated.

6.1 Introduction

Globalisation of international markets has created a high level of competition that requires great agility, rapid changes in production styles and fast configuration of manufacturing systems. As the bridge between computer-aided design (CAD) and computer-aided manufacturing (CAM), computer-aided process planning (CAPP) plays an important role in the computer integrated manufacturing (CIM) environment. CAPP is related to every stage of the PD process, such as feature recognition, machining operation selection, machine selection, cutting tool selection, fixture selection and design. Manufacturing process planning determines how a product will be manufactured. It is a complex engineering process of selecting and sequencing manufacturing processes and parameters, so that one or more goals such as lower cost and shorter processing time are achieved, while satisfying a set of constraints. In today's global manufacturing market, methods to increase the integration, openness, real-time operation and distribution of process planning to reduce product lead time and cost, and to achieve high quality and productivity, have become paramount.

The increasing needs of CIM systems in various industries require that CAPP systems be extendible and flexible to suit different practical industrial applications. The use of agent technology can break down the rigidity and loss of flexibility and adaptability of CIMs in a dynamic environment, and realise strong cooperation and coordination in the whole system to achieve low cost and short lead-time products. Successful applications of agent technology in support of concurrent design have shown advantages over traditional approaches (Cherkaoui *et al.* 2003, Huang 2004). When a new design for a product is developed via CAD, CAD designers can receive downstream feedback from the process planning phase and the manufacturing shop floor concurrently using agent technology, and an optimal design plan can be achieved. This dramatically reduces PD cost and lead time. Agent technology is regarded as one of the most promising technologies to fulfil the above-mentioned goals for process planning in a distributed manufacturing environment.

Agent technology is not a single, new, emerging technology, but rather the integrated application of multiple technologies. In artificial intelligence (AI), the agent concept grew from early work on blackboards, contract-nets, and actors, to the very popular and widely used term in many research areas today. The agent-based method offers a powerful repertoire of tools, techniques, and metaphors that have the potential to considerably improve the way in which people conceptualise and implement many types of software. Agents are being used in an increasingly wide variety of applications – ranging from comparatively small systems such as personalised email filters, to large, complex mission-critical systems. A number of researchers have attempted to apply agent technology in industrial applications, including process control, manufacturing, air traffic control and industrial robotics.

The properties of agents meet the requirements of the next generation manufacturing systems (Shen and Norrie 1999), which need to be open, dynamic, agile, scalable and fault tolerant. Nowadays, agent technology is recognised as a promising approach for intelligent manufacturing systems. The unique features of agent technology have shown great potential in developing distributed manufacturing systems. It leads to applications in manufacturing areas, such as manufacturing planning, scheduling and control, supply chain management, and enterprise integration. In the traditional manufacturing environment, information systems mimic organisational structures, utilising a top-down, command-and-control structure. Communicating decisions and information down through the organisation is time consuming – making it impossible to respond and adapt quickly to the external environment. Traditional manufacturing relies on schedules as a means of forecasting what needs to be produced. One of the problems with traditional schedulers is that they try to anticipate and plan for every possible change that may occur. However, in real industrial cases, the range of scenarios and possible combinations of parameters is extremely large owing to the complex manufacturing environment. In short, traditional manufacturing facilities have shortcomings that affect their ability to compete in today's constantly changing marketplace. The traditional approaches limit the expandability and reconfiguration capabilities of manufacturing systems. These shortcomings lead to problems such as reduced productivity, increased costs, and missed market opportunities. To remain competitive in today's marketplace, manufacturing systems have to be upgraded to meet today's increasingly dynamic and sophisticated manufacturing requirements.

This chapter proposes an agent-based process planning system (ABPPS) for sheet metal parts, and investigates how agent technology can be applied in sheet metal process planning. The aim is to use agent technology to facilitate flexibility and distribution of process planning. In the agent-based approach, a complex process planning task is divided into small subtasks. Each subtask is performed by a defined agent. The process planning tasks are carried out by various defined agents working cooperatively in a distributed environment. The structure of the system is investigated and tested and agents for sheet metal process planning tasks are defined. The essential point in agent technology is the power of multiple cooperating agents, however, one should also understand the functionality of the individual agent to be able to benefit from agent technology.

6.2 Literature Review

In modern manufacturing, reducing the process planning and machining time has become very important as the industry moves towards small batch production. The emergence of agent technology in process planning is an exciting and promising research area that promises to increase production efficiency and competitive ability in fierce markets. This section gives an overview of process planning and agent technology in process planning. It also discusses current process planning problems and demonstrates how they can be solved using agent technology. Three agent-based approaches for process planning are presented, and key issues in developing an agent-based process planning system are investigated.

6.2.1 Agent Technology in Process Planning

Instead of being one large and standalone expert system, agent technology has emerged as a solution for distributed artificial intelligence and has attracted the attention of researchers in the area of process planning. Owing to the difficulties in solving process planning problems using traditional centralised problem solving methodology, a multi-agent system (MAS) approach is used as an attempt to solve planning and scheduling problems. As a distributed problem-solving paradigm, MAS breaks complex problems into small and manageable subproblems to be solved by individual agents cooperatively. The agent-based approach is recognised as an effective way to realise adaptiveness and dynamism in process planning. Agents are ideally suited to systems that require communication between many data sources. The agents can easily interface with various planning databases to provide optimal process plans. In addition, the attributes of agents make the use of agent technology in the development of CAPP/CAM systems particularly useful (Allen *et al.* 2001). Agents can be programmed to accomplish various planning activities such as machine type, cutting tool and cutting parameter selecion to machine specific geometries or features. These can then be integrated to build up a collaborative CAPP system.

Although agent technology has shown great potential in collaborative process planning, it is still somewhat underdeveloped compared with traditional CAPP research. Until now, there has been no mature system, tool, or module developed that could lead to a novel approach for developing agent-based process planning systems. There have been a number of agent approaches in recent years. For example, a cooperative CAPP system was proposed by Zhao and Wu (1999). This system aimed to meet five major requirements of a distributed CAPP system: autonomy, flexibility, interoperability, modularity, and scalability. The CoCAPP system attacks process planning problems by distributing them to special process planning agents. These agents coordinate and cooperate with each other via a commonly shared language with the expectation of reaching agreements when conflicts occur. Bose (1999) introduced a cooperative distributed problem-solving (CDPS) framework to solve CAPP problems arising from a rigid hierarchical structure of tasks. CDPS also decomposes a problem into subproblems and distributes them among a network of problem solvers (agents), which cooperate with each other to work out solutions. Three types of agent are included in the process planning system; a job process planning manager, a work centre manager, and a work centre knowledge expert. Agents solve a problem by sharing their expertise, resources, and information, and generate a complete solution to the problem. This proposed system focused on the manufacturing process, but also has the potential to be used for a cooperative scheduling system. Zhang *et al.* (1999) proposed an agent-based adaptive process planning (AAPP) system, which is on top of an object-oriented manufacturing resources modelling (OOMRM) framework. The OOMRM describes the manufacturing resources capability and the capacity in an object-oriented manner. The man–machine integrated process planning platform is implemented based on the OOMRM and an experienced manufacturing engineer is able to map out a more reasonable and flexible machining process. Five agents with distributed process knowledge are developed in the AAPP to perform tasks such as part information classification, manufacturing resources mapping, process planning, human planning, and machining parameter, by working together in a cooperative manner. The coordination between agents is based on the contract net protocol. An integrated, distributed and cooperative process planning system (IDCPPS) formed on the basis of concurrent engineering and using a cooperatively agent-based approach, was proposed by Chan *et al.* (2001) to integrate design, process planning and scheduling. The process planning agents are at three levels; initial planning level, decision-making level and detailed planning level. Sluga *et al.* (1998) introduced a virtual work system as an essential building block for decision-making in a distributed manufacturing environment. Shih and Srihari (1995) proposed a distributed AI-based framework for process planning. The approach decomposes the entire production control task into several subtasks, each of which is implemented by an intelligent agent. The agents can reach a solution for the problem by working collaboratively. Research has also been carried out by Park and Baik (1999) and Sun *et al.* (2001) to develop methods and technology for cooperative agent-based process planning. Denkena *et al.* (2002) proposed a multi-agent architecture to determine operating routes and schedules. The proposed multi-agent planning architecture builds a flexible, reliable and fault-tolerant information logistics to enable supply chains, temporal logistic networks or virtual enterprises.

In addition to the approaches identified above, there is also research using agent technology in process planning, which does not fall into the above-mentioned categories. Wang and Shen (2003) presented a new distributed process planning methodology by integrating machining feature-based planning, function block-based control, and agent-based distributed decision-making. The proposed methodology is suitable for dynamic, reconfigurable and distributed manufacturing environments. An agent-based approach was adopted for intelligent decision-making involved in distributed process planning. A two-level architecture for supervisory planning and operation planning was presented.

Kornienk *et al.* (2004) used autonomous agents to solve the assignment problem, which is often encountered in manufacturing. Generally, the assignment problem can be classified into scheduling, resource allocation and planning of operation order. This is a classical NP-hard problem. The agent-oriented approach addressed the lowest level of manufacturing architecture, where low-level jobs are assigned to available machines. The aim of the research was to generate this assignment via agents that represent different factory departments as well as processing elements. The approach addressed how agent-based process planning was used, and when production orders should be assigned to available machines after taking into account the technological, and organisational restrictions as well as optimisation criteria. Another example demonstrated an agent-based optimisation approach for manufacturing process planning (Deshpande and Cagan 2004).

6.2.2 Issues in Developing Agent-based Process Planning Systems

The general issues related to agent-based manufacturing systems were discussed in Shen and Norrie (1999). They include agent technology for enterprise integration and supply chain management, agent encapsulation, multi-agent organisation, dynamic system reconfiguration, learning in agent-based manufacturing systems, design and manufacturability assessments, distributed dynamic scheduling, planning, scheduling and execution, factory control architectures, and tools and standards. Most of these issues also exist in ABPPS. In this section, the key issues in developing an ABPPS are identified. They are: 1. architecture issues, which include agent architecture issues and system architecture issues; 2. communication issues, which include standards and protocols; 3. application issues in support of collaborative process planning.

To develop an ABPPS, the first step is to identify tasks involved in the system, and then divide and assign these tasks to agents. The decomposition of a high-level task into subtasks or subsubtasks is not an easy job (Zhang and Xie 2007). The task decomposition should satisfy some criteria (Gasser and Hill 1990) and task allocation should apply some allocation techniques. Currently, task decomposition is often manually done by human beings. Automatic task decomposition needs to be investigated.

Before assigning tasks to agents, agent partition and the classification of different types of agents are needed for this application. The agents in an ABPPS need to be classified and should follow the defined principles:

1. Depending on the different functions agents perform in an ABPPS, agents can be grouped into feature recognition agents, machining operation agents, tool selection agents, *etc.*
2. Some agents may be developed by encapsulating existing tools and databases to provide specific services.
3. The size of an agent should be neither too small or too large. If an agent is too small, it will need to define more interface messages and will be more difficult to integrate. If it is too large, it will be insufficiently configurable and difficult to change.

These principles for a complex multi-ABPPS are too abstract and simple. More detailed principles of division should be established.

6.2.2.1 Agent-oriented Analysis and Design

At present, extensions of object-oriented or knowledge-oriented manufacturing analysis and design methods are used to develop agent-based manufacturing systems. The agent-oriented design and analysis method is extended from object-oriented methods. Until now, there has been no widely accepted method for agent-oriented analysis and design for ABPPS, or for agent-based manufacturing systems. Most existing agent-oriented analysis and design methods are still under development. Therefore, agent-oriented analysis and design methods for process planning need to be investigated. Therefore, the investigation of agent technology and object-oriented technology integration for a collaborative process planning needs to be further researched.

6.2.2.2 Platforms Supporting Applications

Multi-agent systems can be adequately developed using the usual object-oriented languages. Platforms have also been developed or are developing for multi-agent systems, such as AgentBuilder (www.agentbuilder.com), concordia (Concordia 2001), JatLite, IBMAglets, FIPA-OS and JADE. Most of these platforms use the Java language because of its various features for implementing distributed systems. These platforms have their own characteristics in applications and are still undergoing improvement. Therefore, how an appropriate and effective platform can be selected for the development and implementation of an ABPPS needs to be investigated.

6.2.2.3 Implementation

The development of an agent-based application requires many infrastructure services, such as standards of implementation and mechanisms for establishing communication among agents. In addition, although there are commercially and academically available tools or frameworks to implement multi-agent systems, tools and platforms should be based on standards. However, there are no widely accepted agent-oriented standards for agent-based systems development. Distributed component object model (DCOM)/component object model (COM) and common object request broker architecture (CORBA), which have been widely accepted for developing agent-based manufacturing systems, will continue as intermediate standards for object-oriented implementations until agent-oriented standards are available for system implementation. More effort is required in the research of agent-oriented implementation standards for coordinative process planning.

In addition to the issues discussed above, other problems include agent encapsulation, decision schemes of individual agents, and theatrical investigation of methodology for agent-based process planning.

6.3 An Agent-based Sheet Metal Process Planning System

The sheet metal industry has focused on the automation of punching, cutting, shearing and nesting processes for the development of sheet metal parts. According to the literature, there is currently no research being carried out to apply agent technology in sheet metal process planning. This chapter investigates an ABPPS for sheet metal parts and the structure of the ABPPS is discussed. In this system, the process planning of sheet metal parts is divided into several planning tasks, which are completed by several defined agents. These agents collaborate and cooperate with each other to work out the final process plan.

6.3.1 Agent-oriented Analysis and Modelling

Agent technology has been applied in many areas in recent years, and the complexity of applications has also increased gradually in these areas. Therefore, it has become increasingly necessary to develop a set of methodologies to guide agent-based applications. The agent-oriented (AO) methodology has the potential to develop agent-based applications, and without a doubt, is the foundation for developing ABPPS. Originating from the object-oriented (OO) methodology, AO is already regarded as another important methodology or technique in the software development area. Therefore, it is worthwhile to investigate how OO and AO methodologies are used to design systems.

6.3.1.1 Object-oriented Methodology

Most existing agent-oriented analysis and design methods are extensions of object-oriented analysis and design methods or are extensions of knowledge engineering method. There are three steps to developing an object-oriented system: object-oriented analysis (OOA); object-oriented design (OOD); and object-oriented programming (OOP).

The idea of OOA is to build models for software requirements by adopting object-oriented concepts and methods. In this step, customers' requirements are analysed so that they can be made precise and consistent. The main steps in OOA include objects recognition, recognition of their attributes, behaviours, and their categories and structures. Problem statements are carefully analysed, and the system analyst must work closely with the requester (customer) to understand the problem. At this stage, a preliminary design model which is concise, consistent and clear is built.

OOD is closely related to OOA. Normally, methods of OOD are based on OOA models. The model is now subdivided into logical and physical subsystems based on the proposed architecture. It is augmented and complemented by more design details. A design model based on the analysis models is developed. The goal in the design stage is to transfer the OOA model to a model that is easy to implement.

OOP codes the OOD models in a form which computers can accept. Compared with traditional programming languages, object-oriented languages (such as Smalltalk, C++, Java,) can generate programs quickly and naturally.

Figure 6.1 shows a typical object-oriented development process (Katayama 2002). As can be seen from the figure, the three steps are:

1. Analyze the requirements of the system, define objects and classes, regard an object as an abstract unit, and define relationships among objects, which include static and dynamic relationships.
2. Design the structure and contents of objects, such as attributes, methods and so on, and develop the object models.
3. Develop object models by writing codes in OO languages.

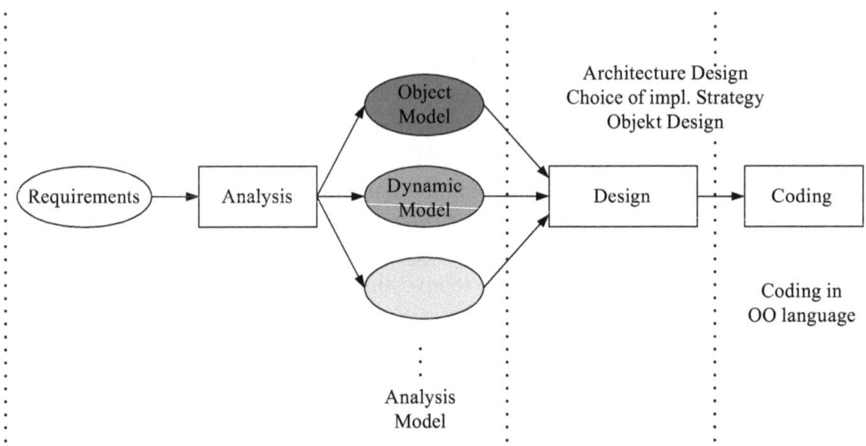

Figure 6.1 Typical object-oriented development process

6.3.1.2 Agent-oriented Methodology

Agent-oriented methods regard an agent as a basic unit and abstraction. An agent goes through the whole process of analysis, design and system realisation. Analysis and design can be thought of as a process of developing increasingly detailed models of the system to be constructed. These two phases are discussed below.

Agent-oriented Analysis

The objective of the analysis stage is to develop an understanding of the system and its structure. This understanding is reflected in the system's organisation. We view an organisation as a collection of roles that stand in certain relationships to one another, and that take part in systematic, institutionalised patterns of interactions with other roles (Wooldridge and Jennings 2000). Let us take the human society as an example to facilitate agent-oriented analysis. The concept of MAS originated from simulation of the organisational structure of human society. Therefore, we start analysing an agent-based application system by analysing the organisation of human society. Normally, "people" are the units of human society. Every person plays a different role. Roles determine the organisational structure and relationships among people. Similarly, we may consider the agent-based application area as a human society. The organisation of this agent-based application is comprised of many entities, which play their own roles. These roles are then further analysed. The analysis of roles includes four steps:

1. Define different roles involved in the system, and then further categorise roles according to their similarity; analyse roles and determine the criteria for defining key roles.
2. Analyse responsibilities that must be completed by key roles: a role is created in order to do something. That is, a role has a certain functionality. This functionality is represented by an attribute known as the role's responsibilities. The knowledge, capability and related behaviours of roles are then determined according to their responsibilities.
3. Analyse relationships among key roles, such as peer-to-peer relationships or hierarchy relationships; the relationships among them determine the communication and control mechanism.
4. Analyse the contact process among key roles and determine the form of information exchange (protocols), information content, etc, so that the requirements for communication functionality can be determined; analyse the cooperation process among them, so that the cooperation mechanism can be determined.

Agent-oriented Design

Agent-oriented modelling views the agent as an abstract unit; modelling methods are based on results at the analysis stage. It also includes four steps:

1. Abstract key roles to agents. When defining an agent, the size of the agent cannot be too small or too large. If the agent is too small, it will need to define more interface messages and will be more difficult to integrate. If it is too large, it will be insufficiently configurable and difficult to change.
2. Design agents' architectures based on their knowledge, capacity and behaviours; determine the knowledge representation, internal rules and decision-making methods.
3. Design relationship models among agents, such as cooperation or competition. Due to the autonomy of an agent, its roles may change, resulting in changing relationships among agents. Therefore, the relationship model is dynamic.
4. Determine communication mechanism, protocols and communication language (message syntax and semantics) and coordination mechanism.

6.3.2 The Structure of Agent-based Process Planning System

Process planning for sheet metal parts is a very complex and difficult problem. It involves activities such as nesting, cutting and bending, and different types of machines for sheet metal machining, including punching machines, cutting machines and unfolding machines. The complexity the process planning of a job and the vast amount of information that a system would need to process to make planning decisions makes process planning an appropriate candidate to be designed as a multi-agent planning system.

Using agent-oriented analysis and design methods, sheet metal process planning is divided into three phases:

1. task identification – this analyses and assigns process planning tasks for given sheet metal parts;
2. process planning – this includes generating a nesting plan, a tool path plan, and a bending plan;
3. resource selection – this involves selecting machining methods, machines, machining tools and work holding devices.

Three main managers, each in charge of three key roles, and a supervisor that monitors and supervises the process status of all agents, are defined. In order to meet the requirements of a modern process planning system for sheet metal parts, this proposed system aims to design a distributed and collaborative environment for sheet metal parts process planning, and make process planning for sheet metal parts more flexible, autonomous and efficient.

This ABPPS is based on the Java Agent DEvelopment (JADE) Framework. The JADE-based architecture of the prototype ABPPS is shown in Figure 6.2. JADE provides an organised environment to run a collection of agents in the ABPPS. These agents may run on different computers connected to the network. They perform different system functions and cooperate with each other. All agents work together to solve the entire process planning problem cooperatively.

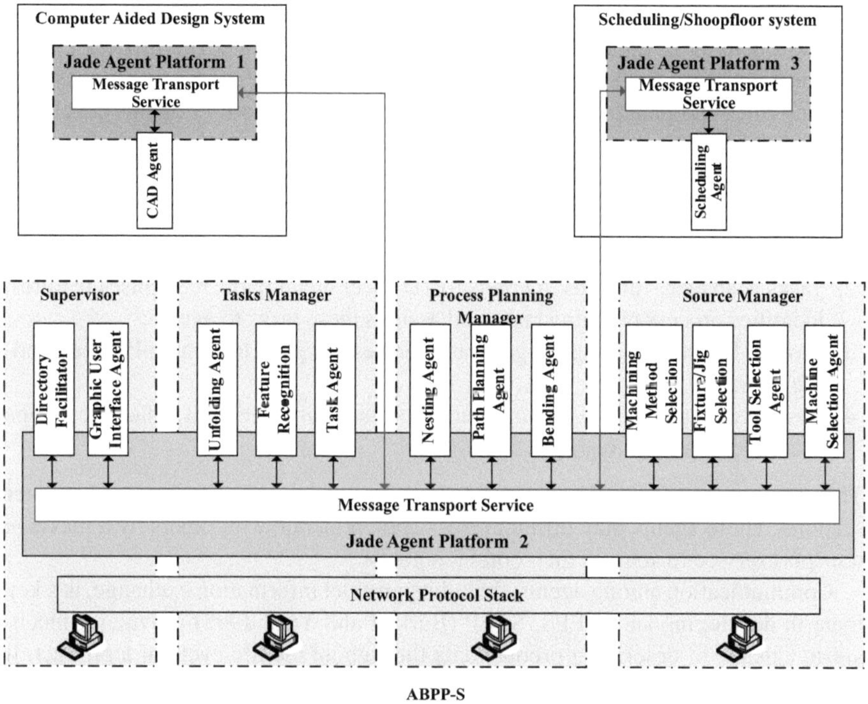

Figure 6.2 Agent-based process planning system for sheet metal parts

As shown in this figure, three agent platforms are developed based on three different systems; the CAD system, the scheduling system, and the ABPPS system. These three platforms run in individual computers in a distributed environment. The CAD agent and the scheduling agent are located in two agent platforms separately. The ABPPS communicates with the CAD system and scheduling system via a message transport service channel, which uses mechanisms and protocols to perform communicating among platforms. A CAD agent is defined in the CAD system, a scheduling agent in the scheduling system, and a number of other agents are defined in the ABPPS.

The ABPPS system receives product design information from the CAD agent of the CAD system and production constraints from the scheduling agent of the scheduling/shop floor system. The ABPPS and the CAD system work concurrently. ABPPS generates the process plans according to the product design information and gives feedback of design suggestions to the CAD agent. This means the process planning system serves the CAD system at any time, and *vice versa*. The scheduling system is responsible for providing information about the processing potential on the shop floor. The information provides resource constraints and capabilities to the ABPPS system when process plans for a part need to be generated.

In ABPPS, process knowledge is distributed into different intelligent agents and process planning is fulfilled by communication among them. The process knowledge is knowledge used in performing a planning task such as nesting algorithms, bending rules and path planning algorithms. In total, there are 12 agents developed to carry out various process planning tasks. In the prototype system, these agents are grouped into four coordinated clusters based on their type of task. These four clusters have been defined as:

1. Supervisor – used to register agents and monitor all agents' process status;
2. Tasks Manager – unfolds 3D parts to 2D sheet metal parts, recognises features, identifies process planning tasks and assigns these tasks to agents;
3. Process Planning Manager – generates the nesting plan, tool path plan and bending plan;
4. Resource Manager – used to determine machining methods, machines to be used, tools and fixtures/jigs.

These managers have their own defined agents, which will be discussed in later sections. These agents play different roles and cooperate with others by a message transport service to achieve their common goals.

Communication among agents, including product information exchange, is a key issue in developing an ABPPS. STEP (Burkett and Yang 1995) provides a mechanism capable of describing product data throughout the life cycle of a product. It has become a commonly accepted standard and is commercially available. The proposed ABPPS adopts STEP as the product modelling standard. It can enhance integration and communication among the CAD system, the ABPPS and the scheduling system.

6.3.2.1 Supervisor

The supervisor consists of a directory facilitator agent and a graphical user interface agent.

Directory Facilitator Agent (DF). This agent is developed in JADE. If one agent wants to be recognised by other agents and provide services for other agents, it must register first with a DF agent. A DF agent is responsible for the registry of agents and investigates their abilities to provide a "yellow pages" service by means of which an agent can find other agents providing the services it requires in order to achieve its goals. A yellow pages service allows agents to publish one or more services they provide so that other agents can find and exploit them, as described in Figure 6.3 (Caire 2003). Agent1 (A1), Agent2 (A2) and Agent3 (A3) publish their services to the DF. As we can see from this figure, A1 provides service X and service Y, A2 provides service Z, and A3 provides services W, K and H. Agent4 (A4), Agent5 (A5) and Agent6 (A6) search for required services on the yellow pages. A6 exploits the required services provided by A3. In addition, this multi-agent architecture is dynamic, which allows new agents to join and old agents to quit.

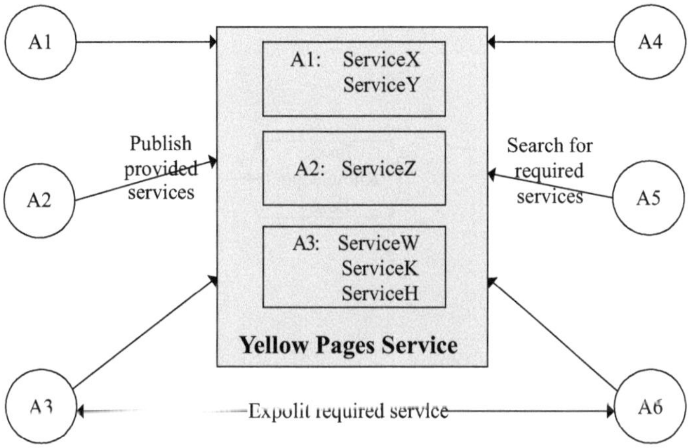

Figure 6.3 Yellow pages services

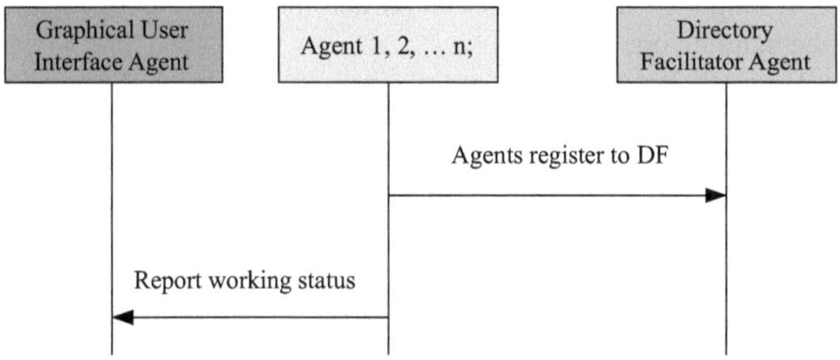

Figure 6.4 Functionality of directory facilitator (DF) agent and graphical user interface (GUI) agent

Graphical User Interface Agent (GUI). This agent is built to monitor the work status of all agents in the ABPPS system. The current working status of all agents can be checked at any time. Agents report their working status after finishing a job. From this, one can check what job a particular agent is doing, and which agent is doing a particular job. Figure 6.4 shows the functionality of DF and GUI. Agents (1, 2,..., *n*) of this system register their services with DF. After performing tasks, these agents regularly report their working status to GUI.

6.3.2.2 Tasks Manager

This includes an unfolding agent, a feature recognition agent and a task agent.

Unfolding Agent. The unfolding agent receives 3D product models of sheet metal parts from the CAD agent, and automatically converts 3D solid bent sheet metal parts into 2D sheet metal parts. Some possible changes are taken into account during the bending operations by this unfolding agent, such as a change in angle formation

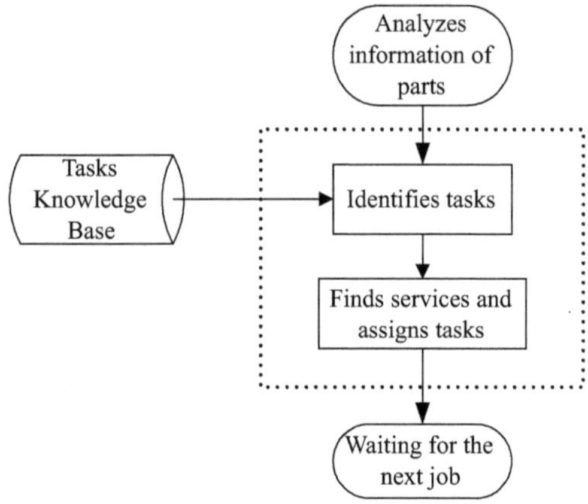

Figure 6.5 Behaviour of the task agent

and change in length of the bent section. It automatically calculates the changes caused by the bending operations.

Feature Recognition Agent. This agent maps product information from the unfolding agent into a set of machining features, such as cutting features, punching features and bending features. Then, it analyses the relationships between these features. It also checks how many different parts are to be recognised. Finally, it sends its analysis results to the task agent.

Task Agent. Figure 6.5 shows the behaviour of the task agent. It receives results of part features analysis from the feature recognition agent. It also checks how many different parts are involved in the process planning. It identifies particular tasks needed to be completed for process planning. Then, based on the services provided by the directory facilitator agent, the task agent finds suitable agents and assigns different tasks to them. It then waits for the next set of jobs.

6.3.2.3 Process Planning Manager

The process planning manager plays a fundamental role in the system. It includes a nesting agent, a path planning agent and a bending agent.

Nesting Agent. The nesting agent is in charge of nesting different shapes of sheet metal parts in one sheet to reduce set-up times and material wastage. The main purpose of the agent is to find the optimum layout of 2D parts of different shapes and sizes within the available sheet, and to reduce waste and maximise the number parts to be nested. It recognises and stores all the geometric information of the part features, then generates a scheme, usually optimal, to layout the components on the sheet metal to achieve a certain objective, which is usually maximum material utilisation ratio. The nesting agent contains knowledge of nesting algorithms within a knowledge base, and includes genetic algorithms, heuristic algorithms, and bottom-left algorithms. An optimal layout of components can be generated by inte-

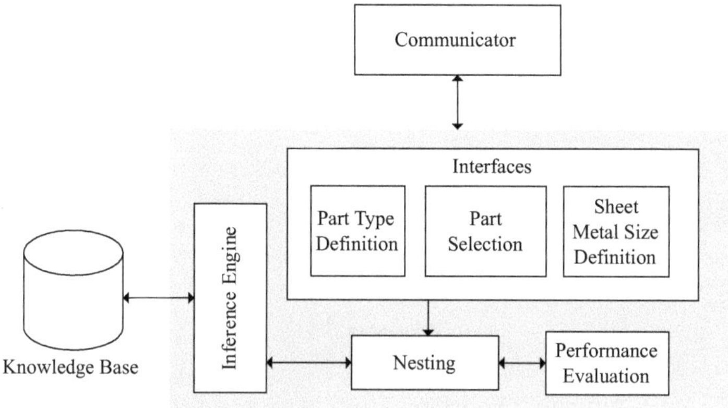

Figure 6.6 Structure of the nesting agent

grating appropriate nesting algorithms. The structure of the nesting agent is shown in Figure 6.6. The nesting agent includes the following four modules.

1. A communicator controls interactions with other agents by performing communication mechanisms, including intercommunications, negotiations, coordination and cooperation.
2. A local knowledge base contains the domain knowledge for nesting, such as genetic algorithms, heuristic algorithms, bottom-left algorithms, *etc.*, and geometric information of parts to be nested. The knowledge base is built based on previous experience and experiments with nesting algorithms under various nesting scenarios.
3. An inference engine is the generic control mechanism that applies the knowledge present in the knowledge base to the task-specific data to deduce conclusions. In this case, the inference engine navigates through and manipulates the knowledge to deduce nesting results in an organised manner by executing an inference mechanism. Two methods of inference are often used, forward and backward chaining (Navin 1997).
4. Interfaces include a part type definition interface, a part selection interface, and a sheet metal size definition interface. The part type definition interface determines single part types or multiple part types to be nested. The part selection interface is defined for selecting the data file of parts that have to be nested on the sheet, and determining the number of each type of part. The sheet size definition interface defines the size of the sheet metal and gaps between parts when nesting. A screen shot of the interfaces is shown in Figure 6.7: a multi-part type is selected by the part type definition interface; three parts (named 01-1-6, 01-2-1, 02-6-3) and their numbers (5, 6, 5 separately) to be nested are chosen by the part selection interface; sheet size definition defines the sheet size as 4 m long and 3.5 m wide.

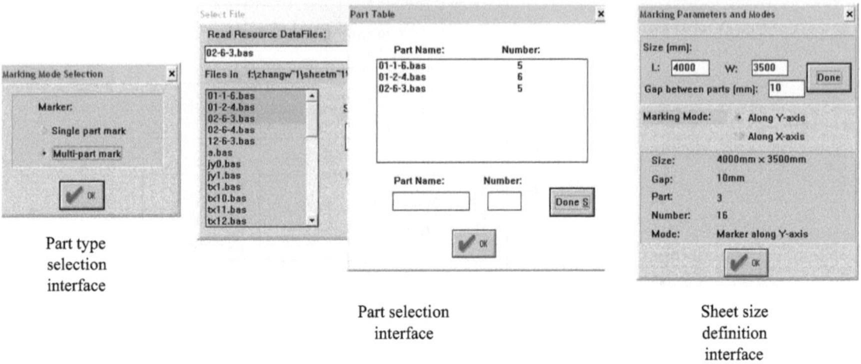

Figure 6.7 Schematic diagram of interfaces of the nesting agent

After all parameters have been set using the interfaces, the nesting agent starts nesting using the algorithms stored in its knowledge base. Within the nesting module there is also an interface for users to rotate, copy and delete parts to improve the utilisation of sheets. When nesting is finished, an evaluation is performed. The performance evaluation module calculates the usage ratio of the sheet. This is reported back to the task agent. If the task agent is not satisfied with the result, it will ask the nesting agent to repeat the nesting, until a satisfactory result is reached.

Path Planning Agent. The path planning agent optimises the machining path after the nesting task is completed. It receives notification from the nesting agent once a nesting task is completed. The agent also communicates with the machining method selection agent to check what machining operations have been chosen for machining. Then, the path planning agent calculates the optimal path using algorithms, including a nearest neighbour algorithm, a shortest insert algorithm, an optimum weight spanning-trees algorithm (Xie *et al.* 2001), or an algorithm for a combined punch–laser machine (Wang and Xie 2005). The results are stored in the knowledge base of the path planning agent. These algorithms are used to save machining time and reduce tool travel distance. The structure of the path planning agent is shown in Figure 6.8, and is comprised of four modules.

1. A start point definition module. This module is designed to define the start cutting or punching point for a feature. This requires industrial experience and should be considered case by case.
2. Auxiliary cutting path insertion module. This module is designed to add cut-in and cut-out auxiliary path to improve the cutting quality and is called up only when a cutting process is selected. The length and shape of the auxiliary path depend on the cutting speed, the shape of the feature, and the sheet material.
3. Path planning and simulation module. This module is designed to generate the tool path using path optimisation algorithms contained in its knowledge base. It can also simulate the machining operation.

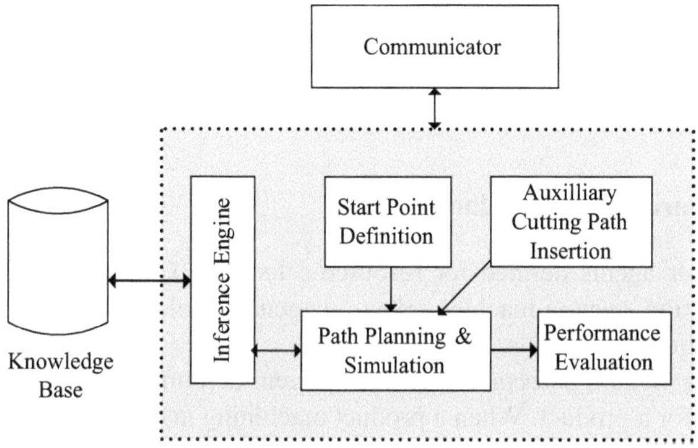

Figure 6.8 Structure of the path planning agent

4. The performance evaluation module evaluates the machining time and machining efficiency of the tool path. This is reported back to the task agent; if the task agent is not satisfied with the result, it can ask the path planning agent to further optimise the machining path.

Bending Agent. If there are parts which need to be bent in the process planning, a bending agent is called to determine the sequence of bending operations to minimize cost. It generates a bending sequence based on bending rules defined in its knowledge base. The bending agent has a similar architecture to the nesting agent. It also includes a communicator, knowledge base and an inference engine. The structure of the bending agent is shown in Figure 6.9.

1. The knowledge base contains the precedence rules and constraints for determining the bending sequence. For example, outside bends should be made before inside bends; internal tab bends, tab bends, and short bends should be bent first.
2. The bending sequence module uses bending rules stored in the knowledge base to arrange bending sequences for different parts (de Vin *et al.* 1994, Wang and Bourne 1997).

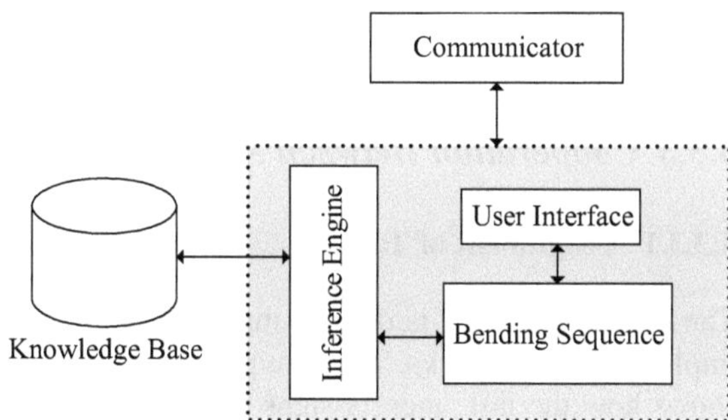

Figure 6.9 Structure of the bending agent

3. The user interface is designed to resolve conflicts when the precedence rules and constraints do not work well. Industrial experience and knowledge are required because the conflict resolution methods vary with products and machines.

6.3.2.4 Resource Selection Manager

There are four agents defined for resource selection. They include a machining method selection agent, a machine selection agent, a tool selection agent and a fixture/jig selection agent.

Machining Method Selection Agent. This agent determines the features machining methods for a product. When a product machining task is given, the machining method agent utilises its knowledge to analyse the features to generate an optimal machining method based on constraints such as cost, lead time and machine capability.

Machine Selection Agent. This selects machines for different machining methods. It receives machining methods data and chooses machines by checking the available machines and their capabilities. The machine agent organises the available machines on the shop floor and knows the ability of each machine and its machining parameters, such as power, speed, *etc.* These machines include cutting machines, punching machines, combined machines performing both punching and laser-cutting operations, shearing machines, bending machines, and welding machines. There are rules defined for machine selection. For instance, if both cutting and punching features exist in one sheet, a combined machine is a better choice than separate cutting and punching machines because multi-machinery requirements influence not only the manufacturing efficiency, but also the manufacturing quality and cost. If a combined machine is not adequate for fabrication, other machines are chosen based on real applications. The machine selection agent has knowledge of all the machines available in the company.

Tool Selection Agent and Fixture/Jig Selection Agent. These are defined to select tools and fixtures/jigs for machining. The agents know the capabilities of tools and fixtures, such as their type, material, working conditions, machine requirements and state. When the machining method selection agent and the machine selection agent require tools or fixtures, tool selection and fixture/jig selection agents determine the appropriate ones to use.

6.3.3 Cooperation Between Agents

6.3.3.1 Assignment of Tasks

The process planning is divided into two major subtasks, assignment of tasks and implementation of tasks. They are performed by particular task agents. Figure 6.10 shows how the task agent assigns tasks to task implementation agents. There are four steps:

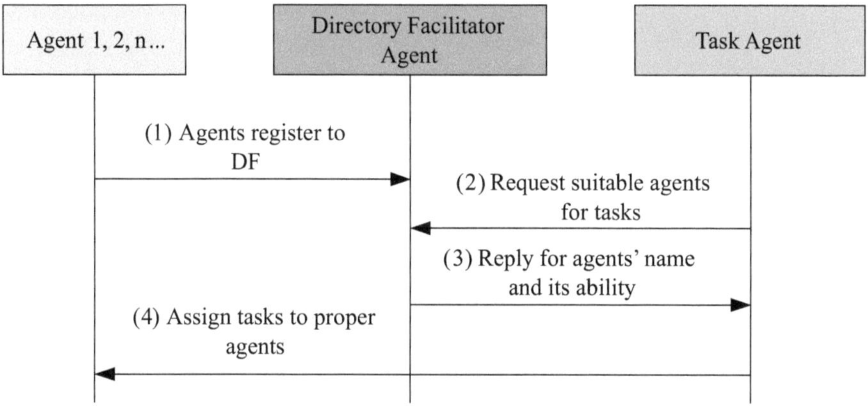

Figure 6.10 How the task agent assigns tasks

1. Agents register with the directory facilitator agent, so that other agents can find the required services provided by the registered agents.
2. The task agent analyses parts and part features and generates subtasks. It sends requests to the directory facilitator agent inquiring about the abilities of agents.
3. After the directory facilitator agent accepts requests to retrieve information from the task agent, it checks the registered agents, chooses a qualified agent, and sends this information to the task agent.
4. The task agent receives the reply from the directory facilitator agent, ensures the information is what it needs, and awards tasks to the task implementation agents.

6.3.3.2 Tasks Implementation

After a task is assigned to a suitable agent, the agent implements the task. Figure 6.11 shows the process of the generation of detailed planning and cooperation between agents. As can be seen from the figure, there are two different types of planning, initial planning and detailed planning. Some agents – such as the fixture/jig selection agent, the tool selection agent, the machine selection agent and the machining method selection agent – need to do initial planning first. When they receive sufficient information from other agents, they will perform the detailed planning. Nesting and bending agents can perform detailed planning directly because they have adequate information from the task agent. The path planning agent performs detailed planning once results have been received from the nesting agent and machining method selection agent.

In the task implementation process, agents need to cooperate with other related agents to obtain the necessary information for final decision-making. There are three phases of operation in terms of timing order.

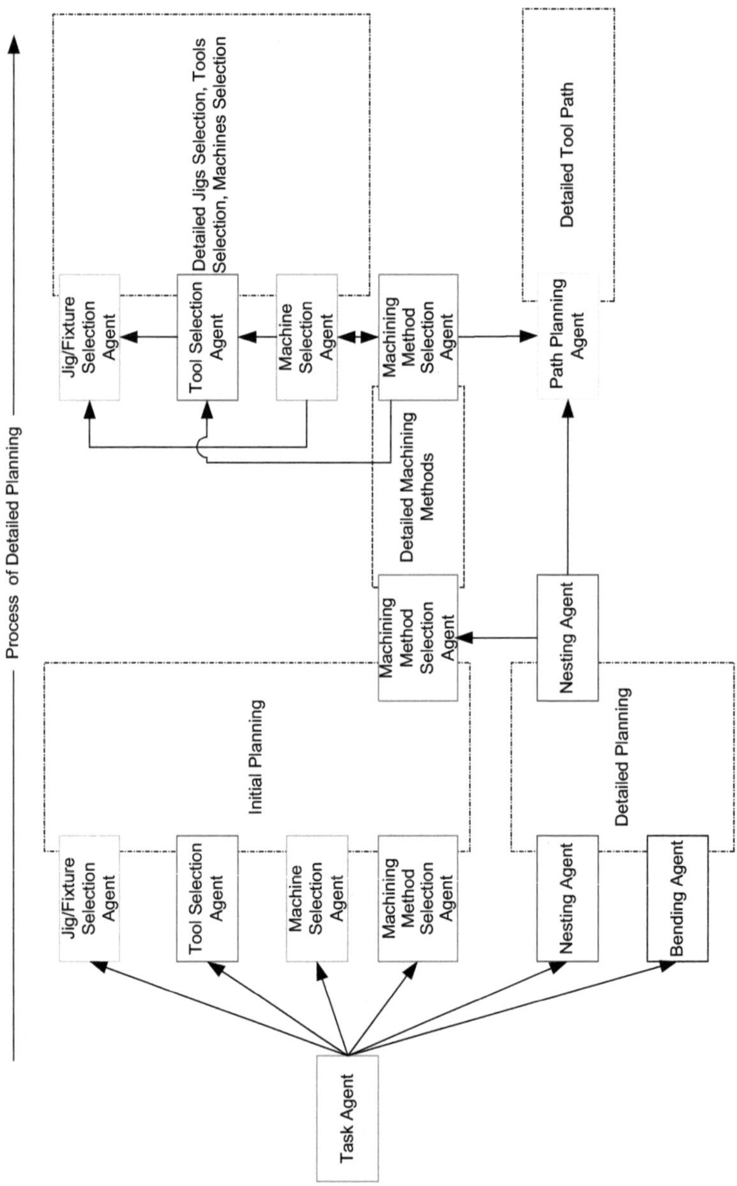

Figure 6.11 Detailed process planning

Early Phase

When agents located in the resource manager receive assigned tasks and relevant information, initial planning is first carried out; this involves only agents within the resource manager, including the machining method selection agent, the machine selection agent, the tool selection agent and the fixture/jig selection agent. Once these agents receive tasks, they execute initial planning where a few possible plans for the assigned tasks are developed. The machining method selection agent generates a plan detailing the machining operations that can be adopted for the features, *e.g.*, cutting or punching. This plan is based only on the features information obtained from the task agent. The machine selection agent, tool selection agent and fixture/jig selection agent operate in a similar manner to the machining method selection agent. They provide alternative cutting, punching and bending machines, tools and fixtures. The nesting and bending agents do not have to do initial planning because they receive tasks from the task agent with sufficient information to make final plans.

Intermediate Phase

The nesting agent sends nesting results to the machining method selection agent once a nesting task is completed. The machining method selection agent then calculates the final machining methods for each feature based on the nesting layout. Here, the following question could arise: Why does the nesting agent send nesting layout results to the machining method selection agent to help it make the final decision? The answer is because the initial machining methods plan is based only on the features themselves; it does not consider the final layout of multiple parts. The final machining methods are determined by the machining method selection agent by following defined rules. For example, if there are holes in several parts of one sheet, the initial machining plan for these holes could be cutting or punching. However, if the nesting layout is taken into account by the machining method selection agent, the final machining operation could be just cutting or just punching or an efficient mix of the methods.

Final Phase

The path planning agent becomes involved only after nesting is completed and machining methods are selected. The path planning agent then determines an optimal tool path for the machining methods adopted.

The machining method selection agent informs the machine selection agent of the machining methods selected. The machine selection agent then chooses suitable machines. The results of the machining methods and machines selection are then sent to the tool selection agent. Tools are chosen by the tool selection agent based on to the machining methods and on machine constraints.

The fixture/jig selection agent receives information from the machine selection agent and the tool selection agent to enable the selection of fixtures/jigs. It checks the available fixtures, taking into account any constraints imposed by machines or tools requirements. The final fixtures/jigs are then determined.

All the tasks implementation agents report their work status to the graphical user interface agent after finishing their tasks.

6.4 Conclusion

Currently, agent technology is a hot research topic in the area of artificial intelligence. A multi-agent system is capable of performing complicated tasks through cooperative multiple agents. It has distinct advantages when dealing with complex and distributed problems. The process planning of sheet metal parts is such a problem. This chapter proposes an agent-based process planning system (ABPPS), for solving the complex issues involved in the development of sheet metal products. Based on the agent-oriented analysis of the roles involved in sheet metal parts process planning, an ABPPS framework is proposed for sheet metal parts. The system architecture is based on decentralised, collaborative planning agents that attempt to solve the subproblems decomposed from the traditional hierarchically structured process planning operations. There are three clusters of agents defined: an unfolding agent, feature recognition agent and task agent in one cluster; a nesting agent, path planning agent and bending agent in a second cluster; and a machining method selection agent, machine selection agent, tool selection agent and fixture/jig selection agent in a third cluster. The cooperation of agents is also discussed in detail. The main contributions of this chapter include:

1. A prototype ABPPS has been developed.
2. A multi-agent system framework is proposed and the architecture of the system is discussed. Based on an analysis of task decomposition for process planning of sheet metal parts, agents and their structures are defined and implemented in the system.
3. A case study was carried out to demonstrate the feasibility of the ABPPS and its workings. Three typical agents, a task agent, a nesting agent and a path planning agent, were defined. The case study proved that the three agents could perform the basic process planning tasks by interacting with each other.

References

Allen, R.D., Harding J.A. and Newman S.T., 2001, The application of STEP-NC using agent-based process planning, Wolfson School of Mechanical Engineering, Loughborough University, Loughborough, UK.

Bose, U., 1999, A cooperative problem solving framework for computer-aided process planning, *Proceedings of the 32nd Hawaii International Conference on System Sciences*, 05–08 Jan 1999, Maui, HI, USA, p. 9.

Burkett, W.C. and Yang, Y., 1995, The STEP integration information architecture, *Engineering with Computers*, **11**, 136–144.

Caire, G., 2003, JADE Programming for Beginners, JADE Tutorial, http://jade.cselt.it/.

Chan, F.T.S., Zhang, J. and Li, P., 2001, Modelling of integrated, distributed and cooperative process planning system using an agent-based approach, *Proceedings of the Institution of Mechanical Engineering, Part B Journal of Engineering Manufacture*, **215**, 1437–1451.

Cherkaoui, S., Desrochers, A. and Habhouba, D., 2003, A multi-agent architecture for automated product technical specifications verification in CAD environments, Journal of Integrated Design and Process Science, **7**(2), 21–38.

Concordia Web Site, 2001, http://www.meitca.com/HSL/Projects/Concordia.

Deshpande, S. and Cagan, J., 2004, An agent based optimization approach to manufacturing process planning, *Journal of Mechanical Design*, **126**, 46–55.

Denkena, B., Tönshoff, H.K., Zwick, M., and Woelk, P.O., 2002, Process planning and scheduling with multiagent systems, in *Knowledge and Technology Integration in Production and Services– Balancing Knowledge and Technology in Product and Service Life Cycle*, Kluwer Academic Publishers, pp. 339–348.

Gasser, L. and Hill, R.W., 1990, Coordinated problem solvers, *Annual Reviews of Computer Science*, **4**, 203–253.

Huang, C.-C., 2004, A multi-agent approach to collaborative design of modular products, *Concurrent Engineering*, **12**, 39–47.

Katayama, T., 2002, http://www.soi.wide.ad.jp/class/20010030/slides/05/15.html.

Kornienk, S., Kornienk, O. and Priese, J., 2004, Application of multi-agent planning to the assignment problem, *Computers in Industry*, **54**, 273–290.

Navin, K., 1997, An expert systems application in electromagnetic compatibility, Master's thesis, University of Missouri-Rolla.

Park, H.G. and Baik, J.M., 1999, Enhancing manufacturing PD through learning agent system over internet, *Computers & Industrial Engineering*, **37**, 117–120.

Shen, W., Wang, L. and Hao, Q., 2004, Agent-based integration of manufacturing process planning and scheduling: a review, *Proceedings of FAIM 2004*, Toronto, Canada, pp. 906–914.

Shih, W. and Srihari, K., 1995, Distributed artificial intelligence in manufacturing systems control, *Computers & Industrial Engineering*, **29**, 199–203.

Sluga, A., Butala, P. and Bervar, G., 1998, A multi-agent approach to process planning and fabrication in distributed manufacturing, *Computers & Industrial Engineering*, **35**, 455–458.

Sun, J., Zhang, Y.F. and Nee, A.Y.C., 2001, A distributed multi-agent environment for product design and manufacturing planning, *International Journal of Production Research*, **39**, 625–645.

de Vin, L.J., de Vries, J., Streppel, A.H., Klaassen, E.J.W. and Kals, H.J.J., 1994, The Generation of Bending Sequences in a CAPP System for Sheet-Metal Components, *Journal of Materials Processing Technology*, **41**, 331–339.

Wang, C.H. and Bourne, D.A., 1997, Design and manufacturing of sheet metal parts: using features to aid process planning and resolve manufacturability problems, *Robotics and Computer-Integrated Manufacturing*, **13**, 281–294.

Wang, L.H. and Shen, W., 2003, DPP: an agent-based approach for distributed process planning, *Journal of Intelligent Manufacturing*, **14**, 429–439.

Wang, G. and Xie, S.Q., 2005. Optimal process planning for a combined punch and laser cutting machine using ant algorithms, *International Journal of Production Research*, **43**(11), 2195–2216.

Wooldridge, M. and Jennings, N.R., 2000, The Gaia methodology for agent-oriented analysis and design, *Autonomous Agents and Multi-Agent Systems*, **3**, 285–312.

Xie, S.Q., Tu, Y.L., Liu, J.Q. and Zhou, Z.D., 2001, Integrated and concurrent approach for compound sheet metal cutting and punching, *International Journal of Production Research*, **39**(6), 1095–1112. ISSN 0020-7453 print /ISSN 1366-588X.

Zhang, W.J. and Xie, S.Q., 2007, Agent technology for collaborative process planning: a review, *The International Journal of Advanced Manufacturing Technology*, **32**(3–4), 315–325.

Zhang, Y., Feng, C., Wang, X., Tian, W. and Wu, R., 1999, Object oriented manufacturing resource modeling for adaptive process planning, *International Journal of Production Research*, **37**, 4179–4195.

Zhao, F.L. and Wu, S.Y., 1999, A cooperative framework for process planning, *International Journal of Computer Integrated Manufacturing*, **12**, 168–178.

Part III
Product Modelling and Integration

Part III of the book introduces one of the fast growing research topics in product design and manufacturing – product modelling and integration. With the development of STEP, this research area has progressed rapidly in order to achieve seamless product integration and information sharing. This part introduces our research work in the development of a generic product modelling framework for modelling customised products.

To model OKP products, Chapter 7 proposes a generic product modelling framework (GPMF) that aims to provide an infrastructure for modelling various types of product information. The outcome of the GPMF is a set of data models that is defined to model product information at different stages of its development processes. This chapter discusses the structure of the GPMF and its main components.

Chapter 8 investigates the EXPRESS data model (EDM) that is developed based on the STEP-based modelling environment and the "five-phase" modelling method. The structure of the EDM will be discussed. The elements of the EDM schemas and the relationships between these schemas are represented by both EXPRESS language and EXPRESS-G diagrams.

Chapter 9 further develops the product modelling methods into a GPMF for efficient information exchange and sharing. This framework consists of four functional components: an EXPRESS data model, namely EDM, a STEP-based modelling environment, a "five-phase" modelling method, and three EDM data exchange and sharing methods. The focus of this research is placed on the modelling methodologies and the definition of schemas for various product development activities such as manufacturing, inspection, etc., and the integration of the schemas with other resources defined within STEP. There are a total of 25 schemas defined to ensure that the proposed GPMF is compatible, and can be used to model various types of products. These aspects, to the best of our knowledge, are not reported extensively in the literature.

Chapter 10 targets product management issues and provides a solution to achieve seamless information integration, which is an important issue in supporting integrated and concurrent product development. This chapter also uses sheet metal as an example and explores the definition and structure of an information framework. This framework aims to build an information bridge to fill the gap between the sheet metal part design, process planning and manufacturing systems. It is based on the principles of zero thickness and zero bend radius, which are used to abstract the geometry entities of sheet metal parts in order to facilitate part modelling and information modelling. In this chapter, a tree-based step-structure information modelling methodology for sheet metal parts is proposed and a case study is given.

Chapter 7
Generic Product Modelling Framework

Abstract In response to the rapidly changing manufacturing environment, product modelling technology has been widely applied to provide the essential information for supporting product development (PD) processes. The traditional product modelling technologies are unable to support the information exchange and sharing at the various stages of PD processes that could be taking place among different departments in a company or even among different companies in a distributed manufacturing environment. This has caused many problems such as information loss, data format incompatibility and reduced efficiency and effectiveness of product data applications. This has consequently created bottlenecks for the integration of PD processes. This chapter presents a generic product modelling framework (GPMF) to overcome the problems of information exchange and sharing in today's manufacturing environment. This framework uses the Standard for the Exchange Product Model Data (STEP) as a foundation. It consists of four functional components: an EXPRESS Data Model (EDM) for presenting the structure of product data; a STEP-based modelling environment used to build up the EDM; a "five-phase" modelling method which is proposed to model the EDM; and three EDM data exchange and sharing methods which implement the EDM in the modelling product.

7.1 Introduction

The market for manufactured goods has changed dramatically in recent years. The first change has been a shift from a seller-oriented market to a customer-oriented market (Xie and Tu 2005). To obtain greater market share, many manufacturing companies have to satisfy the customer's demanding requirements by agilely and rapidly developing new products. Secondly, the manufacturing industry has been moving towards globalisation (Tu *et al.* 2003). As more new firms join the industry, competition becomes more intensive. To survive in this competitive environment, manufacturing companies must utilise state-of-the-art technologies to improve all aspects of the product development (PD) process, from product design to manufacturing to recycling. Finally, as products increase in complexity, more areas of expert knowledge are required. This results in the creation of a culture of cooperation between manufacturing companies in PD processes.

These changes within the new manufacturing environment are closely related to PD processes. Due to this, efficiency enhancement of PD processes has become an issue for manufacturing enterprises. It is essential to find globally optimised concurrent PD processes which can shorten the product life cycle, reduce development costs and lead time, achieve high quality and productivity, respond quickly and effectively to customer demand, and bring all the processes together to meet the requirements of the customer-oriented market.

A key for success lies in the method of managing and utilising product information and knowledge within the production processes. Product modelling technology is one of the key and indispensable technologies in the PD process and can be used to support data exchange and sharing in PD processes. One of the major functions of product modelling technology is to provide a well-organised information structure to model product data and information. PD team members can use an information model to store, exchange and manipulate product data. At the same time, this well-structured product data can be used to support further developments of similar products. For example, a data model of a product can support future PD by making reusable data and information available. They can be used to improve both product innovation and new PD. These benefits place product modelling at the heart of PD activities. Subsequently, this determines engineering productivity and eventual industrial competitiveness (Krause *et al.* 1993, Nambiar *et al.* 2009).

Product modelling technology does not have a long history. Murphy (1950) was the first to define the term "model" as a device "which is so related to a physical system that observations of the model may be used to predict accurately the performance of a physical system in the desired respect". The concept of modelling was formed with the development of computer aided technologies. Data, structure, interface, algorithm, and even an entire system can be modelled. The models are the abstract specifications of the domain functions that perform certain operations. In PD processes, a product can be seen as a functional unit with particular materials, a fixed form and other designated features. The product and its related information are naturally regarded as modelling objects. The model developers combine the modelling concepts to build a product model and use it in its development processes. In each phase of the PD process, the product model is a prerequisite for representing the product data, and sharing and exchanging information among different PD stages.

The explicit description of a product model first appeared with the introduction of geometric models employed by various computer aided design (CAD) systems. The primary purpose of the early product models was to represent the product structure and its two-dimensional (2D) or three-dimensional (3D) shape data. In today's concurrent and integrated PD environment, the method of modelling objects has been expanded to other PD processes. A complete product model consists of useful data and information of a product all the way through its entire life cycle, *e.g.*, information about geometry, structure, function, assembly, materials and product usage.

The information required by specific PD activities, such as process planning, cost analysis and simulation, can be obtained from the complete product model.

7.2 Literature Review

Product modelling is a complex process. It involves a wide range of research issues such as product data standards, modelling resources, language and methods, and the implementations of product models. In recent years, considerable effort has been put into the above mentioned issues.

New modelling methodologies have been developed in the last few decades due to the fact that product modelling has become more and more important. Krause categorised different product models into five basic types (Krause *et al.* 1993) and these five product models correspond with the five basic modelling methodologies:

1. structure based product modelling methodology;
2. geometry based product modelling methodology;
3. feature based product modelling methodology;
4. knowledge based product modelling methodology;
5. integrated product modelling methodology.

Much research work has been carried out using the above product modelling methods. For example, Yang and Li (1992) used a CSG method to construct machining form features. This feature-based workpiece model can support the integration of CAD and computer aided process planning (CAPP). Gu (1992) presented a five-level product modelling language (PML), which included a form-feature level. Salustri (1996) presented a formal theory for knowledge-based product model representation called Axiomatic Information Model for Design (AIM-D). Song *et al.* (1999) developed an integrated product model that was established based on generalised features. This integrated product model was integrated with an expert system for product manufacturability evaluation and a concurrent design controller to form a real-time concurrent product and processes design system for mechanical parts.

With the development of STEP, effort has been placed on modelling products using STEP. Gu and Chan (1995) developed a STEP-based generic product modelling (GPM) system that was designed and implemented according to the generic resources of STEP. The system can be used to integrate manufacturing activities, such as process planning and inspection planning, in the concurrent engineering environment. Li *et al.* (1996) developed a feature-based parametric product modelling system, which employed a product model based on the STEP and managed by an object-oriented database. This system was suitable for application in a computer integrated manufacturing (CIM) environment. Usher (1996) presented a STEP-based object-oriented product model based on STEP AP 224. This model was proposed to support CAPP analysis. Ming *et al.* (1998) presented a STEP-based part information model for process planning purposes. Their model consists of a process planning information model and a production resource information model. Tang *et al.* (2001) presented a STEP-based die and product integrated information model (DPIIM), in which integrated resources of STEP were used to model six EXPRESS schemas. These models could support the concurrent development of stamp and die products. Zha and Du (2002) presented a product data exchange using STEP (PDES)/STEP-based assembly model for concurrent integrated design and assembly planning. Xu *et al.* (2005) presented a comprehensive review of using STEP-NC for the integration of CAD, CAPP, CAM and CNC. The issues and challenges for the development of a STEP-NC-compliant process chain are discussed and supported by a product and manufacturing model.

It can be concluded that the STEP-based modelling method has become the core of product modelling processes to organise product data in a standardised representation. This greatly enhances the capability of data exchanging and sharing in the integrated manufacturing environment.

Conventional product modelling technologies have significantly enhanced the performance of PD processes. Geometrics models are used to model product geometric information, and can support CAD systems in the exchange and sharing of product data. However, conventional modelling technologies cannot meet the requirements of current PD processes due to the new challenges from the ever-changing manufacturing environment. To the authors' knowledge, no generic modelling framework has been proposed for efficient information exchange and sharing among different computer aided systems. The limitations of conventional product modelling technologies have become the main hurdles for the development of modern manufacturing systems.

7.2.1 Problems of Integration

Conventional product modelling technologies have normally been used to model products to support the integration of one or two systems such as a CAD or CAPP system. They cannot be directly used in the integration of systems that are employed in other stages of the PD process such as computer aided manufacture (CAM) or computer aided engineering (CAE). Hence, more work should be carried out to establish seamless product data exchange and sharing across the entire PD processes (Tu *et al.* 2003, Xie and Tu 2005, Xu *et al.* 2005). For instance, the CAD-based geometric product model does not define all the necessary information for the downstream manufacturing processes.

On the other hand, for downstream systems such as CAPP and CAM, there can be some situations when the geometric information provided by a CAD-based product model is converted to product models of CAPP or CAM systems (Lee 1999). Information loss may result from incompatible formats of product models in different systems. This has become a major problem when attempting to integrate these systems. The representation format of product data can impede the fluent information flow between systems, and can lead to high development costs as a result of unnecessary costly rework and redesign. Consequently, an inefficient PD process will influence the competitiveness of the manufacturing company. This requires developing some new kinds of product modelling methods and technologies.

7.2.2 Problems of Cooperation

Complicated products are usually developed by combining the strengths of several manufacturing companies. Data exchange and sharing between these companies should be very efficient and effective. Most companies structure their products using different modelling methods. Hence, it has become an issue for them to cooperate with each other in support of the development of a particular product. Normally, an extra data conversion process is carried out, but this is very inefficient. Conflicts over the model structures may sometimes cause loss of information which cannot be converted. Therefore, a non-compatible product model is a great barrier to effective performance where cooperative effort is required. A suitable modelling method is required to build up the high compatibility product model.

7.3 A Generic Product Modelling Framework

This chapter proposes a generic product modelling framework (GPMF) that attempts to provide an infrastructure for modelling various types of product information. The outcome of the GPMF is a set of data models defined to model product information at different stages of the development processes. This section will discuss the structure of the GPMF and its main components.

Figure 7.1 shows the structure of the GPMF developed based on STEP. It consists of four functional components including an EXPRESS data model – EDM, a STEP-based modelling environment, a 'five-phase' modelling method, and three EDM data exchange and sharing methods.

The EXPRESS data model (EDM) is the core of the modelling framework GPMF. The EDM defines a complete product data structure and uses the standardised data format. It consists of 11 defined EXPRESS schemas and STEP AP 203. Each schema utilises either STEP resources or STEP-based compatible resources defined by our research group to model a particular type of product information. These schemas are presented in this chapter.

The STEP-based modelling environment is built up for the GPMF. Within the environment, the modelling language, EXPRESS, and its graphical representation method, EXPRESS-G, are used to model product structure. STEP generic resources are used to model product information that is defined by STEP, and STEP

AP 203 is used to model product geometric information. There are also new modelling resources defined for modelling product information that is not covered in STEP.

The 'five-phase' modelling method is proposed to build up the EDM. It defines a formal approach to logically organise all the tasks of building up the EDM in the modelling processes.

There are also three EDM data exchange and sharing methods used in the GPMF. As shown in Figure 7.1, product data can be exchanged and shared through either exchange files, or working forms, or database management systems. The product models defined within the GPMF can be exchanged or shared using one of these methods. The three methods are easily integrated into any application software environment, which makes it easy to implement the product models defined by the GPMF in applications. The following sections discuss the details of these components used in GPMF.

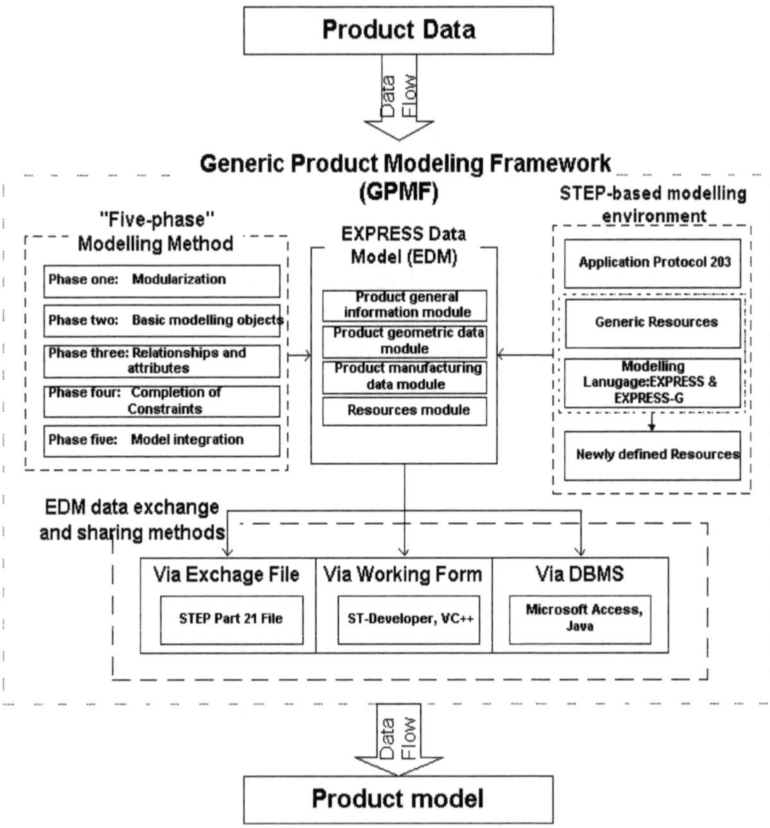

Figure 7.1 Structure of STEP-based GPMF

7.3.1 STEP-based Modelling Environment

The main purpose of the framework is to establish a STEP-based modelling environment, where data exchanges are through a STEP-compliant product data model. EXPRESS data modules and modelling methods are used to develop STEP-compliant data models for the integration of systems such as CAD, CAPP, and CAM. The following sections detail the methods and resources used to develop such an environment.

7.3.1.1 Modelling Language: EXPRESS and EXPRESS-G

EXPRESS modelling language is a data description language. It consists of language elements that allow unambiguous data definition and specification of constraints on the data defined (ISO 1994b). EXPRESS language is one part of STEP defined in Part 11 (ISO 1994a). All the resources in STEP including generic resources and APs, are presented as EXPRESS schema. In the GPMF, EXPRESS modelling language is used to build up the EDM, which represents the structure of product data. Figure 7.2 shows the main elements of an EDM. The data in the EDM is represented by one or more SCHEMAs, which "group together the modelling objects with related meaning and purpose" (Kahn *et al.* 2001). The most important EXPRESS language element is the ENTITY data type, which defines the objects of interest in the domain being modelled. The ENTITY is characterised by its attributes and constraints. The attribute is of three types, including the explicit attribute, the derived attribute (DERIVE), and the inversed attribute (INVERSE). The constraints are applied to the specific entity or other data types by using the WHERE and UNIQUE rules, or are applied in the SCHEMA using the RULE constraint. The FUNCTION is defined to support constraint definition. The relationships among entities are defined by schema interfaces including the USE FROM

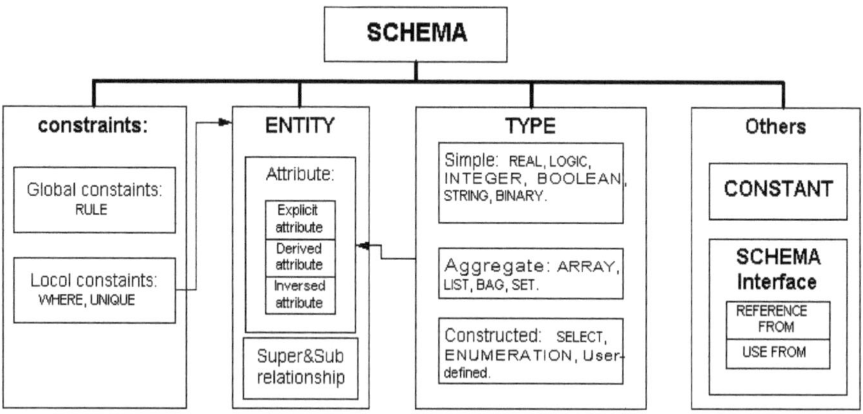

Figure 7.2 Main elements of EXPRESS data model

and the REFERENCE FROM. EXPRESS language also supports various kinds of data types, including simple types (*e.g.*, INTEGER, STRING, REAL, *etc.*), aggregation types (*e.g.*, SET, LIST, BAG, and ARRAY), and constructed types (*e.g.*, user defined types, SELECT and ENUMERATION).

EXPRESS is an object-oriented language and has a mechanism for presenting the specialisation and generalisation of the entities. It is possible to create a hierarchy of entity types in which each node in a given path is a specialisation of the nodes above and a generalisation of the nodes below. There is no restriction on the number of levels in the super/sub-type hierarchy, thus a very complex object can be modelled. In such a hierarchy, "the more general objects are called the super-types of their specializations; the less general objects are called sub-types of their generalizations". The sub-type inherits all attributes of the super-type. EXPRESS supports single and multiple inheritance as well as complex relationships (ONEOF, AND, ANDOR) between the sub-types of a super-type.

The EXPRESS elements that EXPRESS-G does not support are: FUNCTION definition, RULE definition, and local constraints definition (WHERE and UNIQUE). EXPRESS-G is represented by graphic symbols to form a diagram. The notation has three kinds of symbols (ISO 1994):

1. definition: symbols denoting simple data types, names of data types, constructed data types, and schema declarations;
2. relationship: symbols describing relationships that exist among the definitions;
3. composition: symbols enabling a diagram to be displayed on more than one page.

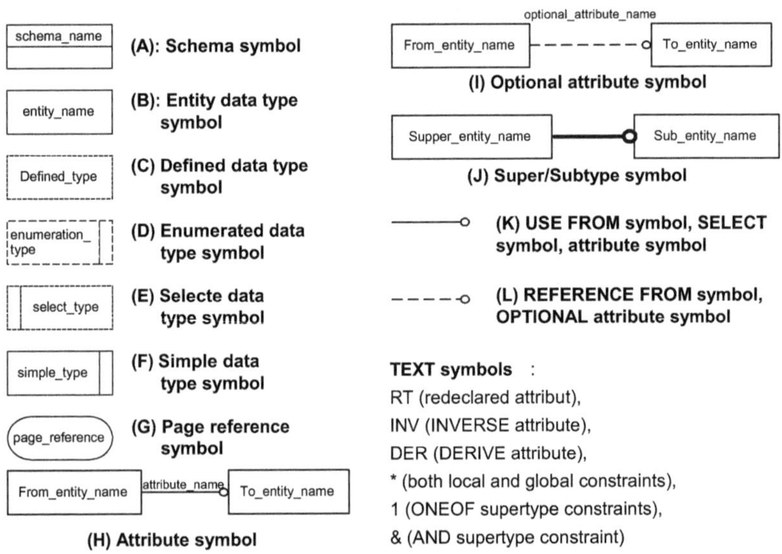

Figure 7.3 Symbols and text symbols of EXPRESS-G notations

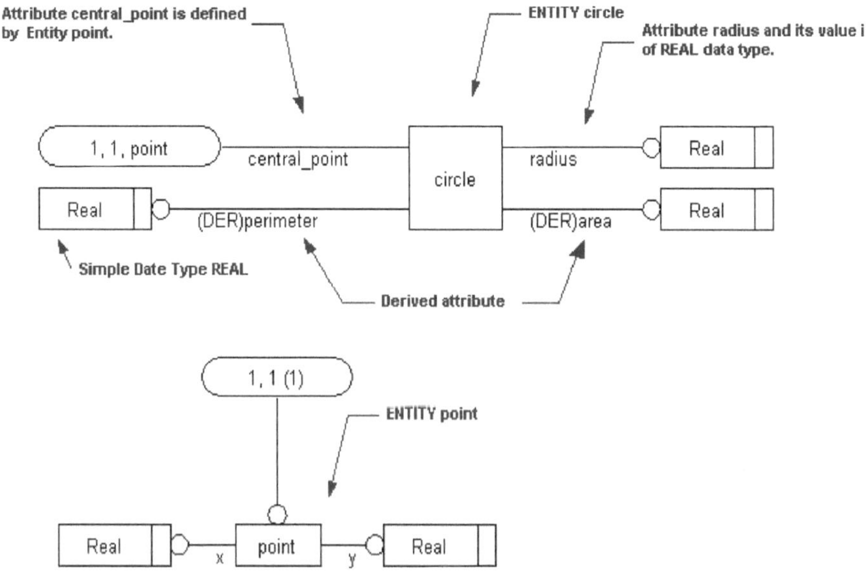

Figure 7.4 EXPRESS-G diagrams of circle model

Figure 7.3 shows a summary of EXPRSS-G definition symbols, relationship symbols, and text symbols. They are used to create diagrams to represent product data structure.

Two EXPRESS-G diagrams, as shown in Figure 7.4, represent an object "circle". The symbols of EXPRESS-G notation: entity, attribute, simple data type are all present in the circle model. Two EXPRESS-G diagrams explicitly represent the components and the relationships of the modelling objects. The first diagram defines the entity *circle*. The four attributes and their attribute value type are presented. However, the detailed definition of two derived attributes *perimeter* and *area* cannot be explicitly represented. The second diagram defines the entity *point*, which has two attributes *x* and *y* indicating the *x* axis position and the *y* axis position using REAL. The *point* is utilised by the attribute *circle.central_point*[1]. The relationships between the *circle* and the *point* are defined.

7.3.1.2 Generic Resources and STEP AP 203

Generic resources can be directly utilised to define the basic data objects in an EDM. For example, entities defined in product_definition_schema in STEP Part 41 (ISO 2000) are used to present the product general information in the EDM; STEP Part 45 (ISO 1998) is used to define the EXPRESS schema that presents the data structure for material information. The generic resources are defined in STEP. The use of generic resources enhances the compatibility of the GPMF.

[1] The *circle.central_point* means the entity *central_point* in the schema circle.

STEP AP 203 (ISO 1994b) is used to model product geometric information in the EDM. It is employed by different CAD systems as the data model to structure product geometric data. Thus, the product geometric data module of EDM can be integrated with CAD systems based on STEP AP 203.

7.3.1.3 New Modelling Resources

STEP has already defined many modelling resources to support product modelling. However, it is still under development. There are still not sufficient resources defined to present information of different types in PD processes. Hence, two kinds of new modelling resources are defined and used in the GPMF to supplement STEP-defined modelling resources.

One new modelling resource is called the hybrid modelling resource. This is modified from STEP generic resources. For instance, the document_schema in Part 41 (ISO 1998) is modified to generate a data model for document information. The method_definition_schema and process_property_schema in Part 49 (Lee 1999) are modified to define the data model in the EDM for process information.

There are also new STEP-compatible modelling resources defined by the authors. These resources are supplements to STEP-defined modelling resources to model product data which are not covered in STEP. Examples of the schemas of these resources are discussed in Section 7.3.4.

7.3.2 'Five-phase' Modelling Methodology

A "five-phase" modelling method is proposed for the development of the EDM. This methodology standardises the modelling of different types of product information. It is composed of five phases: 1. modularisation; 2. basic modelling objects; 3. relationships and attributes; 4. completion of constraints; and 5. module integration. The five steps are discussed in the following subsections.

7.3.2.1 Phase One: Modularisation

This phase defines and modularises an EDM. The main tasks include: 1. analysing the product modelling objective; 2. classifying the product data; 3. modularising the EDM.

Product modelling objective analysis is very complex and contains large quantities of different types of data. Modelling these data without analysis and structuring could result in data loss or repetition. Consequently, the EDM will not be able to correctly represent a product. Thus, the EDM needs to be modularised according to the product data classification.

In the modularisation phase, the EDM is divided into four modules. These include a product general information module, a product geometric data module, a product manufacturing data module, and a resources module developed to present the sharable basic modelling objects extracted from the other modules.

7.3.2.2 Phase Two: Basic Modelling Objects

The principle objective of this phase is to define the basic modelling objects and the general structure of each EDM module. The phase starts by analysing the structure of each EDM module. The fundamental elements of that module are identified then defined as basic modelling objects. A way to structure these basic modelling objects is then analysed and applied in the EDM. For example, in the product manufacturing data module, the product assembly information is normally defined by four basic modelling objects including an assembly product object, a product component object, a subassembly component object, and an object called connector, used to represent the connections between the other objects. In this phase, these objects are named as *assembly_product*, *part*, *subassembly*, and *connector*, respectively.

7.3.2.3 Phase Three: Relationships and Attributes

This phase refines and enriches the basic modelling objects defined in the second phase by adding relationships to the defined objects. The following tasks are included in this phase: 1. definition of attributes and relationships between entities; 2. enhancement of the defined entities; 3. definition of new entities; and 4. correctness checking. This phase must be continued until the EDM has reached the desired level of detail for representing the content of objects and their relationships with other objects. For example, the basic modelling object *part* has to be further specified to have seven attributes: id, name, description, of_assembly, super_item, level_in_assembly_hierarchy, and connecting_information. Their value types are defined as identifier, label, text, assembly_product, super_item_type, integer, and connector, respectively.

7.3.2.4 Phase Four: Completion of Constraints

This phase models the constraints of the EDM. It has the following three tasks: 1. definition of constraints of objects and their relationships; 2. addition of global constraints to the model; and 3. model error-checking. Constraints modelling is just as important as objects modelling. The tasks model the objects and their relationships based on requirements. After this phase, a complete EDM module is developed. For example, one local constraint named WR1 is defined in the part. This constraint specifies the value range of level_in_assembly_hierarchy attribute, which means that the part must be in the second or under the second level of the assembly model tree (AMT) (Zha and Du 2002). The error-checking includes grammar errors, missing information, and conflicts of constraints.

7.3.2.5 Phase Five: Model Integration

Phase five integrates the modules defined in the previous four steps to form an EDM. The main tasks involved in this phase include checking how data is represented in each module, checking the inputs and outputs of the four modules, and checking whether EDM is complete, minimally redundant, unambiguous, and error free.

7.3.3 EDM Data Exchange and Sharing Methods

The EDM presents the structure information of product data. To support data exchanging and sharing, the EDM must be mapped to the product model through a proper method. Lee (1999) demonstrated three implementation methods for the information models that are currently used in practice. These three methods were indicated by Wilson (1990) and Loffredo (1998) as the three implementation methods for EXPRESS data model, and are: 1. implementation via an exchange file; 2. implementation via a working form; and 3. implementation via a DBMS. In the GPMF, these three methods are used as EDM data exchange and sharing methods to support GPMF implementation in product modelling processes.

7.3.3.1 Level One: Via Exchange File

This level is the least complex EDM data exchange and sharing method. The product data can be organised into a product model with exchange file format, which is a neutral file with a predefined syntax structure or format. The downstream applications systems can directly read and write this file to exchange and share product data.

The STEP defines a formal exchange file named STEP Part 21 file in Part 21. The STEP Part 21 file can represent product data for different applications without any constraints. Compared with conventional exchange files such as Initial Graphics Exchange Specification (IGES), which supports exchanging product geometric data only, the STEP Part 21 file can be applied to model more product data rather than just geometric product data. The STEP Part 21 file uses the product data structure defined in the EDM, and in the GPMF it is used as a standard file format for data exchange and sharing.

7.3.3.2 Level Two: Via Working Form

This method possesses all the features offerred by level one and also manipulates data based on the EDM using a working form. The working form is an in-memory representation of data structure. The exchanging and sharing of product data is carried out using a number of defined working forms. The original data is not moved around but applied in memory.

STEP defines a standard application programming interface (API) named Standard Data Access Interface (SDAI) for working form applications. The SDAI programming language bindings defined in STEP can be used to generate the application programs based on the EDM. The application programs can model product data and support product data exchange and sharing. A number of working form implementations have been developed, such as ROSE C++ Library and NIST SCL (Wilson 1990). In the GPMF, the ROSE C++ Library, which uses the SDAI C++ binding, is used as the working form to support implementation of this level method.

7.3.3.3 Level Three: Via DBMS

This level method contains all the features offered by level two and makes product data available by using API, which can present data as EDM-defined structures. The DBMS can support manipulating the product database, including data query, data update and data retrieval. Different application systems can access the product database to read and write product data. These product data are exchanged and shared within these systems, as shown in Figure 7.5.

The data structure defined in the EDM should be mapped into the product database. This structure information can support possible DBMS applications to implement this level method. In the GPMF, a DBMS application prototype system is developed as a product data management system.

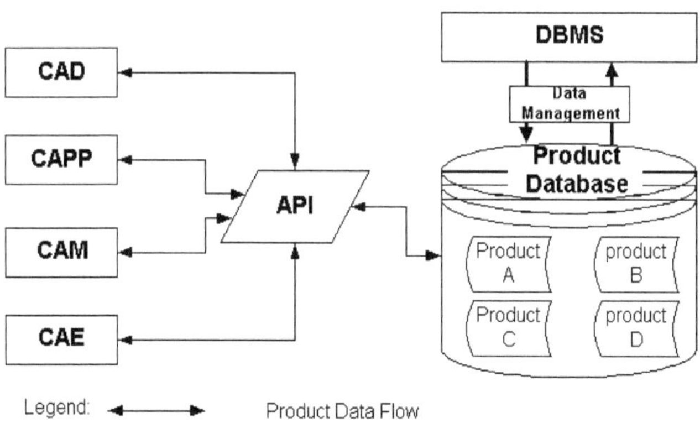

Figure 7.5 Product data exchange and sharing based on DBMS

7.3.4 Conclusion and Future Work

This chapter presented a STEP-based generic product modelling framework (GPMF), to support efficient information exchange and sharing in a distributed manufacturing environment. The overall structure of the GPMF is presented and discussed. The focus was placed on the STEP-based modelling environment, the "five-phase" modelling method, the three EDM data exchange and sharing methods and the definition of three schemas. By using the methods presented, product models are easily created for applications involving PD activities. Case studies are carried out to validate the GPMF. A prototype PDMS is developed to demonstrate the compatibility of the GPMF and the defined schemas. The following conclusions are drawn:

1. The GPMF is a STEP-based product modelling framework. The framework is used to build generic product models in support of efficient information sharing and exchange for integrated or concurrent PD.
2. The EDM is the core of GPFM. It is developed using a "five-phase" modelling method. The five modelling phases ensure that all elements of the EDM are explicitly and precisely defined and modelled. The EDM and its modules include all the data models defined to reduce data loss.
3. The STEP modelling language EXPRESS and EXPRESS-G are employed to build up the EDM. This makes the model computer compatible and helps software take advantage of the EDM definitions without human transcription.
4. STEP generic resources and AP 203 are used as the basic modelling resources to build up the EDM. New modelling resources are also defined by the authors using EXPRESS language. These are compatible with STEP-defined resources and are supplementary resources for the GPMF to model various types of product data. Three schemas are presented as examples to demonstrate how the GPMF works in practice.
5. Three EDM data exchange and sharing methods are used to support information exchange and sharing. They are integrated with different application software tools to simplify the implementation of EDM in applications.
6. GPMF provides a well-established mechanism to support the integration of various manufacturing systems. However, owing to time limitations, the GPMF has not been applied in an actual integrated manufacturing environment to support the integration of CAD, CAPP and CAM. Future work in this area has great potential and is not limited to the following two areas.
7. The first area for future work is the further development of the prototype PDMS system. This includes the input/output interfaces for the integration of various computer aided systems, such as CAD, CAPP and CAM. The system needs to provide a standard interface to transfer product data in a proper format to the end user.
8. The second area for future work is the exploration of the possibility of extending the proposed GPMF to support Web/Internet-based manufacturing. Integrating web technologies into the modelling framework can improve the information ex-

changing effectiveness through Internet/intranet. This can enable the proposed GPMF for the distributed PD environment to support web-based PD activities. A possible implementation method involves mapping between XML and EXPRESS, which is defined in STEP Part 28 (ISO 2003).

References

Gu, P. H., 1992, PML: product modelling language, Computers in Industry, 18, 265–277.

Gu, P. H. and Chan, K., 1995, Product modelling using STEP, Computer-Aided design, 27, 163–179.

ISO, 1994a, Industrial automation systems and integration: Product data representation and exchange: Part 11: Description methods: The EXPRESS language reference manual, Reference number: ISO 10303-11:1994(E). 1st edition, Switzerland.

ISO, 1994b, Industrial automation systems and integration: Product data representation and exchange: Part 203: Application protocol: Configuration controlled 3D designs of mechanical parts and assemblies, Reference number: ISO 10303-203:1994(E). 1st edition, Switzerland.

ISO, 1998, Industrial automation systems and integration: Product data representation and exchange: Part 45: Integrated generic resource: Materails, Reference number: ISO 10303-45:1998(E). 2nd edition, Switzerland.

ISO, 2000, Industrial automation systems and integration: Product data representation and exchange: Part 41: Integrated generic resource: Fundamentals of product description and support, Reference number: ISO 10303-41:2000(E). 2nd edition, Switzerland.

ISO, 2003, Industrial automation systems and integration: Product data representation and exchange: Part 28: Implementation methods: XML representations of EXPRESS schemas and data, Reference number: ISO 10303-281:2003(E). 1st edition, Switzerland.

Kahn, H., Filer, N., Williams, A. and Whitaker, N., 2001, A generic framework for transforming EXPRESS information models, Computer-Aided Design, 33, 501–510.

Krause, F. L., Kimura, F., Kjellberg, T., Lu, S. C. Y., Van derWolf, A. C. H., Ating, L., ElMaraghy, H. A., Eversheim, W., Iwata, K., Suh, N. P., Tipnis, V. A. and Weck, M., 1993, Product modelling, CIRP Annals: Manufacturing Technology, 42, 695–706.

Lee, Y. T., 1999, An overview of information modelling for manufacturing system integration, NISTIR 6382, National Institute of Standard and Technology, Gaithersburg, MD, USA.

Li, H. L., Han, J. H., Dong, J. X. and Wang, Y., 1996, Feature-based, parametric modelling system for CAD/CAPP/CAM integrated system, Industrial Technology, 1996. (ICIT '96), Proceedings of The IEEE International Conference on 2–6 December, Shanghai, China, pp. 329–333.

Loffredo, D., 1998, Efficient Database Implementation of EXPRESS information models, PHD Thesis, Rensselaer Polytechnic Institute, Tory, New York.

Ming, X. G., Mak, K. L. and Yan, J. Q., 1998, A PDES/STEP-based information model for computer aided process planning, Robotics and Computer Integrated Manufacturing, 14, 347–361.

Murphy, G., 1950, Similitude in Engineering, Ronald Press Company, New York.

Nambiar, A. N., Judd, R. P. and Greenblatt, J., 2009, Data exchange among different modules in a manufacturing or service framework. International Journal of Product Development, 8(2), 122–140.

Salustri, F. A., 1996, A formal theory for knowledge-based product model representation, in Knowledge-Intensive CAD II: Proc. IFIP WG5.2 Workshop, eds S. Finger, T. Tomiyama and M. Matyla, Chapman & Hall, London, 59–78.

Song, Y. Y., Chu, X. P. and Cai, F. Z., 1999, Real-time concurrent product and process design system for mechanical parts, High Technology Letters, 5, 74–80.

Tang, D., Zheng, L., Li, Z. and Chin, K. S., 2001, STEP-based product modelling for concurrent stamped part and die development, Computers in Industry, 46, 75–94.

Tu, Y. L., Fung, R. Y. K., Tang, J. F. and Kam, J. J., 2003, Computer aided customer interface for rapid PD, The International Journal of Advanced Manufacturing Technology, 21, 743–753.
Usher, J. M., 1996, STEP-based object-oriented product model for process planning, Computers and Industrial Engineering, 31, 185–188.
Wilson, R. R., 1990, Information modeling and PDES/STEP, Technical Report 90017, Rensselaer Design Research Center, Pensselaer Polytechnic Institute, Troy, New York.
Xie, S. Q. and Tu, P. L., 2005, Rapid one-of-a-kind PD, International Journal of Advanced Manufacturing Technology, 27, 421–430.
Xu, X, Wang, H., Mao, J., Newman, S. T., Kramer, T. R., Proctor, F.M. and Michaloski, J. L., 2005, STEP-compliant NC research: the search for intelligent CAD/CAPP/CAM/CNC integration, International Journal of Production Research, 43, 3703–3743.
Yang, J. G. and Li, B. Z., 1992, Research on transformation of workpiece information in CAD/ CAPP, Zhongguo Fangzhi Daxue Xuebao/ Journal of China Textile University, 18, 1–7.
Zha, X. F. and Du, H., 2002, A PDES/STEP-based model and system for concurrent integrated design and assembly planning, Computer-Aided Design, 34, 1087–1110.

Chapter 8
EXPRESS Data Model

Abstract This chapter presents the EXPRESS Data Model (EDM) that has been developed based on the STEP-based modelling environment and the "five-phase" modelling method. The structure of the EDM is discussed and the elements of the EDM – schemas and the relationships between schemas – are presented using EXPRESS language (see Appendix A.3) and EXPRESS-G diagrams.

8.1 Structure of EDM

Figure 8.1 presents the structure of the EDM which consists of four modules. The product general information module, product geometric data module and product manufacturing data module are defined and based on the results of classifying product data in the first modelling phase as discussed in Section 7.3.1. The resources module is developed by grouping the shareable basic modelling objects to support the development of other modules.

The product general information module represents product data that are not directly related with product manufacturing, such as the product identity, product property, and the relationship between products. In this module, the "product_definition_schema" is defined to support the modelling of this aspect of the product data.

The product geometric data module supports the modelling of product geometric data, such as shape information and dimension information. This module is key for integrating different computer aided systems, such as CAD/CAPP/CAM. In the product geometric data module, STEP AP 203 (ISO 1994) is directly utilised to represent and exchange product 3D geometric information. This chapter will not discuss how this AP is implemented; detailed information can be obtained from STEP Part 203 (ISO 1994).

The product manufacturing data module is the core of the proposed EDM, and consists of nine EXPRESSS schemas to support the modelling of different manufacturing data.

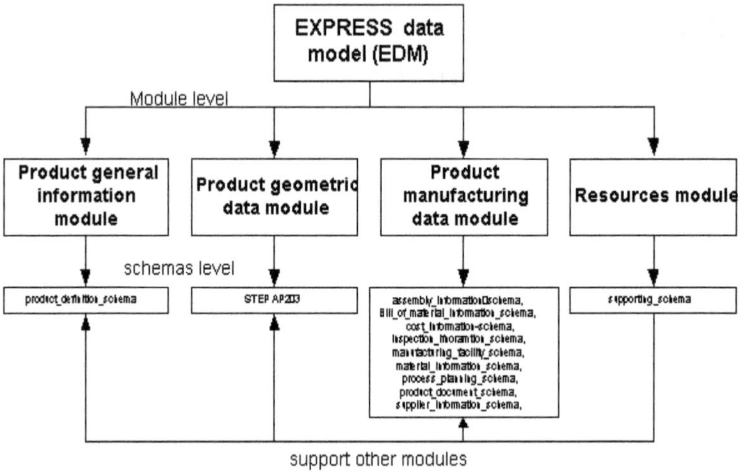

Figure 8.1 Structure of the EDM

- The *supplier_informaton_schema* is defined to model the supplier information. The supplier is an essential component of the manufacturing industry. Manufacturing companies need information about suppliers and their products to arrange the manufacturing processes.
- The *manufacturing_facility_schema* is defined to model the facility information. The manufacturing facilities, including machines, tools and fixtures are directly involved in product manufacturing processes. They are the key resources that determine product manufacturing processes and influence the product quality.
- The *product_document_schema* is defined for the documentation aspect. Documents are defined as industry standards and documented criteria that regulate and guide the manufacture of a product.
- The *bill_of_material_information_schema* is defined to represent Bill of Material (BOM) information. The BOM, also called "part list", is a list of product components or resources for manufacturing or assembling the product. The BOM is used to determine the product cost and as an effective tracking source during the manufacturing and assembly processes.
- The *material_information_schema* is defined to represent material information. The material is the basic element for a product and directly influences the selection of proper manufacturing facilities and manufacturing processes.
- The *process_planning_information_schema* is defined to model process data. The manufacturing process information is the detailed description of actual manufacturing activities. It is essential to consider this aspect of information in the design stage in order to optimise manufacturing processes.
- The *assembly_information_schema* is defined for two aspects of product data. Assembly product is one of the most important types of products. To develop an assembly product, the components information and the assembly method are required.

- The *inspection_information_schema* is defined to model inspection information. Inspection processes are essential parts of the PD process used to control product quality. The inspection results can assist in indicating problems with a product and its production processes.
- The *cost_information_schema* is defined for the cost information occurring in PD processes. Cost is one of the most important items of information for a manufacturing company. Managing cost information can help the firm to increase its competitive ability.

The resources module defines basic modelling objects that are shared by the other modules. All these basic modelling objects are grouped into the *supporting_schema.* The resources in the supporting_schema are presented as EXPRESS ENTITY, the constructed TYPE, and FUNTION. Through the EXPRESS schema interface, the resources in this schema are used by other schemas to structure an effective and efficient data model representation.

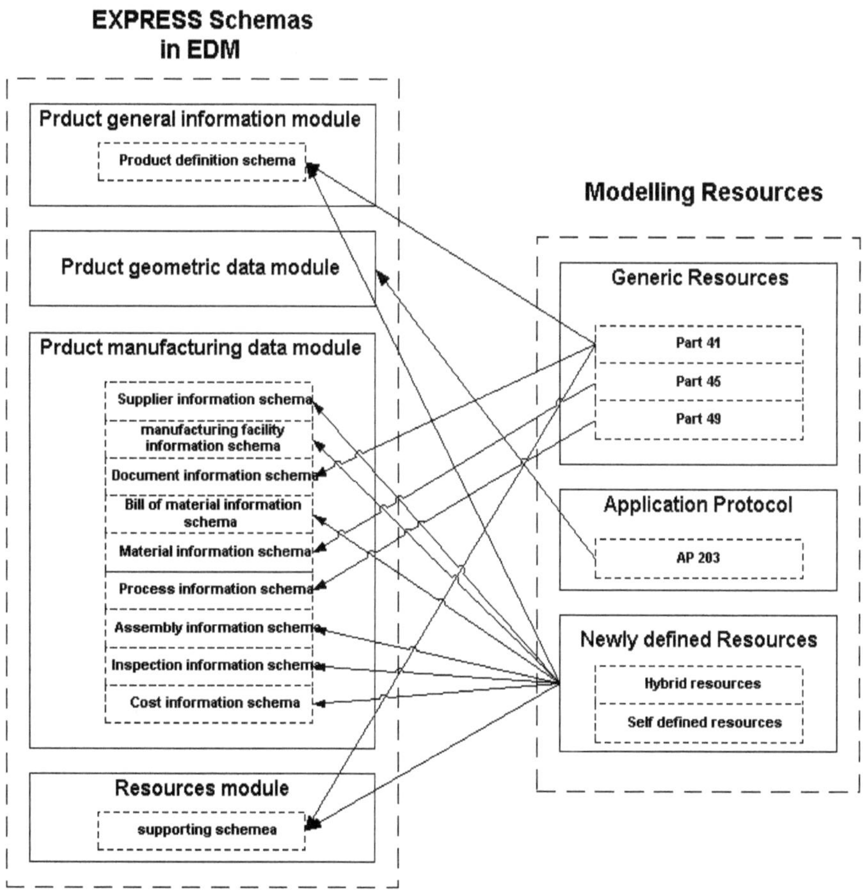

Figure 8.2 Relationships between data model schemas and modelling resources

The four modules of the EDM are developed by applying STEP generic resources including Part 41 (ISO 2000), Part 45 (ISO 1998a), and Part 49 (ISO 1998b), STEP AP 203 (ISO 1994) as well as the newly defined STEP compatible modelling resources. The relationships among EDM schemas and these resources are presented in Figure 8.2.

8.2 Product General Information Module

The product general information module has one schema, named *product_definition _schema*. This schema is modified from the product_definition_schema in STEP Part 41 (ISO 2000). It refers to the resources defined in the supporting_schema. Figure 8.3 shows the structure of this schema.

This schema consists of five entities – product, product_relationship, product_category, product_category_relationship, and product_property.

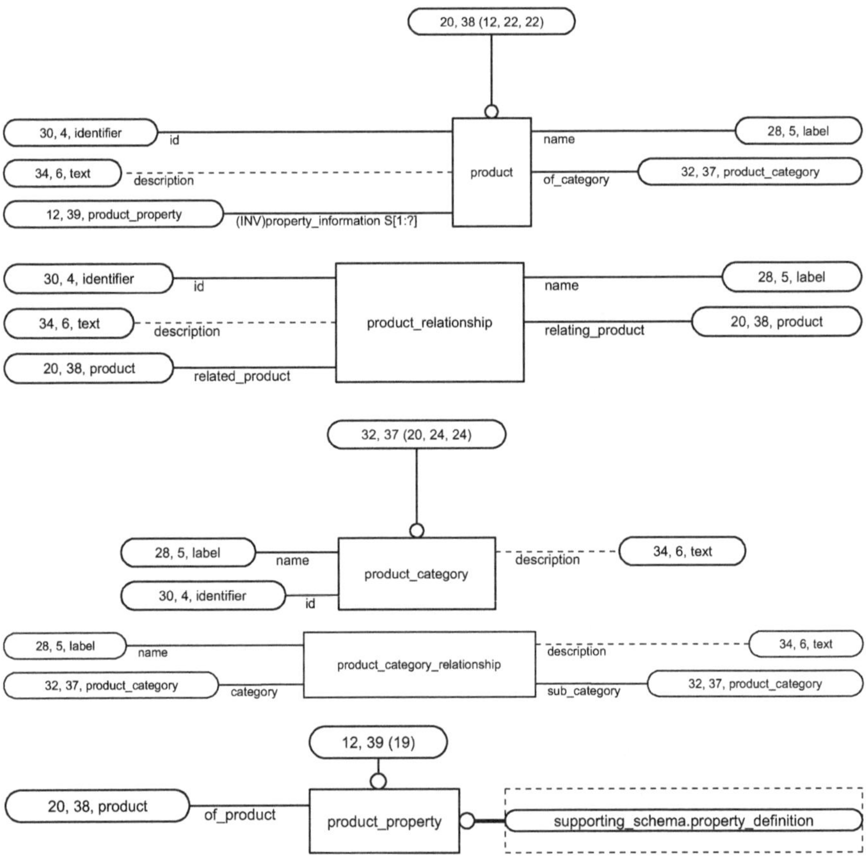

Figure 8.3 EXPRESS-G diagrams of the product_defintion_schema

1. *Product.* The product has five attributes: id, name, description, of_category, and property_information. The id, name, and description represent the id identification, the name information, and the text description information of a product.[1] The of_category utilises an instance of product_category to define a product's category. The inversed attribute, property_information, utilises a set of product_property instances to define all the properties of a product.
2. *Product_relationship.* The product_relationship represents relationship information between two products, such as "innovation version" relationship and "substitution" relationship. It has five attributes: id, name, description, relating_product, and related_product. The relating_product defines one instance of product, which is a part of this relationship. The related_product defines an instance of product as well, which is the other part of the relationship.
3. *Product category.* The product_category defines category information about a kind of product. For example, the instances of product "car", "bus" and "truck" belong to the same product_category named "automobile". The product_category has three attributes: id, name, and description.
4. *Product_category_relationship.* The product_category_relationship hierarchically relates one product category with another and provides a description of the relationship. This entity has four attributes: name, description, category, and sub_category. The category employs an instance of product_category to define the parent of the sub_category. The sub_category also utilises another instance of product_cateogory to define the child of the category. For example, an instance of product_category called "piping part" is the parent category of a sub_category such as "cold water pipe". The product_category_relationship has a local constraint. A function named acyclic_product_category_relationship, which is defined in this schema, is applied to ensure there is no repeated value of category and sub_category.
5. *Product_property.* The product_property is a self-defined entity. This entity is a subtype of supporting_information_schema.property_definition.[2] All the attributes of property_definition are inherited by the product_property. In addition, the product_property has an individual attribute called of_product which describes what product this product_property is defined for.

8.3 Product Manufacturing Information Module

The manufacturing information module is defined specifically to model manufacturing information in order to reuse the information to support the computer aided systems in other PD processes. In this module, there are nine schemas and they are developed by generic resources as well as newly defined resources.

[1] According to other entities, attributes id, name, and description present id identification, name information, and text description, respectively, of that entity.

[2] The supporting_information_schema. property_definition means the entity property_definition that is defined in the supporting_information_schema.

8.3.1 Supplier Information Schema

The supplier_information_schema is a self-defined schema. This schema refers to the resources defined in five other schemas: supporting_schema, product_definition_schema, manufacturing_facility_schema, assembly_information_schema and material_information_schema. The structure of the supplier_information_schema is presented in Figure 8.4.

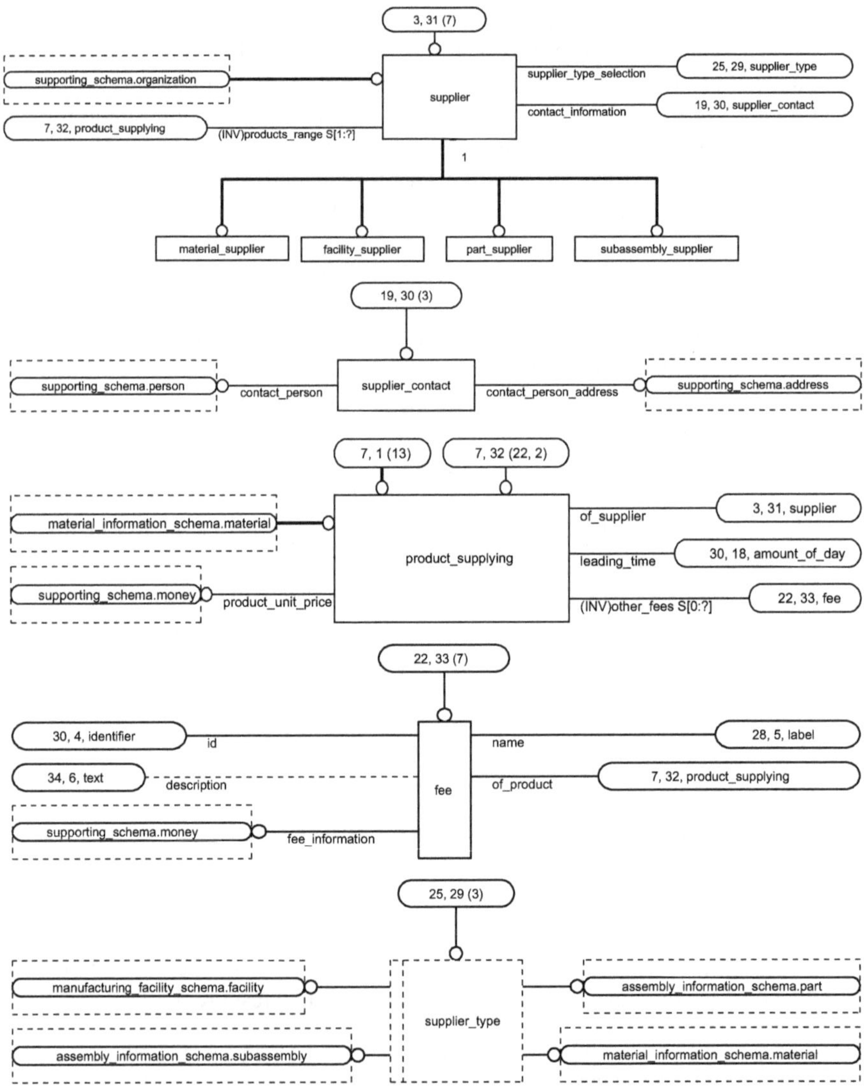

Figure 8.4 EXPRESS-G diagrams of the supplier_information_schema

There are four entities in the supplier_information_schema: supplier, supplier_contact, product_supplying, and fee.

1. *Supplier.* The supplier identifies a supplier and presents its information. This entity is a subtype of supporting_schema.organisation and inherits all the attributes of the organisation. The supplier has three individual attributes: supplier_type_selection, product_range, and contact_information. The supplier_type_selection uses the TYPE supplier_type to categorise a supplier. The inversed attribute, product_range, presents a set of product instances that the supplier provides. The contact_information utilises an instance of supplier_contact to denote the information.
   The supplier is the super-entity of four other entities: material_supplier, facility_supplier, part_supplier, and subassembly_supplier. They indicate four particular kinds of suppliers respectively. Each sub-entity has a constraint to confirm that the supplier_type_selection attribute value is matched with the actual supplier type of these sub-entities.
2. *Supplier_contact.* The supplier_contact represents the contact information of a supplier and has two attributes. The contact_person attribute is defined by the supporting_schema.person. The contact_person_address attribute is defined by the supporting_schema.address.
3. *Product_supplying.* The product_supplying represents the information of the product provided by supplier and has two super-entities: product_definition_schema.product and material_information_schema.material. It inherits the attributes of either product or material. The product_supplying includes three individual attributes: leading_time, product_price, and other_fees. These model the product leading time, the product selling price, and other expenses respectively. The leading_time is defined by the supporting_schema.amount_of_day. The product_unit_price is defined by the supporting_schema.money. The other_fees is defined as a particular expense by a fee, such as delivery fee, stocking fee.
4. *Fee.* The fee defines the information about an expense. It has four attributes: id, name, description, of_product and fee_information. The of_product links the fee with an instance of product_supplying. The fee_information is defined by the supporting_schema.money.

The supplier_infomation_schema has a SELECT TYPE named supplier_type. It presents the type of a supplier as one of the following: manufacturing_facility_schema.facility, assembly_information_schema.part, assembly_information_schema.subassembly, or material_information_schema.material.

8.3.2 Manufacturing Facility Schema

The manufacturing_facility_schema is a self-defined schema. The manufacturing_facility_schema refers to the resources defined in the suppor-ting_schema, the product_definition_schema and the product_document_schema. Figure 8.5 shows the structure of this schema.

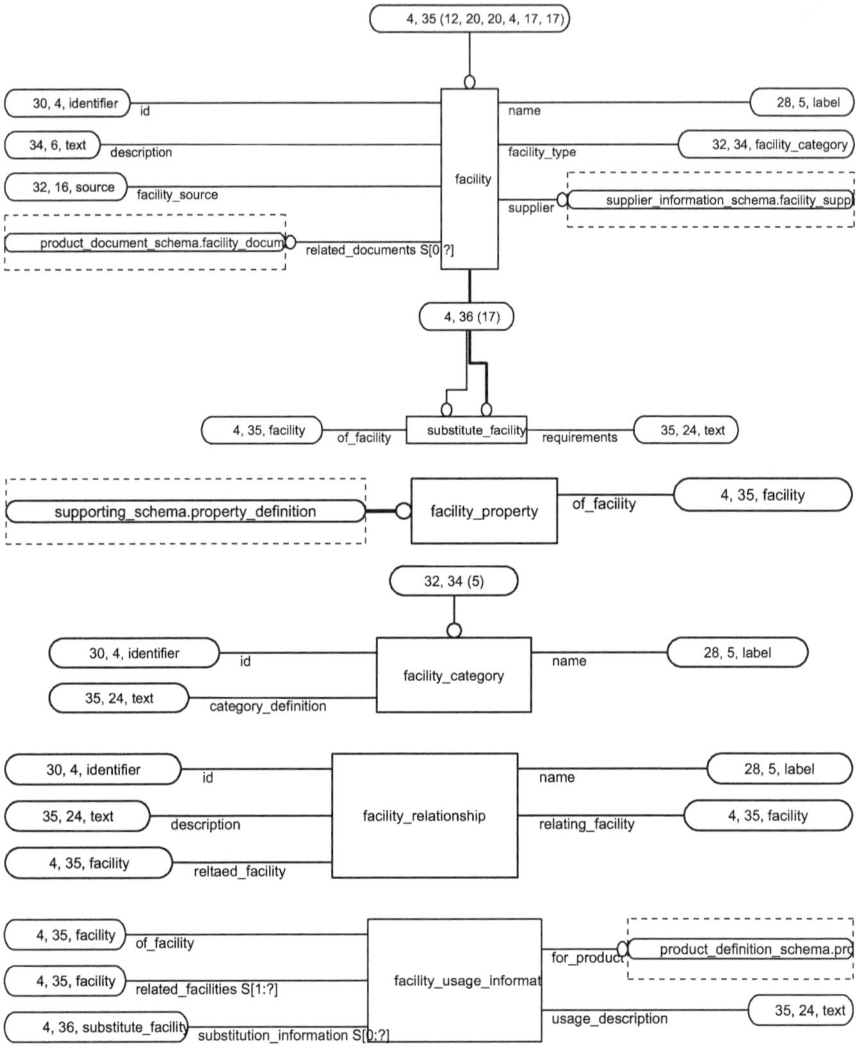

Figure 8.5 EXPRESS-G diagrams of the manufacturing_facility_schema

This schema has six entities: facility, facility_property, facility_category, facility_relationship, substitute_facility, and facility_usage_information.

1. *Facility.* The facility is the core of the manufacturing_facility_schema, and contains the information of an instance of facility utilised in the manufacturing process. It has seven attributes: id, name, description, facility_type, facility_source, supplier, and related_documents. The facility_type is defined by the facility_category. The facility_source is defined by TYPE supporting_schema.source. The supplier is defined by the supplier_information_schema.facility_supplier.

The related_documents is defined by a set of instances of product_document_schema.facility_document.

2. *Facility_property.* The facility_property presents one characteristic of a facility instance, for example "hardness" is a facility_property of an instance facility named "cutter". The facility_property is the sub-entity of supporting_schema.property_definition, and inherits all attributes of the super-entity. In addition, the facility_property has one individual attribute named of_facility to link an instance facility with an instance of facility_property.
3. *Facility_category.* The facility_category defines information about different categories of facilities, such as "milling machine category", "drilling machine category", *etc.* It has three attributes: id, name, and category_definition. The category_definition is the text definition description about an instance of facility_cateogory.
4. *Facility_relationship.* The facility_relationship is used to support an association or relationship between two different instances of facility. It has five attributes: id, name, description, relating_facility and related_facility. The related_facility consists of an instance of facility, which is related with another facility instances stored in the relating_facility. The relationship between these two instances is presented by the facility_relationship.
5. *Substitute_facility.* The substitute_facility is used to provide information to support substitution or replacement of an original facility within the manufacturing processes. It is the sub-entity of facility and inherits all attributes of its super-entity. The substitute_facility has two individual attributes, of_facility and requirements. The of_facility associates an instance of substitute_facility with an instance facility, this means the two instances are able to replace each other. The requirements use text descriptions to present the constraints for carrying out this replacement.
6. *Facility_usage_information.* The facility_usage_information collects and presents information about how a facility instance can be implemented in an actual manufacturing process. This entity has five attributes: of_facility, for_product, related_facilities, usage_description, and substitution_information. The of_facility links an instance of facility_usage_information to an instance of facility. The for_product defines what product_definition_schema.product the facility instances are used for. The related_facilities defines one or more instances of facility employed to manufacture the product defined in the for_product. The usage_description uses text to describe the utilisation details of this facility. The substitute_information uses a set of instances of substitute_facility to present the replacement information.

8.3.3 Product Document Schema

The product_document_schema defines the standards or documents used in the PD processes. The document_schema in STEP Part 41 (ISO 2000) is used to develop this schema. The product_document_schema refers to the resources defined in the supporting_schema, the supplier_information_schema, and the manufacturing_facility_schema. Figure 8.6 shows the structure of this schema.

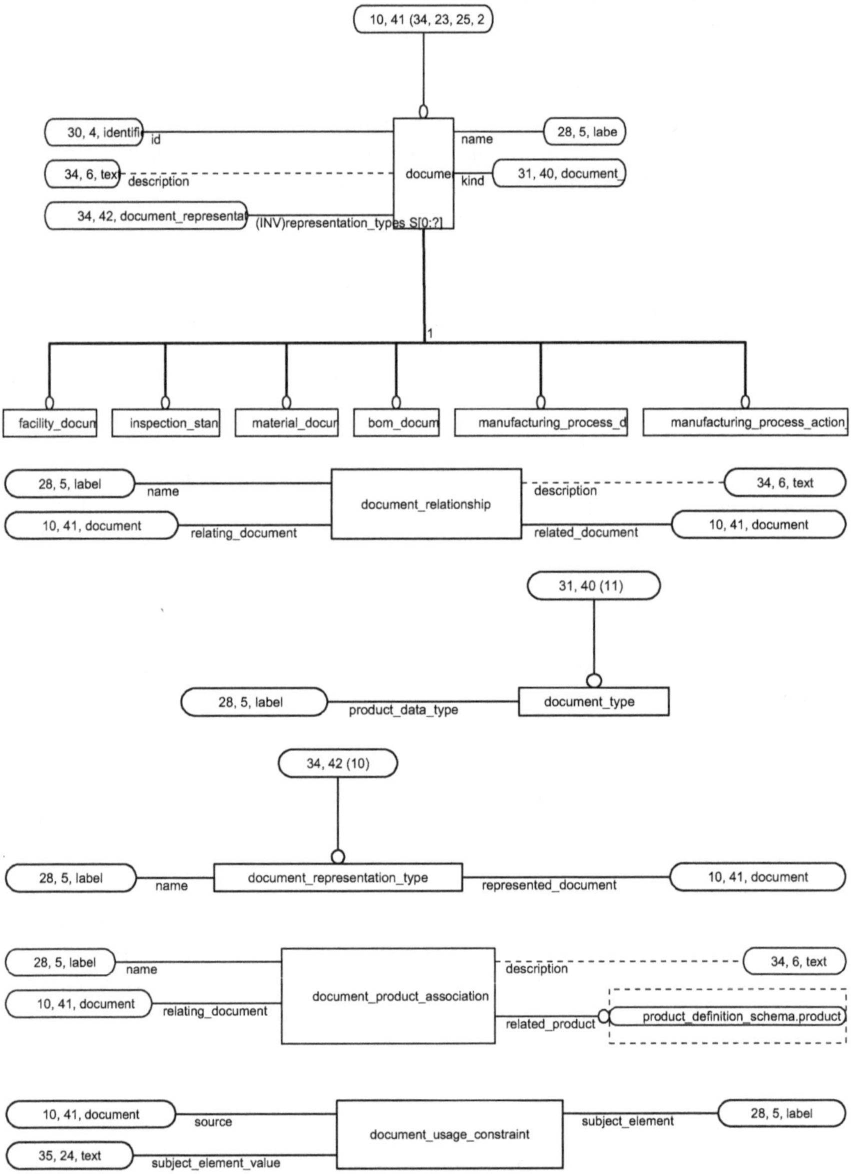

Figure 8.6 EXPRESS-G diagrams of the product_document_schema

There are twelve entities defined in the product_document_schema. The document, document_relationship, document_type, document_representation_type, document_product_association, and document_usage_constraint are defined by utilising the document_schema in STEP Part 41 (ISO 2000). The facility_document, inspection_standard, material_document, bom_document, manufacturing_process_document, and manufacturing_process_action_document are sub-entities of document and are used for modelling six specific kinds of documents.

8.3.4 Bill of Material Information Schema

The bill_of_material_information_schema is a self-defined schema to model the BOM information. It refers to the resources in the supporting_schema, the assembly_information_schema, the product_definition_schema, the material_information_schema, and the supplier_information_schema. Figure 8.7 shows the structure of this schema.

There are four entities in the bill_of_material_information_schema: bom_item, bill_of_meterail_list, external_ording_information, and internal_ordering_information.

1. *Bom_item.* The bom_item defines the information of one element listed on the BOM, this can be an individual part, a subassembly component, or a kind of material used for the product. The bom_item has eight attributes: id, name description, of_bom_list, quantity, item_type, item_source, and item_ordering_information. The of_bom_list links an instance of bom_item to an instance of bill_of_material_list. The quantity defines how many of these instances of bom_item are needed in the manufacturing processes. The item_type utilises the SELECT TYPE bom_item_type to define whether the bom_item is an instance of assembly_information_schema.part, an instance of assembly_information_schema.subassembly, or an instance of material_information_schema.material. The item_source utilises the supporting_schema.source to identify whether the bom_item is purchased, self-made or unknown. The derived attribute, item_ordering_information, calls the function order_type_selection to provide the order information based on the order type of bom_item. If the value of item_source is "bought", the item_ordeing_information value is defined by the external_ordering_information. If the value of item_source is "made", the item_ordering_information value is defined by the internal_ordering_information. Otherwise, the value is defined by the supporting_schema.unknown.
2. *Bill_of_meterial_list.* The bill_of_material_list is the representation of a BOM. It has three attributes: id, of_product and bom_list. The of_product indicates what product this instance of bill_of_material_list is used for. The inversed attribute bom_list lists all the instances of bom_item in this BOM.
3. *External_ordering_information.* The external_ordering_information presents the information of an external order. It has six attributes: id, supplier_identifier, order_product, order_quantity, person_in_charge, and delivery_time. The sup-

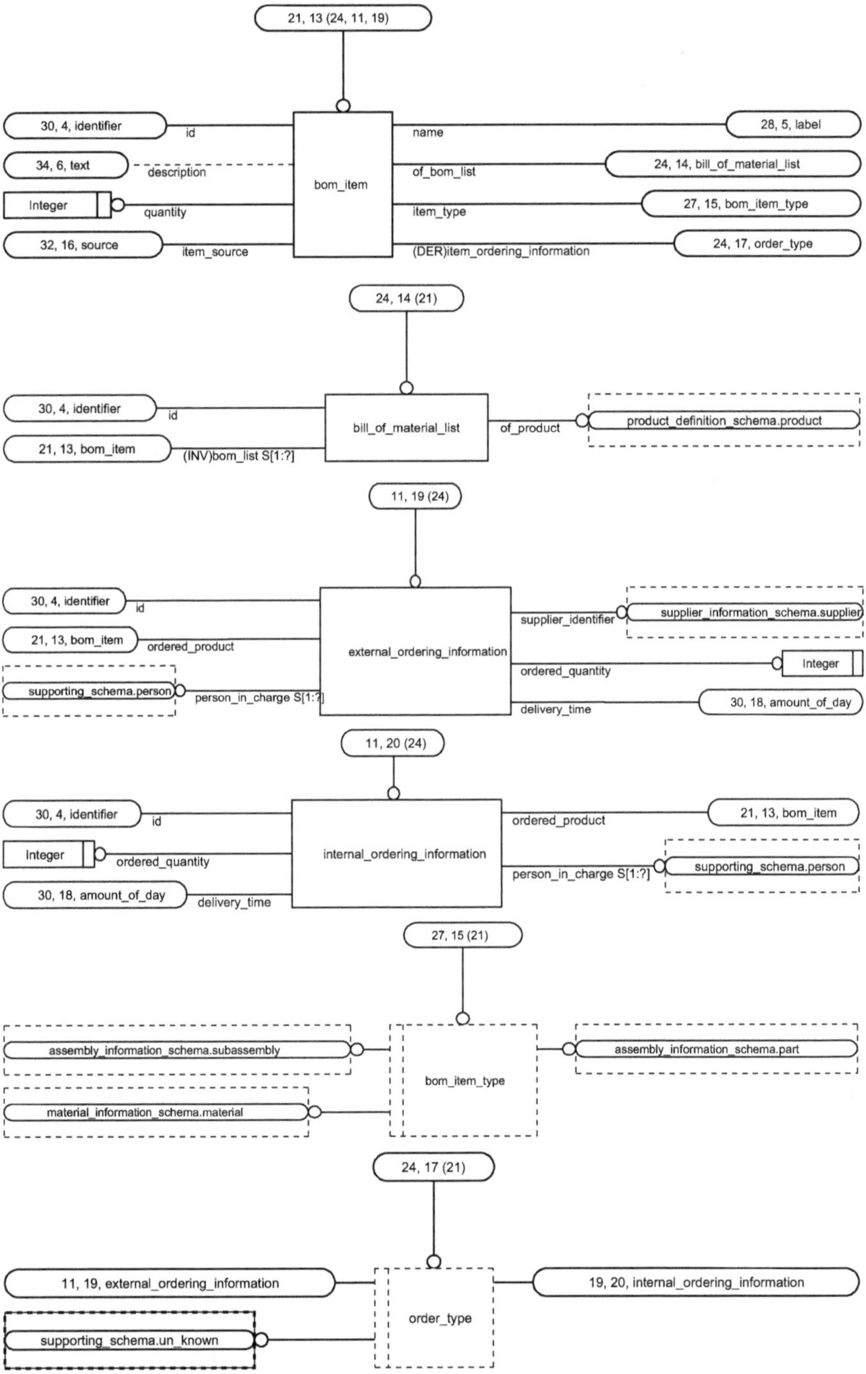

Figure 8.7 EXPRESS-G diagrams of the bill_of_material_information_schema

plier_identifier uses supplier_information_schema.supplier to represent the supplier information about this order. The ordered_product defines what bom_item is associated with this external order. The ordered_quantity represents the how many ordered_product are needed in this order. The person_in_charge uses a set of instances of supporting_schema.person to define the people who are in charge of this order. The delivery_time is defined by supporting_schema.amount_of_day.

4. *Internal_ordering_information.* The internal_ordering_information defines the information of an internal order. Its five attributes, id, order_product ordered_quantity, person in charge, and delivery_time are defined in the same way as they are applied in the external_ordering_information.

There are also two SELECT TYPEs in this schema: bom_item_type and order_type. The bom_item_type defines the type of an instance *bom_item.* The order_type defines the type of an instance order as either an external order, an internal order, or an unknown. The function order_type_selection defines the way the type of order is analysed.

8.3.5 Material Information Schema

The material_information_schema presents the material information. It refers to the resources defined in the supporting_schema, the supplier_information_schema, and the manufacturing_facility_schema. STEP Part 45 (ISO 1998a) is used to define the definitions in this schema. The structure of this schema is shown in Figure 8.8.

There are four entities in the material_information_schema: material, material_fabrication_information, material_property and substitute_material.

1. *Material.* The material presents information related to a kind of material. It has seven attributes: id, name, description, supplier, characteristics, substitution_select, and fabrication_information. The supplier uses supplier_information_schema.material_supplier to indicate information of a set of material suppliers. The inversed attribute, characteristics, presents one or more specific material characters defined in the instances of material_property. The inversed attribute, fabrication_information, collects information about how this kind of material is manufactured, and represents it by using the material_fabrication_information.
2. *Material_fabrication_information.* The material_fabrication_information describes how an instance of material is fabricated. It has three attributes: of_material, relating_manufacturing_facility, and fabrication_requirements. The of_material associates the material_fabrication_information with an instance of material. The relating_manufacturing_facility presents a set of instances of manufacturing_facility_schema.facility to fabricate the material defined in the of_material. The fabrication_requirements utilises a set of texts to describe the requirements of the material's fabrication process.

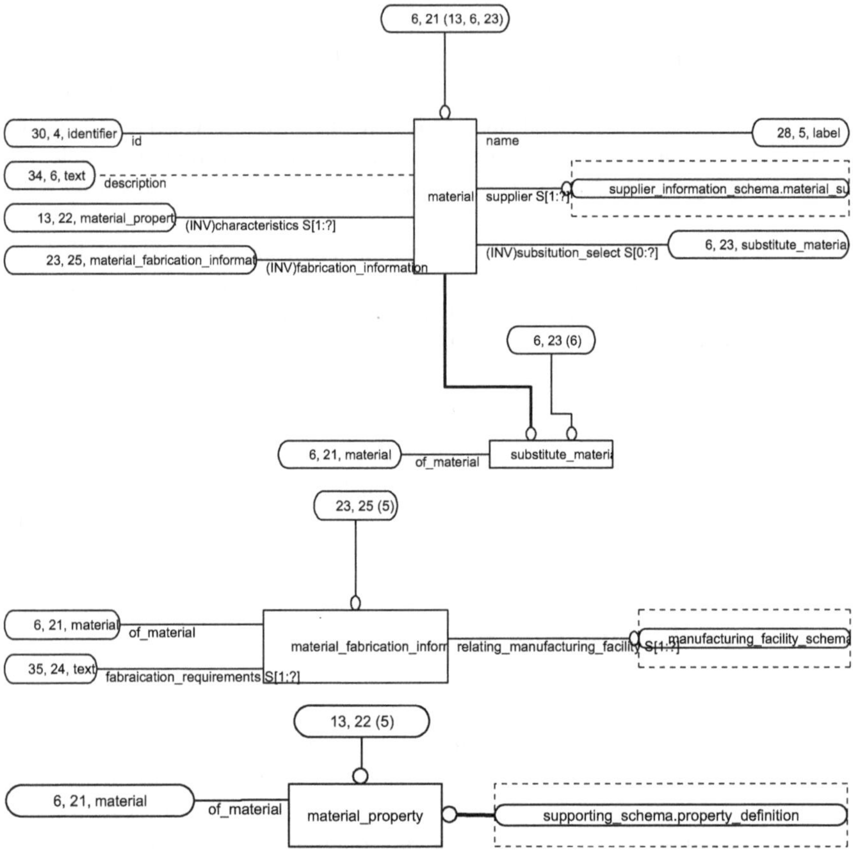

Figure 8.8 EXPRESS-G diagrams of the material_information_schema

3. *Material_property.* The material_property defines a characteristic of a material. It is the sub-entity of supporting_schema.property_definition and it inherits all attributes of the super-entity. The material_property has an individual attribute of_material, which links the material_property to an instance of material.
4. *Substitute_material.* The substitute_material is a sub-type of the material and inherits all attributes of the super-entity. The substitute_material links to an instance material through its individual attribute of_material.

8.3.6 Assembly Information Schema

The assembly_information_schema defines the information of the assembly products and the assembling processes. An assembly product can be constructed and based on a hierarchical assembly model tree (AMT) presented in Figure 8.9. The

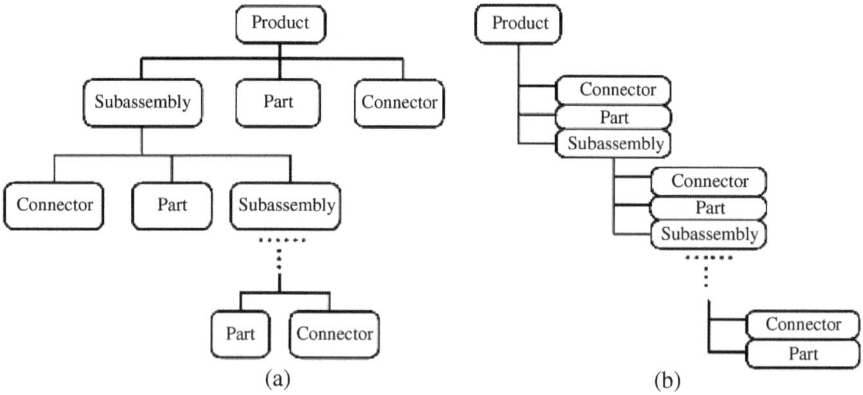

Figure 8.9 Hierarchical structure of assembly product model tree (Zha and Du 2002), (a) product assembly structure; (b) assembly modelling

root of AMT is the assembly product and has three nodes including part component, subassembly component and connector that are in hierarchy level one. The "part" component is the individual component in the AMT and there are no subcomponents under it. The "subassembly" component, which has the same structure as the assembly product, has its own "part" components, "subassembly" components and "connector" components. The "connector" component is the assembling information shared among different components. The AMT will stop when there are only "part" components and "connector" components left.

The assembly_information_schema presents the information of an assembly product by mapping the structure of AMT. It refers to the resources defined in the supporting_schema, and the product_definition_schema. Figure 8.10 shows how the assembly_information_schema represents the AMT structure and models assembly products.

There are five entities in the assembly_information_schema: assembly_product, part, subassembly, connector and connecting_method.

1. *Assembly_product.* The assembly_product defines the assembly product information. It is the sub-entity of product_definition_schema.product. The assembly_product has three individual attributes: sub_parts, sub_assemblies, and connection_information. The inversed attribute sub_parts, applies a set of instances of part to define all the individual "part" components. The inversed attribute sub_assemblies utilises a set of instances of sub_assembly to define all the "assembling" components. The inversed attribute connection_information, utilises a set of instances of connector to present the connection information among "parts" and "subassemblies".

   The assembly_product possesses three local constraints that define the values of level_in_the_assembling_hierarchy attributes of all the instances of part, subassembly, and connector that the assembly_product consists of, which means they all exist in the second level of the AMT.

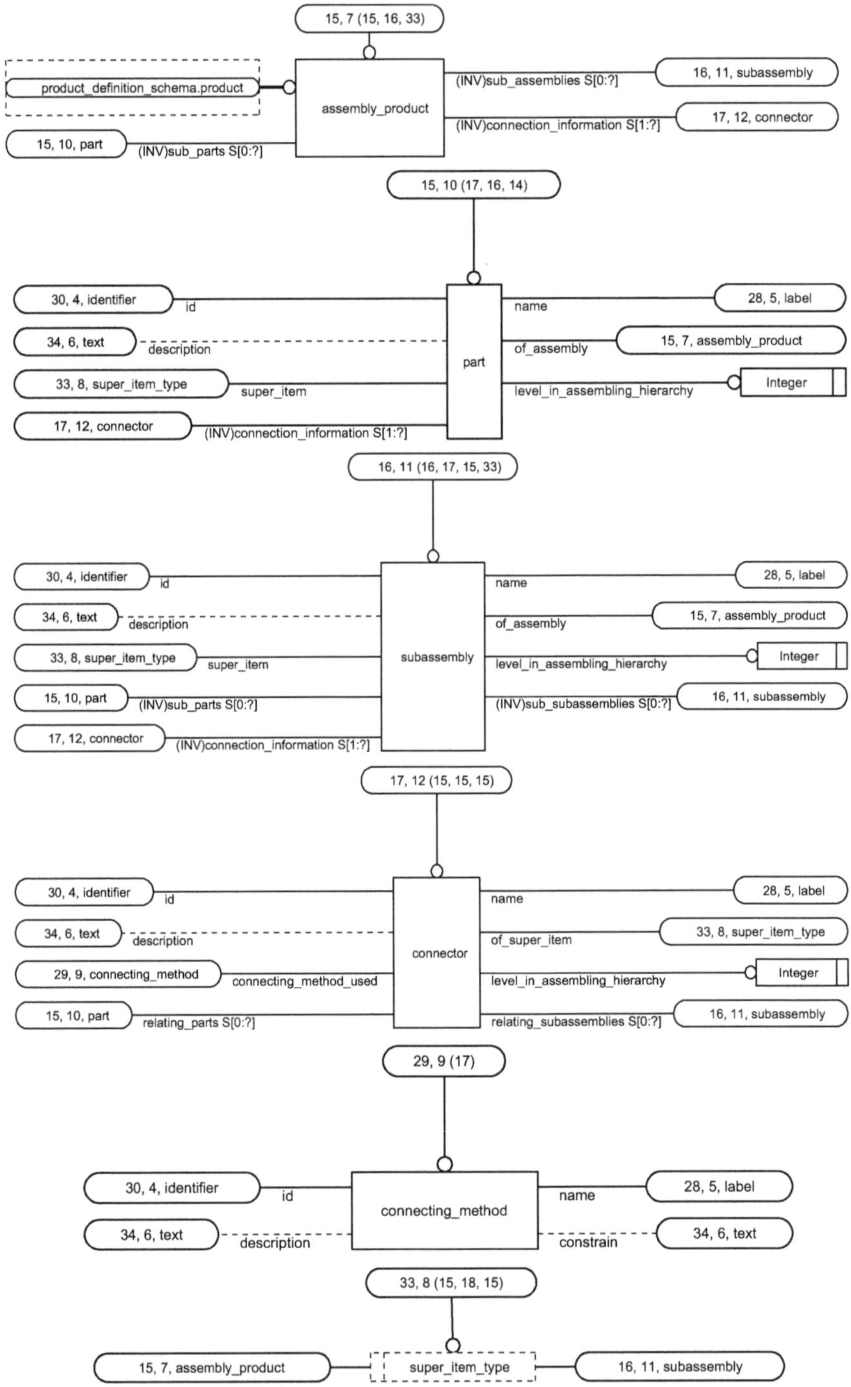

Figure 8.10 EXPRESS-G diagrams of the assembly_information_schema

2. *Part.* The part defines an individual component named "part" in the AMT. It has six attributes: id, name, description, of_assembly, super_item, level_in_assembling_hierachy, and connection_information. The of_assembly is utilised to link the part with an instance of assembly_product. The super_item utilises a SELECT TYPE super_item_type to define the upper level component of the part. The level_in_assembling_hierarchy defines the hierarchical position of the part in the AMT. The inversed attribute part_relating_connector utilises a set of instances of connector to present all the connection information related to the part. The part has one local constraint which defines the value of attribute level_in_the assembling_hierarchy as greater than or equal to two.
3. *Subassembly.* The subassembly defines the "subassembly" component in the AMT, which consists of its own sub-components. The subassembly has a total of nine attributes. The attributes id, name, description, of_assembly, super_item and level_in_the_assembling_hierarchy, have similar functions as they have in part. The inversed attribute sub_parts defines a set of instances of part within the subassembly. The inversed attribute sub_subassembly defines the "subassembly" components within the subassembly. The connection_information defines information on how an instance of subassembly can be constructed using the connector.
   The subassembly entity has five constraints. The constraint "WR1" defines that the subassembly must contain either "part" components or "sub-subassembly" components. The constraint "WR2" defines the value of level_in_assembling_ hierarchy of subassembly as greater than or equal to two. The other three constraints define that the attribute level_in_assembling_hierarchy of "parts", "sub-subassemblies" and "connectors" must have the same values, which equals the value of the subassembly.level_in_assembling_hierarchy[3] plus one.
4. *Connector.* The connector defines information about the connection between part and part, subassembly and subassembly, or part and subassembly.
   There are eight attributes of connector: id, name, description, of_super_item, connecting_method_used, level_in_assembling_hierarchy, relating_parts, and relating_subassemblies. The of_super_item utilises SELECT TYPE super_item _type to define what component the connector is used to construct. The connecting_method_used utilises an instance of connecting_method to define the connecting methodology. The level_in_assembly_hierarchy indicates the level information related to where this entity connector belongs in the assembly model tree. The relating_parts and relating_subassemblies define the "parts" and "subassemblies" components that are associated by this connector.
   The connector has three constraints. The constraints "WR1" and "WR2" define the connector and all components related to it to have the same values of level_in_assembling_hierarchy. The constraint "WR3" defines the value of level_in_assembling_hierarchy of connector as greater than or equal to two.

[3] The subassembly.level_in_assembling_hierarchy means the level_in_assembling_hierarchy attribute of the subassembly.

5. *Connecting_method.* The connecting_method defines information about a particular method for assembling different components of an assembly product, such as gluing method, and welding method. This entity has three attributes: name, description, and constraint. The constraint describes what requirements are needed to apply this connecting_method to connect assembly components.

The assembly_information_schema has one SELECT TYPE, named super_item _type, which defines the super-level component as either an assembly_product or a subassembly.

8.3.7 Process Planning Schema

The process_planning_schema defines the manufacturing process information of a product. The resources of this schema is defined by using the definition of STEP part 49 (ISO 1998b) and the action_schema in STEP Part 41 (ISO 2000). It also refers to the resources defined in supporting_schema and product_document_ schema. Figure 8.11 shows the structure of this schema.

There are fifteen entities defined in the schema. The action, action_method, action_method_relationship, action_relationship, versioned_action_request, and action_resources are modified from the *action_schema* defined in STEP Part 41 (ISO 2000). Others, which are defined based on STEP Part 49 (ISO 1998b), are presented as follows:

1. *Manufacturing_process_action.* The manufacturing_process_action defines information of one step in the manufacturing process. It is the sub-entity of action and inherits all attributes of its super-entity. The manufacturing_process_action is the super-entity of surface_finish.
   The manufacturing_process_action has seven individual attributes. The type_of _process defines the type of the manufacturing process by using TYPE process_method. The desired_result utilises the versioned_action_request to present the desired output of the manufacturing_process_action. The manufacturing_ process_action_property collects a set of instances of action_propertyies to present the characteristics of the manufacturing_process_action. The manufacturing_process_action_time defines the work time of the manufacturing_process _action. The manufacturing_process_action_resources refers to the action_resources to define a set of necessary resources which supports the implementation of this manufacturing_process_action. The documents utilises the product_document_schema.document to define related documents. The relating_ manufacturing_feature utilises the manufacturing_feature to indicate what manufacturing feature is worked on by this manufacturing_process_action.
2. *Manufacturing_process.* The manufacturing_process defines information to support planning of an entire manufacturing process. It has two attributes. The process_action_list attribute lists all the instances of the manufacturing_process_

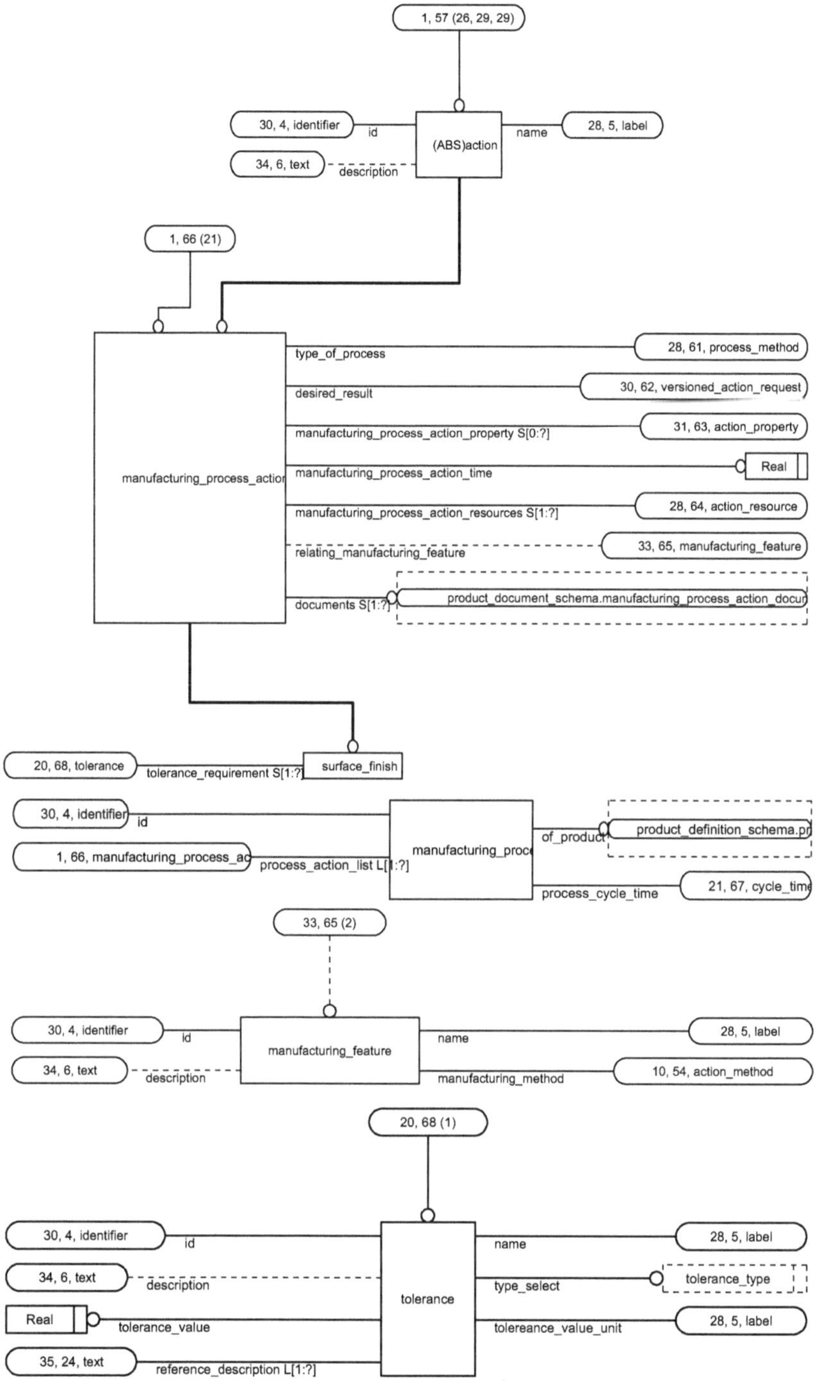

1, 57 (26, 29, 29)
30, 4, identifier
id
(ABS)action
name
28, 5, label
34, 6, text
description
1, 66 (21)
manufacturing_process_actio
type_of_process
28, 61, process_method
desired_result
30, 62, versioned_action_request
manufacturing_process_action_property S[0:?]
31, 63, action_property
manufacturing_process_action_time
Real
manufacturing_process_action_resources S[1:?]
28, 64, action_resource
relating_manufacturing_feature
33, 65, manufacturing_feature
documents S[1:?]
20, 68, tolerance
tolerance_requirement S[1:?]
surface_finish
30, 4, identifier
id
1, 66, manufacturing_process_ac
process_action_list L[1:?]
of_product
process_cycle_time
21, 67, cycle_tim
33, 65 (2)
30, 4, identifier
id
manufacturing_feature
name
28, 5, label
34, 6, text
description
manufacturing_method
10, 54, action_method
20, 68 (1)
30, 4, identifier
id
tolerance
name
28, 5, label
34, 6, text
description
type_select
tolerance_type
Real
tolerance_value
tolereance_value_unit
28, 5, label
35, 24, text
reference_description L[1:?]

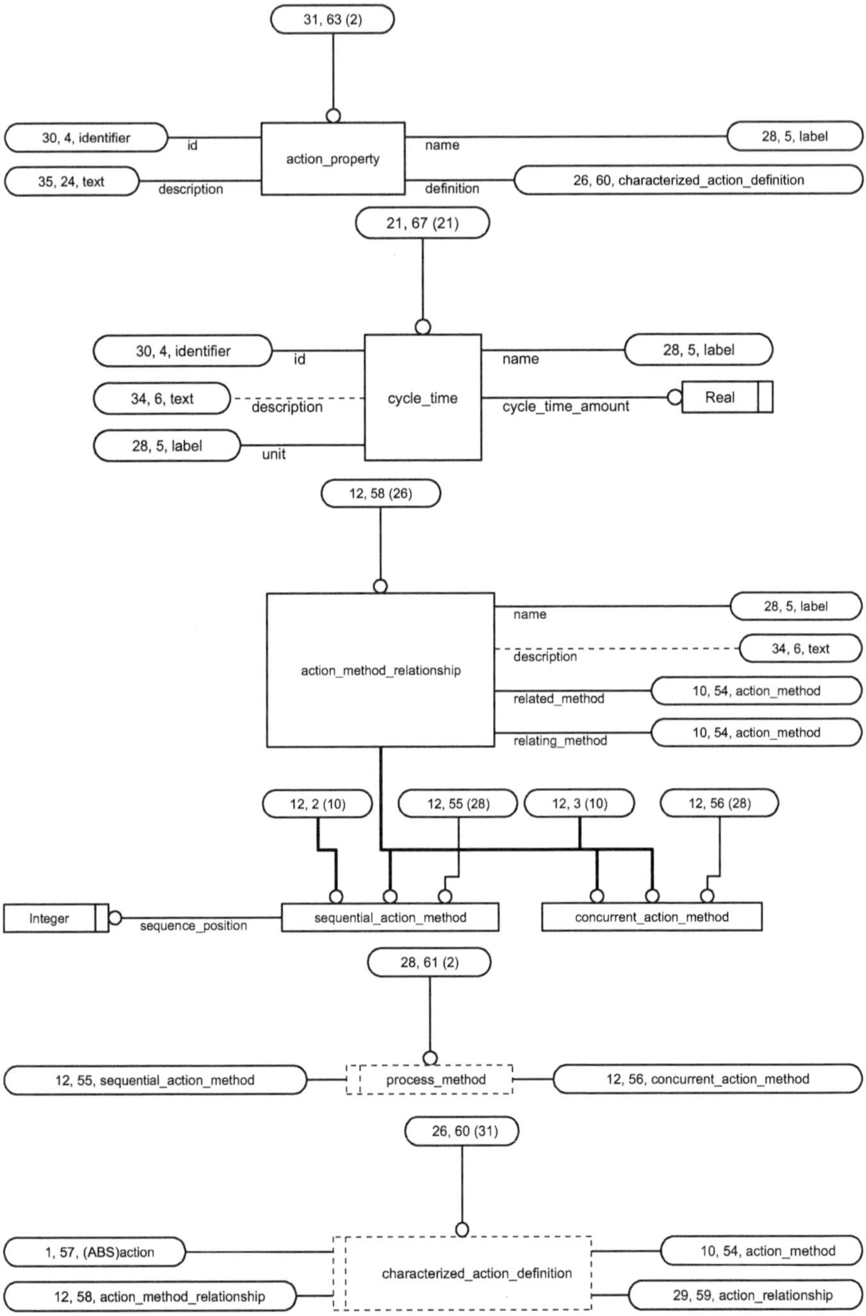

Figure 8.11 EXPRESS-G diagrams of the process_planning_schema

action according to their sequences in the manufacturing process. The process_cycle_
time attribute presents the cycle time of a complete manufacturing process by using the cycle_time.

3. *Surface_finish.* The surface_finish defines information of the surface finishing action in the manufacturing process. It is the sub-entity of manufacturing_process_action and inherits all attributes of its super-entity.
   The surface_finish has one individual attribute: tolerance_requirement. This attribute uses the tolerance to define a set of tolerance information required for the surface_finish.
4. *Manufacturing_feature.* The manufacturing_feature presents information about a manufacturing feature. It has four attributes: id, name, description and manufacturing_method. The manufacturing_method uses the action_method to define the manufacturing method utilised for the manufacturing_feature.
5. *Tolerance.* The tolerance presents tolerance information. It has seven attributes: id, name, description, type_select, tolerance_value, tolerance_value_unit, and reference_description. The type_select presents the type information about an instance of tolerance. It is one of the elements in ENUMERATION TYPE tolerance_type.
6. *Action_property.* The action_property is the description of the behavior, capabilities, or performance measures that are pertinent to a process, an action or a potential action that affects a process; it has three attributes. The definition attribute characterises the property as one value of SELECT TYPE characterised_action_definition.
7. *Cycle_time.* The cycle_time defines the manufacturing process cycle time. It has five attributes: id, name, description, cycle_time_amount and unit.
8. *Sequential_action_method.* The sequential_action_method presents information about the sequential manufacturing method. It is the sub-entity of action_method and inherits all attributes of its super-entity. The individual attribute, sequential_number, indicates the sequence information of an action.
9. *Concurrent_action_method.* The concurrent_action_method presents a concurrent manufacturing processing approach. It is the sub-entity of action_method and inherits all attributes of its super-entity.
   Three SELECT TYPEs are defined, these are tolerance_type, process_method, and characterized_action_definition.
10. *Tolerance_type.* The SELECT TYPE tolerance_type defines the type of a tolerance, which includes the straightness, flatness, circularity, cylindricity, profile of a line, profile of a surface, all around profile, angularity, perpendicularity, parallelism, position, concentricity or coaxiality, symmetry, circular runout, and total runout.
11. *Process_method.* The SELECT type process_method defines two kinds of manufacturing methods: sequential process method and concurrent process method, which are defined by the sequential_action_method and the concurrent_action_method respectively.

12. *Characterised_action_definition.* The SELECT type characterized_action_definition identifies either an action, action_method, action_method_relationship, or action_relationship.

8.3.8 Inspection Information Schema

The inspection_information_schema is a self-defined schema which presents the inspection information in the PD process. The inspection_information_schema refers to resources defined in the supporting_schema, the product_definition_schema, and the product_document_schema. The structure of this schema is shown in Figure 8.12.

This schema defines four entities: product_inspection, inspection_test, inspection_pass_requirement and result_value.

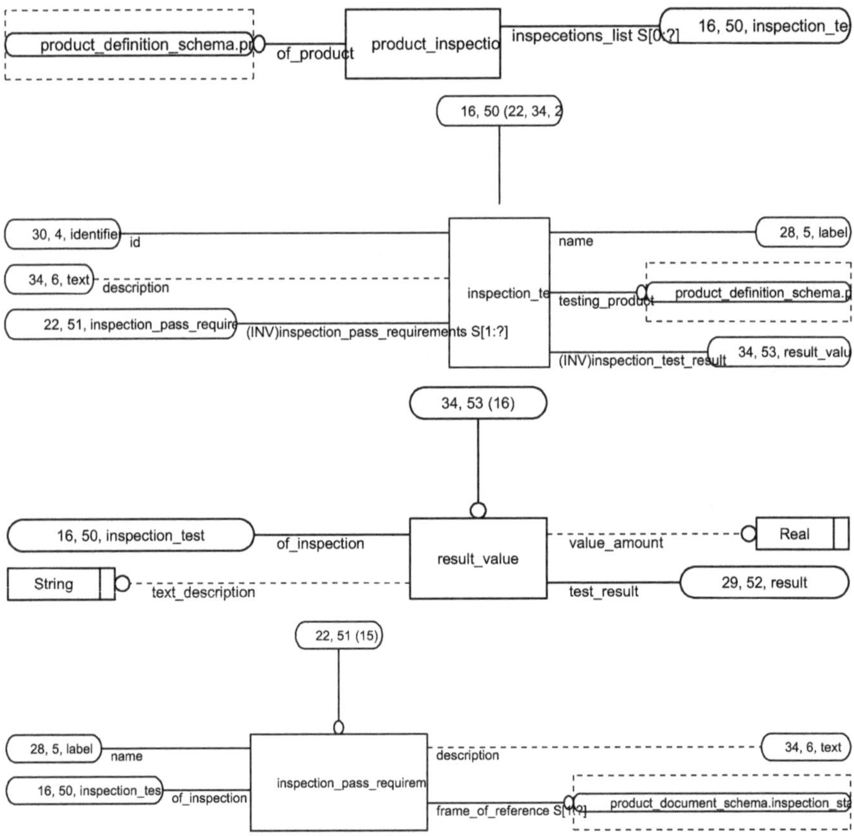

Figure 8.12 EXPRESS-G diagrams of the inspection_information_schema

1. *Product_inspections*. The product_inspections presents information about all the inspections of a product. It has two attributes: of_product and inspections_list. The of_product defines to product_definition.product that the product_inspections belong to. The inspections_list collects a set of instances of inspection_test to present all the inspections.
   The product_inspections has one constraint WR1 to define that each instance of the inspection_test listed in inspection_list has the same values as test_product. This means they have the values as product_inspections.of_product.
2. *Inspection_test*. The inspection_test presents information of one inspection test.

It has seven attributes: id, name, description, frame_of_reference, test_product, inspection_pass_requirements, and inspection_test_result. The test_product identifies what product has been inspected. The inversed attribute inspection_pass_requirements presents all the criteria needed to pass the inspection_test. The inspection_pass_requirements is defined by a set of instances of inspection_pass_requirement. The inversed attribute inspection_test_result presents the result of an instance of inspection_test, which utilises the result_value to define the attribute.

1. *Result_value*. The result_value presents both the overall and the detailed inspection test results. It has four attributes. The of_inspection attribute uses the inspection_test to indicate which inspection has this test result. The value_amount attribute and text_description attribute present the numerical result and the text description of the test result. The test_result attribute presents the overall inspection results as to whether the product passes or fails the inspection.
2. *Inspection_pass_requirement*. The inspection_pass_requirement defines a specific inspection requirement. It has four attributes: name, description, of_inspection, and frame_of_reference. The of_inspection defines which inspection_test the inspection_pass_requirement works for. The frame_of_reference presents a set of instances of product_document_schema.inspection_standard, which outlines the related documents required by inspection_pass_requirement.

There are also three constructed TYPEs, fail, pass, and result, included in the schema.

8.3.9 Cost Information Schema

The cost_information_schema is a self-defined schema presenting the cost information within the PD processes. Both the individual cost information and the overall cost information of the PD process are involved in this schema. The different costs are classified into two categories: fixed cost and variable cost. The fixed costs are independent of the rate of production of goods (Dieter 2000). For example, the depreciation on capital investment, research and development cost, *etc.* The variable costs change with the production rate (Dieter 2000). For example, the direct labour cost, materials cost, *etc.* The variable cost can be calculated as:

$$\text{Variable Cost} = \text{Cost Rate} * \text{Quantity} \tag{8.1}$$

This schema refers to the resources defined in supporting_schema and pro-duct_ definition_schema. Figure 8.13 shows the structure of the cost_information_sche-ma. Seven entities included in this schema are as follows:

1. *Cost.* The cost defines the generic description of the cost information. It has three attributes: id, name, and description, and is the super-entity of fixed_cost and variable_cost.
2. *Fixed_cost.* The fixed_cost presents the information of a fixed cost for developing the product. It is the sub-entity of cost and inherits all attributes of its super-entity. It has three individual attributes. The cost_type attribute defines the type information of an instance of entity fixed_cost. The value of this attribute is a kind of fixed cost defined in ENUMERATION TYPE fixed_cost_type. The cost_amount attribute presents the amount of a fixed_cost. This attribute value is defined by an instance of supporting_schema.money. This attribute links the fixed_cost to an instance of fixed_cost_association.
3. *Fixed_cost_association.* The fixed_cost_association collects all the fixed_cost information inccurred when developing a product. This entity provides information to support the management of the overall fixed cost.
   The fixed_cost_association has three attributes. The of_product attribute defines what product_definition_schema.product these instances of fixed_cost belong to. The inversed attribute cost_association groups all instances of fixed_cost of a product. The sum represents the subtotal of these costs.
4. *Variable_cost.* The variable_cost presents information of a variable cost during product development. It is a sub-entity of the cost and inherits all attributes of its super-entity.
   The variable_cost has five individual attributes: belonging, rate, quantity, cost_ amount_unit, and cost_amount. The belonging links the variable_cost to an instance of variable_cost_association. The rate utilises the variable_cost_rate to define the parameter “cost rate” in Equation 8.1, while the quantity defines the parameter “quantity”. The derived attribute cost_amount utilises Formula 8.1 to calculate the variable cost amount.
5. *Variable_cost_association.* The variable_cost_association collects all the variable_cost information for developing a particular product. It provides information to support the management of the overall variable cost in the PD process.
   It has three attributes: of_product, sum, and association. The inversed attribute association lists all the instances of variable_cost of a product. The of_product utilises the product_definition_schema.product to define what product these instances of variable_cost belong to. The sum represents the subtotal amount of the variable_cost instances defined in entity association.
6. *Variable_cost_rate.* The variable_cost_rate is the unit cost of an instance of variable_cost. It has four attributes: id, name, rate and unit.

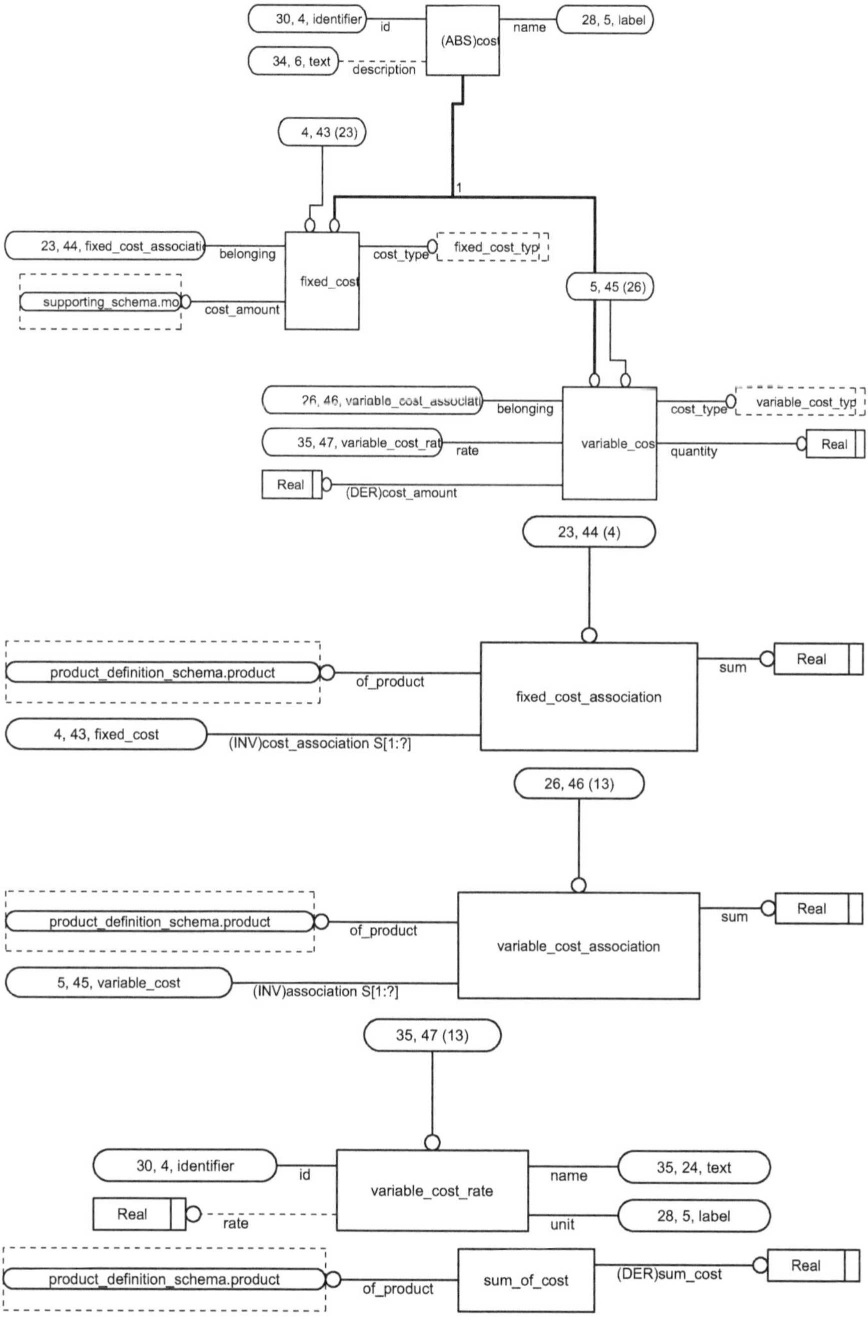

Figure 8.13 EXPRESS-G diagrams of the cost_information_schema

7. *Sum_of_cost.* The sum_of_cost adds all fixed and variable costs to obtain the total cost amount of a product. It has two attributes. The of_product attribute presents what product the instance sum_of_cost is for. The derived attribute sum_cost adds the values of fixed_cost_association.sum and variable_cost_association.sum to get the total cost amount. The two local constraints WR1 AND WR2 guarantee that all costs in the sum calculation process belong to the same product.

There are two ENUMERATION TYPEs defined in the schema: fixed_cost_type and variable_cost_type.

1. *Fixed_cost_type.* The ENUMERATION TYPE fixed_cost_type defines sixteen types of fixed costs: depreciation on capital investment, interest on capital investment and inventory, property taxes, insurance, technical service, product design and development, non-technical service, general supplies, rental of equipment, share of corporate executive staff, legal staff, share of corporate research and development staff, market staff, sales force, delivery and warehouse, and technical service staff (Dieter·2000).
2. *Variable_cost_type.* The ENUMERATION TYPE variable_cost_type defines eight types of variable costs including material, direct label, maintenance costs, power and utilities, quality control staff, royalty payment, packaging and storage costs, and scrap loss and spoilage (Dieter, 2000).

8.4 Resources Module

The resources module includes the schema supporting_schema. The supporting_schema describes the EXPRESS declarations shared and used by other schemas. This schema consists of STEP integrated resources and self-defined resources.

The resources, which are referred from generic resources, include:

1. *Support_resource_schema* (ISO 2000). TYPE identifier, TYPE label, TYPE text, and FUNCTION bag_to_set.
2. *Person_organisation_schema* (ISO 2000): ENTITY address, ENTITY organisation, and ENTITY person.

Other resources are all self-defined resources including: TYPE amount_of_day and amount_of_hour; ENTITY money, unknown, and property_definition. The property_definition is modified from entity property_definition in product_property_definition_schema of STEP Part 41 (ISO 2000). It defines the information required to present a property.

8.5 Conclusion

This chapter discusses the proposed EDM in detail. The EDM is divided into four modules. The 11 schemas defined by modelling resources and STEP AP 203 are involved in these four modules. The relationship between the EDM and the STEP modelling environment is investigated. The EDM contains a large variety of product information and includes not only product geometric data and general information, but also important manufacturing information, such as assembly information, manufacturing process information, inspection information, cost information, *etc.* The EDM provides a mechanism to structure product data systematically and to support product data exchange and sharing.

References

Dieter, G. E., 2000, Engineering design: a materials and processing approach, 3rd edn. McGraw-Hill, Boston, USA.

ISO, 1994, Industrial automation systems and integration: Product data representation and exchange: Part 203: Application protocol: Configuration controlled 3D designs of mechanical parts and assemblies, Reference number: ISO 10303-203:1994(E), 1st edn, Switzerland.

ISO, 1998a, Industrial automation systems and integration: Product data representation and exchange: Part 45: Integrated generic resource: Materails, Reference number: ISO 10303-45:1998(E), 2nd edn, Switzerland.

ISO, 1998b, Industrial automation systems and integration: Product data representation and exchange: Part 49: Integrated generic resources: Process structure and properties, Reference number: ISO 10303-11:1998(E), 1st edn, Switzerland.

ISO, 2000, Industrial automation systems and integration: Product data representation and exchange: Part 41: Integrated generic resource: Fundamentals of product description and support, Reference number: ISO 10303-41:2000(E), 2nd edn, Switzerland.

Zha, X. F. and Du, H., 2002, A PDES/STEP-based model and system for concurrent integrated design and assembly planning, Computer-Aided Design, 34, 1087–1110.

Chapter 9
Generic Product Modelling Framework: Case Study

Abstract The focus of this chapter is placed on the modelling methodologies and the definition of schemas for various activities in a rapid OKP development process such as manufacturing, inspection, etc., and the integration of the schemas with other resources defined within STEP. There are 25 schemas defined to ensure that the proposed generic product modelling framework (GPMF) in Chapter 8 is compatible and can be used in modelling various types of products. These aspects, to the best of our knowledge, have not been reported extensively in the literature. The structure of the framework is discussed in this chapter with the focus placed on the EDM as the core of the framework. Case studies are carried out to validate the proposed GPMF. Two products are chosen from different engineering disciplines. They are modelled into product models according to the GPMF. Each case utilises one of the EDM data exchange and sharing methods and its corresponding software environment to obtain the product models, which are presented as STEP Part 21 exchange files, STEP objects, and a database object. A prototype system called Product Data Management System (PDMS) is developed to test the GPMF. The case studies show that the product models built based on the GPMF are capable of integrating information in product design, manufacturing and assembly, and the GPMF is compatible, comprehensive, and flexible.

9.1 Introduction

The market for manufactured goods has become more and more competitive in recent years. To survive and thrive in this competitive environment, manufacturing companies must use state-of-the-art technologies to improve all aspects of the PD process. It is essential to find globally optimised PD processes that can shorten product life cycle, reduce cost and lead time, and achieve high quality and productivity. This requires product data, information and knowledge to be efficiently managed and used throughout a product life cycle. Product modelling technology is one of the key and indispensable technologies for supporting efficient data exchange and sharing in PD processes.

Product modelling technology is used to provide a well-organised information structure to model product data and information. PD team members can use an information model to store, exchange and manipulate product data, which can be used to support the development of similar products. It can also be used to improve both product innovation and new PD, as well as cut down on unnecessary rework. Subsequently, this determines engineering productivity and eventual industrial competitiveness (Krause *et al.* 1993, Ai *et al.* 2010). The potential benefits of the technology have led to much research effort in this area.

Product modelling technology does not have a long history. Murphy (1950) was the first to define the term "model" as a device, "which is so related to a physical system that observations of the model may be used to predict accurately the performance of a physical system in the desired respect". The concept of modelling was formed with the development of computer aided technologies. Data, structures, interfaces, algorithms, and even the entire system can be modelled. The models are the abstract specifications of the domain functions that perform certain operations. In PD processes, a product can be seen as a functional unit with particular materials, a fixed form and other designated features. The product and its related information are naturally regarded as the modelling objects. The model developers combine the modelling concepts to build a product model and use it in its PD processes. In each phase of the PD processes, the product model is a prerequisite for representing product data, and sharing and exchanging information among different PD stages.

The explicit description of a product model first appeared with the introduction of geometric models employed by various CAD systems. The primary purpose of the early product models was to represent the product structure and its two-dimensional (2D) or three-dimensional (3D) shape data. In today's concurrent and integrated PD environment, the method of modelling objects has been expanded to other PD processes. A complete product model consists of useful data and information of a product all the way through its life cycle, *e.g.*, information about geometry, structure, function, assembly, materials and product usage. The information, required by specific PD activities, such as process planning, cost analysis and simulation, can be obtained from the complete product model.

The conventional product modelling technologies have significantly enhanced the performance of PD processes. Geometric models are used to model product geometric information and can support CAD systems to exchange and share product data. However, the conventional modelling technologies cannot meet the requirements of current PD processes due to the new challenges from the ever-changing manufacturing environment. The limitations of conventional product modelling technologies have become the main barriers to the development of modern manufacturing systems.

This is very inefficient. Sometimes, conflicts over model structures may cause loss of information which cannot be converted. Therefore, a non-compatible product model is a great barrier to effective performance where cooperative effort is required. A suitable modelling method is required to build up the high compatibility product model.

9.2 Literature Review

STEP is currently considered a promising product modelling resource since it can provide a standardised mechanism for product model data representation and exchange. A complete review was presented in Chapter 3. In this section, the effort and research outcomes of applying STEP in product modelling are investigated.

There are some publications on STEP-based product modelling. Gu and Chan (1995) developed a STEP-based generic product modelling (GPM) system designed and implemented according to the generic resources of STEP. The system can therefore be used to integrate manufacturing activities, such as process planning and inspection planning in the concurrent engineering environment. They used an object-oriented approach for building product models to support product design. The focus was placed on the definition of classes and the design of user interfaces with CAD software tools. There were no discussions on the definitions of schemas and on modelling methodologies. Li *et al.* (1996) developed a feature-based parametric product modelling system, which employed a product model based on the STEP and managed by an object-oriented database. This system was suitable for application in a CIM environment. Usher (1996) presented a STEP-based object-oriented product model based on STEP AP 224. This model was proposed to support CAPP analysis. Ming *et al.* (1998) presented a STEP-based part information model for process planning purposes. Their models included a process planning information model and a production resource information model. Tang *et al.* (2001) presented a STEP-based die and product integrated information model (DPIIM), in which integrated resources of STEP were used to model six EXPRESS schemas. These models supported the concurrent development of stamp and die products. Zha and Du (2002) presented a product data exchange using a STEP (PDES)/STEP-based assembly model for concurrent integrated design and assembly planning.

It can be concluded that the STEP-based modelling method has become the core of product modelling processes to organise product data in a standardised representation, which greatly enhances the capability of data exchanging and sharing in the integrated manufacturing environment.

To use the modelling resources defined in STEP, various modelling methods are integrated with STEP to form an integrated product modelling environment.

9.2.1 Geometry-based Modelling Methods

APs are used to build up information models for the integration of STEP with different geometric modelling methods, such as AP204 (ISO 2002) and AP203 (ISO 1994). AP 203 integrates five types of shape representation methods "that include wireframe and surface without topology, wireframe geometry with topology, manifold surfaces with topology, faceted boundary representation, and boundary representation" (ISO 1994) to support configuration controlled 3D design of mechanical parts and assemblies.

Shaharoun *et al.* (1998) utilised STEP to describe geometric data of a particular plastic product. The geometrical descriptions of the product were transferred into a CAD system to assist the design and machining of a suitable mould for the plastic product. Cai *et al.* (2002) proposed a method to build self-defined APs for all kinds of machine parts based on STEP. They implemented this method to develop two APs to represent the geometric data model in a cone gear product for the final drive of an automobile driving axle system.

9.2.2 Feature-based Modelling Methods

For feature-based product modelling, STEP provides a suitable representation method for many different features. For instance, AP224 (ISO 2005) illustrates the mechanical product definition of process plans using *Machining Feature*; AP203 (ISO 1994) is used to represent *Form Feature*. Other APs,, such as AP214 (ISO 2001) and AP118 (ISO 2004), also contain STEP expressions for specific features in particular application areas. The entity defined in STEP can be directly used to represent the target features. Some self-defined features, such as special assembly structures, machining and technique information for particular products, can be structured using EXPRESS modelling language and integrated resources. Both STEP-defined features and self-defined features can optimise the data exchange and sharing capability of the feature-based product modelling method.

Shah and Mathew (1991) integrated STEP and PDES's form feature information model (FFIM) on the ASU Features Testbed. This PDES/STEP-based modelling method can be useful for the static exchange of geometric and topological data. Meng *et al.* (1997) presented a STEP-based feature modelling system, which was based on a user-defined AP development based on AP214. Zhao and Ma (1999) described an object-oriented feature-based aero-engine blade product modelling system. In this modelling system, the design platform used STEP to standardise data modelling and to support information transmission from design platform to analysing system.

9.2.3 Integrated Modelling Methods

There has been much research combining STEP-based product modelling methods with integrated modelling methods. For example, Chin *et al.* (2002) proposed a multiple view methodology for integrated product modelling based on STEP. Song *et al.* (1999) used a STEP-based integrated product model to support the proposed Design for Manufacturing (DFM) system. The aim was to extract the design information of parts from a CAD system to automatically evaluate the manufacturability of those parts. Jasnoch and Haas (1996) developed a collaborative working virtual prototyping environment to integrate existing CAD systems. The underlying product model of this environment was a STEP-based integrated product model.

The purpose of this research is to develop a GPMF to support the integration of various PD activities. The focus of this research is placed on the modelling methodologies and the definition of the structure of the schemas for manufacturing, inspection, etc., and the integration of the schemas with other resources defined within STEP. There are 25 schemas defined to ensure that the proposed GPMF is compatible and can be used to model different types of products. These aspects, to the best of our knowledge, have not been reported extensively in the literature.

9.3 A STEP-enabled Product Modelling Framework

This section proposes a STEP-enabled generic product modelling framework (GPMF) that aims to provide an infrastructure for modelling various types of product information. The outcome of the GPMF is a set of data models defined to model product information at different stages of its development processes. The data models will then be used in PD processes to support efficient information exchange and sharing for the integration of PD systems such as CAD, CAPP and CAM, and product data management systems.

9.3.1 Generic Product Modelling Framework

The GPMF presented in Chapter 7 consists of four functional components including an EXPRESS data model (EDM), a STEP-based modelling environment, a "five-phase" modelling method, and three EDM data exchange and sharing methods. The EDM is the core of the modelling framework GPMF. The EDM defines a complete product data structure and uses a standardised data format. It consists of eleven defined EXPRESS schemas and STEP AP 203. Each schema uses either STEP resources or STEP-based compatible resources defined by our research group to model a particular type of product information.

The STEP-based modelling environment is built up for the GPMF. Within the environment, a modelling language (EXPRESS) and its graphical representation method (EXPRESS-G) are used to model product structure. STEP generic resources are used to model product information defined by STEP. STEP AP 203 is used to model product geometric information, and there are also new modelling resources defined for modelling product information that is not covered in STEP.

The "five-phase" modelling method is proposed to build up the EDM. It defines a formal approach to logically organise all the tasks of building up the EDM in the modelling processes.

There are also three EDM data exchange and sharing methods used in the GPMF. Product data can be exchanged and shared through either exchange files, working forms, or database management systems. The product models defined within the GPMF can be exchanged or shared using one of the methods. These three methods are easily integrated into any application software environment, which makes it easy to implement the product models defined by the GPMF in applications. The following sections discuss the details of these components used in GPMF.

9.3.2 Structure of EDM

The structure of the EDM was presented in Chapter 7. It consists of four modules. The product general information module, product geometric data module and product manufacturing data module are defined and based on the results of classifying product data in the first modelling phase. The resources module is developed by grouping the shareable basic modelling objects to support the development of other modules.

The product general information module represents the product data, which are not directly related with the product manufacturing, such as the product identity, product property, and the relationship between products. In this module, the product_definition_schema is defined to support modelling this aspect of the product data.

The product geometric data module supports modelling product geometric data, such as shape information and dimension information. This module is a key to integrate different computer aided systems, such as CAD/CAPP/CAM. In the product geometric data module, STEP AP 203 (ISO 1994) is directly used for representing and exchanging product 3D geometric information. This chapter will not discuss how to implement this AP; detailed information can be found in STEP Part 203 (ISO 1994).

The product manufacturing data module is the core of the proposed EDM, and consists of nine EXPRESSS schemas to support modelling different manufacturing data.

The supplier_informaton_schema is defined to model the supplier information. The supplier is an essential component of the manufacturing industry. Manufacturing companies need information about suppliers and their products to arrange the manufacturing processes.

The manufacturing_facility_schema is defined to model the facility information. The manufacturing facilities, such as machines, tools and fixtures are directly involved in the product manufacturing processes. They are the key resources determining the product manufacturing processes and influencing the product quality.

The product_document_schema is defined for documentation aspects. The documents are defined as industry standards and documented criteria that regulate and guide the manufacture of a product.

The bill_of_material_information_schema is defined for representing bill of material (BOM) information. The BOM, which is also called a part list, is a list of product components or resources for manufacturing or assembling this product. The

BOM is used to determine the product cost and used as an effective tracking source during the manufacturing and assembling processes.

The material_information_schema is defined to represent the material information. The material is the basic element for a product and it directly influences the selection of proper manufacturing facility and manufacturing processes.

The process_planning_information_schema is defined for modelling process data. The manufacturing process information is the detailed description about the actual manufacturing activities. It is essential to consider this aspect information in design stage, which leads to optimize the manufacturing processes.

The assembly_information_schema is defined for these two aspects product data. Assembly product is one of the most important types of product. To develop an assembly product, components information and assembly method are required.

The inspection_information_schema is defined to model inspection information. The inspection processes are essential parts of PD processes to control product quality. The inspection results can assist in indicating problems with a product and its production processes.

The cost_information_schema is defined for the cost information occurring in PD processes. Cost is of paramount importance to a manufacturing company. Managing cost information can help the firm to increase its competitive ability.

The resources module defines basic modelling objects that are shared by the other modules. All these basic modelling objects are grouped into the supporting_schema. The resources in the schema are presented as EXPRESS ENTITY, the constructed TYPE, and FUNCTION. Through the EXPRESS schema interface, these resources in this schema are utilized by other schemas to structure an effective and efficient data model representation.

The four modules of EDM are developed by applying STEP generic resources including Part 41 (ISO 2000) Part 45 (ISO 1998a), and Part 49 STEP AP 203 (ISO 1998b), as well as the newly defined STEP-compatible modelling resources.

9.4 Case Study

Case studies are conducted to demonstrate the feasibility and the compatibility of the GPMF in modelling products from the different engineering applications. The rationality of the EDM, and the relevant modelling methodologies are also tested. Two sample products of different types are chosen to investigate the above mentioned objectives.

Water Bottle Lid

The water bottle lid product is a simple injection moulded product. The lid has the same large quantity of associated manufacturing data as complex injection products, although it is not geometrically complicated. Figure 9.1 shows a solid 3D model

Figure 9.1 3D model of water bottle lid product

of this product, which is generated from the Pro/ENGINEER® Wildfire® CAD system. The sample lid product is injected moulded on a HUSKY S160 injection machine using high-density polyethylene (HDPE) material. It is manufactured for a plastic water bottle with a 28 mm thread.

Clamp Assembly Product

The clamp assembly product originated from a design project within the Pro/ENGINEER tutorial book (Lamitt 2004). Figure 9.2 shows a solid 3D model of this product, which is generated by the Pro/ENGINEER® Wildfire® CAD system.

Figure 9.3 shows a tree structure model to demonstrate how this product model is assembled. The clamp assembly product consists of two subassembly components: a clamp and a plate. These two sub-assembly components are connected by the arm part. The subassembly clamp includes six components: an arm part, a swivel part, a foot part, a stud part and two ball parts. The subassembly plate includes four components: an arm part, a plate part, a stud part and a flange nut part.

Three EDM data exchange and sharing methods are used to model different aspects of product data for the sample products.

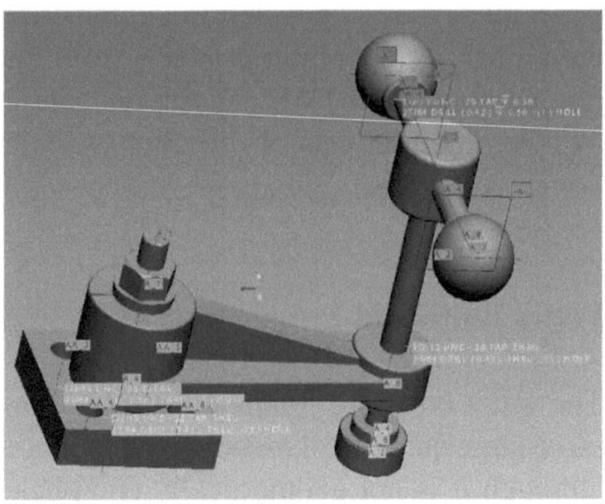

Figure 9.2 3D model of the clamp assembly product

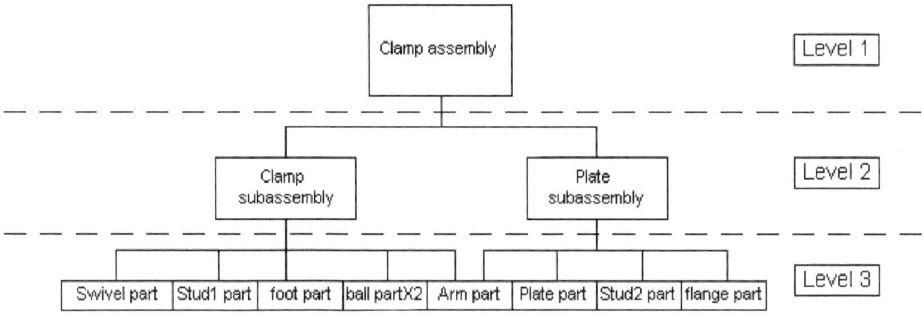

Figure 9.3 Structure of clamp assembly product

1. Product geometric data and product general information of the sample product "water bottle lid" are modelled in STEP Part 21 files. These two products are used to test the product geometric data module and the product_definition_schema.
2. The product inspection data of the sample product "water bottle lid" are modelled as STEP objects[1] by C++ application program. This is to test the inspection_information_schema defined in the EDM.
3. A prototype DBMS application system PDMS is developed to represent product data. The manufacturing process for the "water bottle lid product" and the assembly data for the "clamp assembly product" are modelled. They are used to test the process_planning_schema defined in the EDM.

9.4.1 Modelling Product Geometric Data

The geometric product data of the water bottle lid product are modelled in a Part 21 file through an AP203 output interface defined in Pro/ENGINEER®. The Pro/E ".part" file of the lid product is loaded in the system first; then the geometric data are saved and output through the AP 203 interface in the STEP Part 21 file format.

This Part 21 file can be imported by other application systems to provide the product geometric information. Figure 9.4 shows the windows display results of the lid product when this Part 21 file format product model is loaded in both the Pro/ENGINEER® system and ST-Viewer® in the ST-Developer toolkit. As shown in Figure 9.4, it is apparent that the geometric product data in the product model can be read by both systems to regenerate the 3D geometric models. This is because the two systems have a STEP AP 203 interface to load/output product geometric data. The screenshots of the lid product model in both Pro/E and ST-Viewer are the same, which means that the geometric product data of the product model can be exchanged and shared by the two application systems.

[1] STEP objects are C++ objects with two additional characteristics: 1. the data structure of STEP objects is defined by an EXPRESS information model; 2. they can be written out in the STEP physical file (Part 21) format.

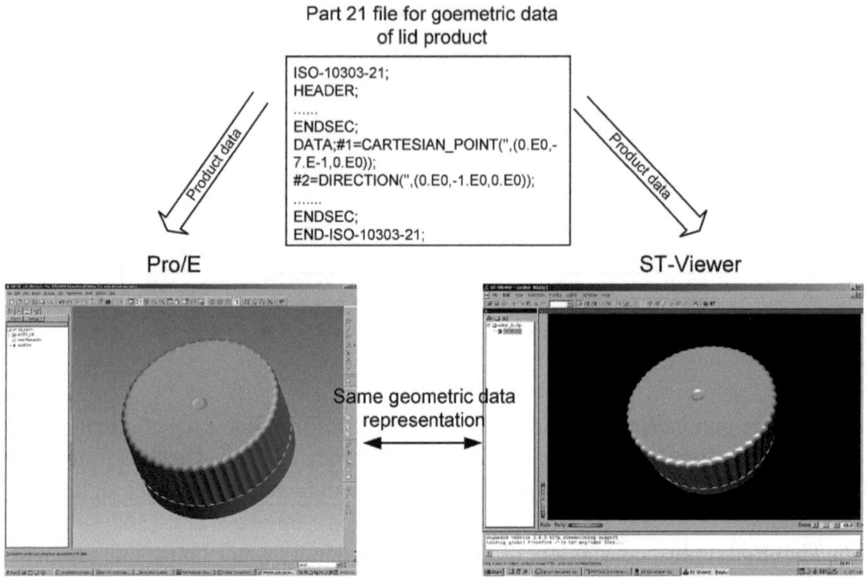

Figure 9.4 Lid product model loaded in Pro/ENGINEER® and ST-Viewer®

9.4.2 Modelling Product General Information

The product general information of the lid product is modelled into a product model with the STEP Part 21 file format. This STEP Part 21 file is based on the structure of the product_definition_schema. When the product model users understand this schema, the product general data can be easily retrieved from the STEP Part 21 file. In this case, the product model for product general information is encoded manually, it is partially presented as follows:

```
ISO-10303-21;
   HEADER;
   ......
   FILE_SCHEMA ((‘PRODUCT_DEFINITION_SCHEMA’,
                                         ‘SUPPORTING_SCHEMA’));
   ENDSEC;
   DATA;
   #10= PRODUCT (‘lid001’, ‘water bottle lid’, ‘Lid diameter is 28mm’, #30);
   #20= PRODUCT (‘lid002’, “water bottle lid’, ‘Lid diameter is 28mm’, #30);
   #30= PRODUCT_CATEGORY (‘cat001’, ‘28mm series’, $);
   #40= PRODUCT_RELATIONSHIP (‘pro-rel001’, ‘substitute product’,
         ‘can substitute each other, but have different sealing design’, #10, #20);
   #50= PRODUCT_PROPERTY (‘property001’, ‘color’, ‘white’, #10);
   #60= PRODUCT_PROPERTY (‘property002’, ‘sealing method’,
         ‘screwing’, #10);
```

ENDSEC;
END-ISO-10303-21;

This file consists of two sections: a header section and a data section. The header section presents the information about this STEP Part 21 file, such as the FILE_SCHEMA declaration regarding how product_definition_schema and supporting_schema in the EDM are utilized to define the product data structure. The data section models product data. In this case, the product general information is modelled from line #10 to #60; each line presents a set of product data for one kind of product general information.

Line #10 refers to the product and models the basic definitions of a lid product. The four data are listed in this line, as shown in Figure 9.5, by matching the sequences of the id, name, description, and category in the product[2]. The id, name, description and the category information of the sample product are represented as "lid001", "water bottle lid", "Lid diameter is 28 mm", and the product data defined in line #30, respectively. Line #20 models another lid product with the id "lid002" using the same presentation as line #10.

Line #30 defines the product data of the category information, which refers to product_category as shown in Figure 9.5. The product data "cat001" and "28 mm series" relate to id and name. The "$" symbol indicates that there are no data of the optional attribute description.

Line #40 demonstrates the relationship between two products defined in line #10 and line #20. The product data in line #40 are sequentially presented as the value of

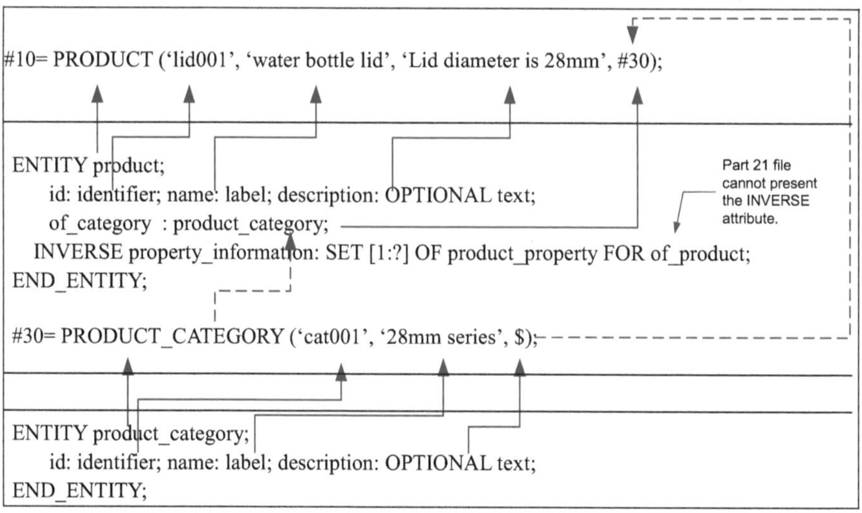

Figure 9.5 Relationships between Part 21 file and EDM schema

[2] Part 21 file can not map the INVERSE attribute. Hence, the product_property information can not be presented in Line #10.

attributes id, name, description, relating_product, and related_product in the product_relationship.

Lines #50 and #60 define two product_property instances named "color" and "sealing method". The product data of these two instances are organised with the same sequence as the attributes id, name, description and of_product[3] are listed in the product_property.

9.4.3 Modelling Product Inspection Information

The inspection data for the lid product has been chosen to be modelled based on the working form ROSE C++ library. Through the integrated development software environment that consists of ST-Developer and Microsoft Visual C++, the inspection data are modelled into STEP objects. This case study is carried out following the above three steps. The product inspection data are modelled into STEP objects and presented as a STEP Part 21 file.

1. The inspection_information_schema is converted to C++ class definitions through EXPRESS Compiler in ST-Developer® as shown in Figure 9.6. Meanwhile, as the product_definition_schema and product_document_schema are

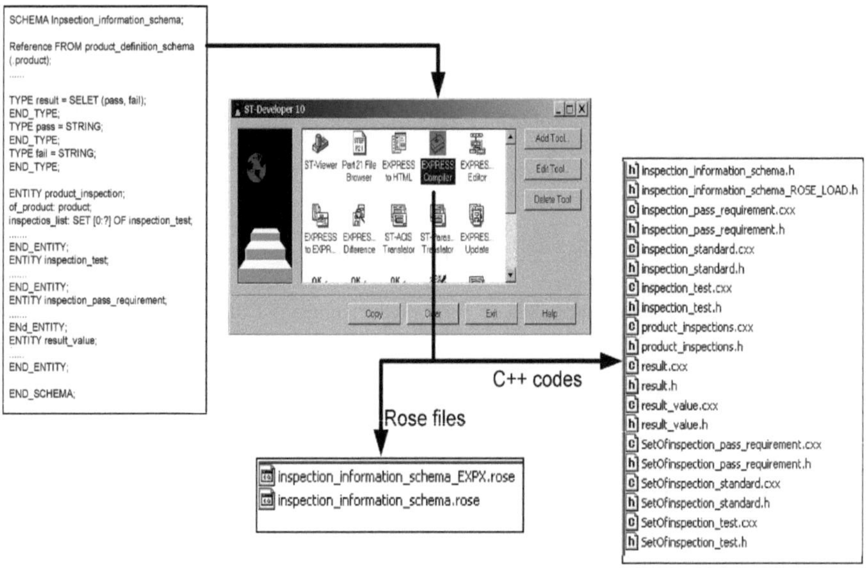

Figure 9.6 Converting inspection_information_schema into C++ classes and ROSE schema file

---

[3] The first three attributes, id, name, and description, are referenced from supporting_schema.**property_definition**.

referenced by the inspection_information_schema, these two schemas are converted to C++ codes as well. The six corresponding ROSE schema files, inspection_information_schema.rose, inspection_information_schema_EXPX.rose, product_definition_schema.rose, product_definition_schema_EXPX.rose, product_document_schema.rose, and product_document_schema_EXPX.rose are generated.

2. The functions and variables in these C++ codes and the data-dictionary in the *.rose files are used to generate the application ROSE C++ program.
3. The STEP objects are displayed in the command window and output in a STEP Part 21 file. The manipulation of these two STEP objects is realised using the functions ROSE.display( ) and ROSE.saveDesign( ) defined in the ROSE Library. Figure 9.6 presents a screen snapshot of ROSE.display( ) results.

There are 19 STEP objects created in this case study, from <0-0> to <0-18>. All these STEP objects are built based on the inspection_information_schema, product_definition_schema and product_document_schema in the EDM.

The STEP objects <0-0> to <0-2> are the three instances of inspection_pass_requirement in the inspection_information_schema. The STEP objects <0-3> to <0-5> define three instances of SetOfinspection_standard[4]. The STEP objects <0-6> to <0-8> present the three instances of inspection_standard in the inspection_information_schema. The STEP object <0-9> defines the product data for an instance of document_type in the product_document_schema. The STEP objects <0-10> to <0-12> present the three instances of result_value in the inspection_information_schema for three inspection tests modelled in STEP objects <0-14>, <0-15>, and <0-16>. An instance of result in the inspection_information_schema is modelled into STEP-object <0-13>.

The STEP objects <0-14>, <0-15>, and <0-16> present three inspection tests: "open torque test", "1.5m falling test" and "airproof test" for the water bottle lid product. They are three instances of inspection_test in the inspection_information_schema. For example, in the STEP object <0-14>, the id, name, and description are valued as "lid-inspection001" and "open torque test", respectively.[5] The test_product is valued by STEP object <0-17>, which stores the definition of the "water bottle lid" product. The frame_of_reference utilises the STEP object <0-3> to define the referring standards. The other two instances inspection_test, STEP objects <0-15> and <0-16> are defined in the same way.

The STEP object <0-17> is an instance of product in product_definition_schema. The of_category utilises STEP objects <0-18> to define the product category data about the product modelled in STEP object <0-17>.

---

[4] The SetOfinspection_standard is the result of converting a set of inspection_standard through the EXPRESS Compiler of ST-Developer

[5] There is no input for description attribute.

9.5 A Prototype Product Data Management System

A prototype PDMS is developed using the third EDM data exchange and sharing method. All the product data is stored and managed by this system. In this case study, two types of product data are modelled to test the proposed GPMF. They include the manufacturing process information of the lid product, and the assembly information of the clamp assembly product.

The prototype PDMS consists of two parts: a product data input/query interface and a product database. These two parts are both developed and based on the EDM. Figure 9.7 shows the structure of such a system, where ODBC is used as a driver for establishing the link between user-interface pages and the created databases.

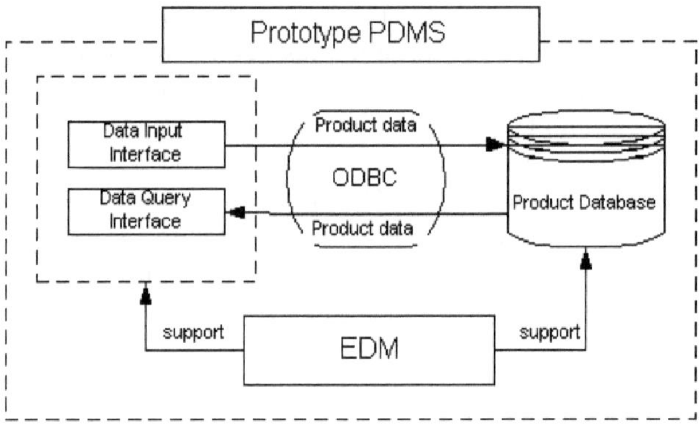

Figure 9.7 Structure of the PDMS

9.5.1 Product Data Input/Query Interface

This interface has two main functions. The first is to provide the interface to input product data into the product database. The second is to manipulate product data, including data querying and data updating, which is based on the support of Microsoft Access DBMS. These two functions are presented in two separated windows: a data input window and a data query window. Their layouts are organised according to the schemas structure of the EDM. Figure 9.8 shows four kinds of product data input windows for: product general information, product inspection information, product manufacturing process information and product assembly information. A data query window is also developed to enable users to search the PDMS. The query field can be chosen from the pop-up menu. The query keyword can be inputted from the blank input field and served for data query after clicking the "submit" button. The query results are presented using a table or a form.

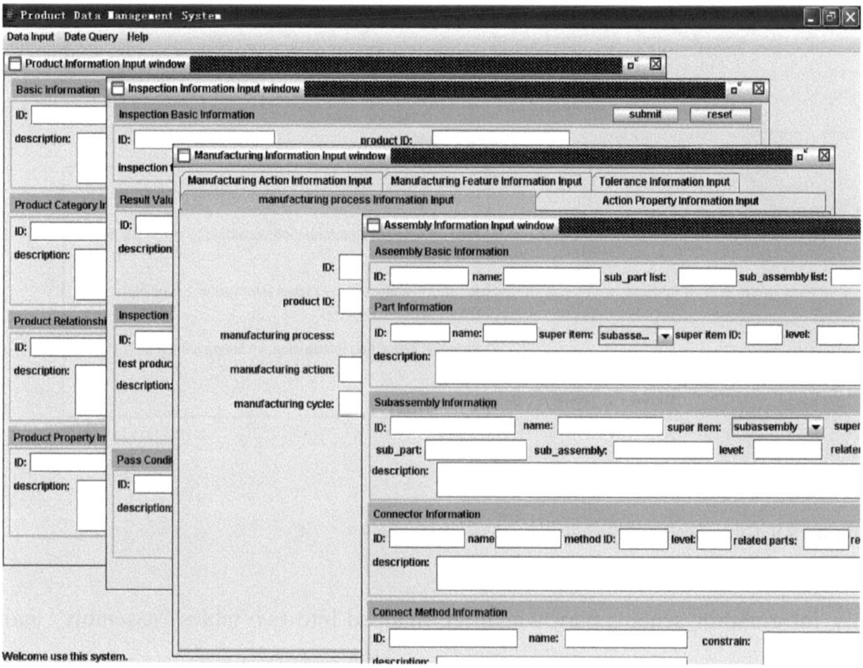

Figure 9.8 Product data input window and query window

9.5.2 Product Database

The main functions of the product database are to store the product data following the structure defined by the EDM, and to use the Microsoft Access DBMS to manipulate data. Example programmes are shown in Appendix A.6. As shown in Figure 9.9, there are 19 tables in the proposed product database and they are categorised into four types: 1. product general information; 2. product inspection information; 3. product assembly information; and 4. product manufacturing process information. These tables are naturally associated with the corresponding schemas: the product_definition_schema, the assembly_information_schema, the inspection_information_schema, and the manufacturing_process_schema.

The structures of the above schemas are mapped into the product database. There are two types of mappings: 1. mapping between EXPRESS ENTITY and the table in the product database; and 2. mapping the relationships between different EXPRESS ENTITYs to the corresponding tables in the product database. For the first kind of mapping, the attributes of an entity are mapped to the corresponding columns of a table. The name and the value type of the input field are mapped from the attribute definition. Figure 9.10a presents the first type of mapping between assembly_information_schema.assembly_product and the table "assembly". The relationships between entities are also presented in between tables. Figure 9.10b shows the relationship between assembly_information_schema.assembl_producty and assem-

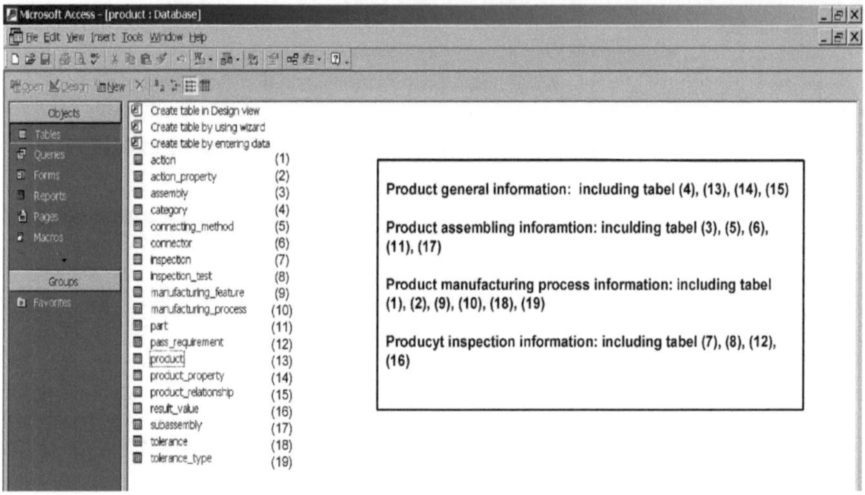

Figure 9.9 Classification of product database data tables

bly_information_schema.part, which are mapped into two tables "assembly" and "part".

9.6 Modelling Product Manufacturing Process Data

The lid product is used to test the process_planning_schema through the prototype PDMS. This test utilises the product manufacturing process data input interface and involves six access tables: action, action_property, manufacturing_process, manufacturing_feature, tolerance and tolerance_type. They are built based on the process_planning_schema.

In this case study, all the data of the manufacturing processes for the water bottle lid are modelled. For example, there are seven steps to manufacture a water bottle lid product. The second manufacturing step, "injection", is modelled through the PDMS, which is shown in Figure 9.11. The product data of this step are input through the manufacturing action information input window (see top of Figure 9.11); they are stored in the product database table "action" (see bottom of Figure 9.11). The product data of this step are demonstrated as: 1. "step002"; 2. "injection process"; 3. "injecting 2.4 * 24 g melted HDPE to the cavity"; 4. "property 001" and "property002", which are 210 degrees and 840 bar, respectively; 5. 0.8 s; 6. "nozzle", "24-cavity mold", "HUSKY S160"; 7. none; and 8. 2. These eight data elements correspond to the eight attributes of manufacturing_process_action. The arrow lines present this mapping relationship. The product data of the other six steps are input into PDMS as presented in the "action" table in Figure 9.11.

Figure 9.12 presents the modelling results of the manufacturing process data for the water bottle lid product. These product data include: an instance of manufactur-

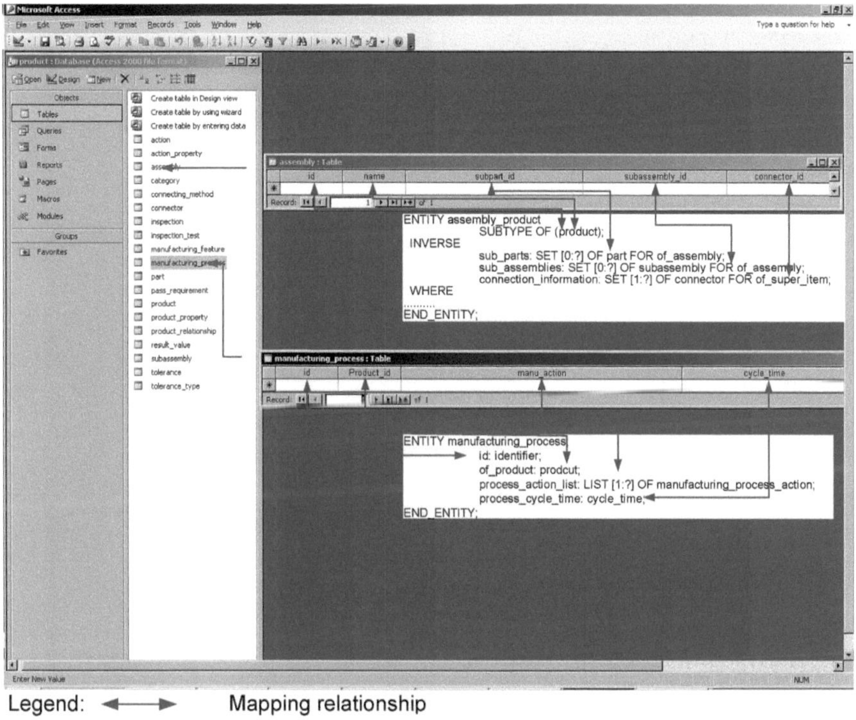

(a)

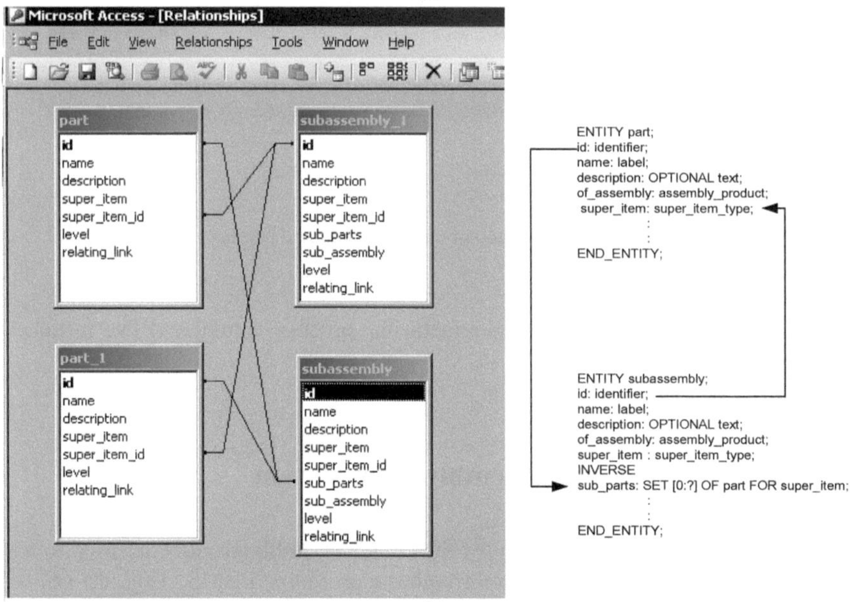

(b)

Figure 9.10 Mappings between product database and the EDM

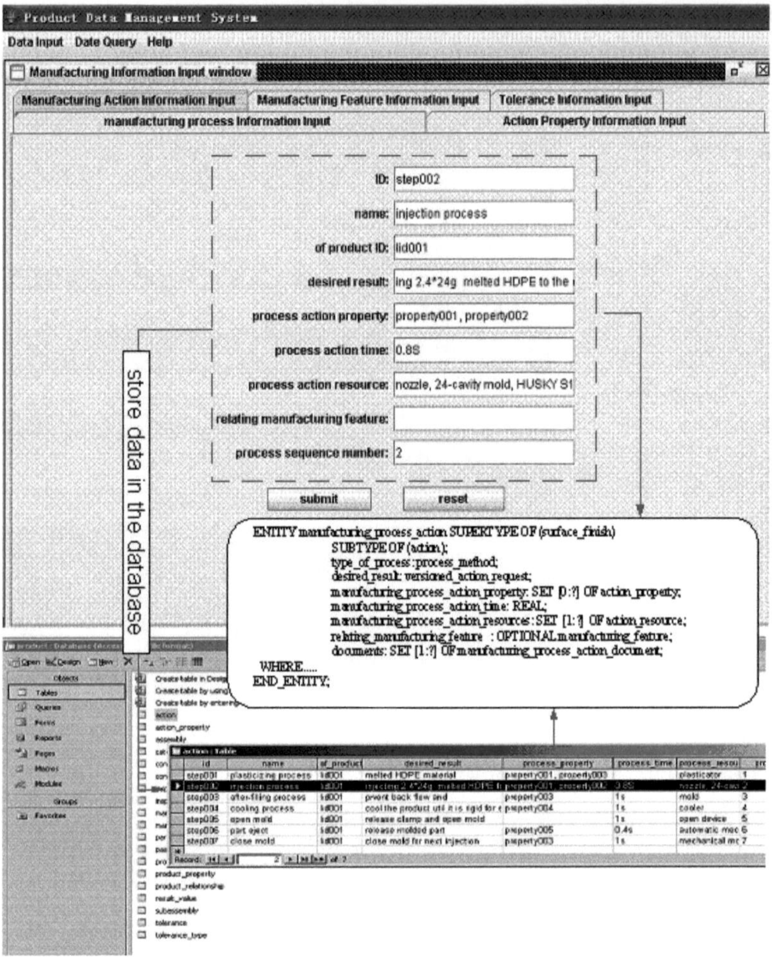

Figure 9.11 Modelling the "injection" step of the water bottle lid product

ing_process, seven instances of manufacturing_process_action, and five instances of action_property.

9.7 Modelling Product Assembly Information

The product assembly information for the clamp assembly product are used to test the assembly_information_schema through the prototype PDMS. This test utilises the product assembly information input interface and it involves five access tables: assembly table, part table, subassembly table, connector table and connecting_method table. These are built up based on the assembly_information_schema.

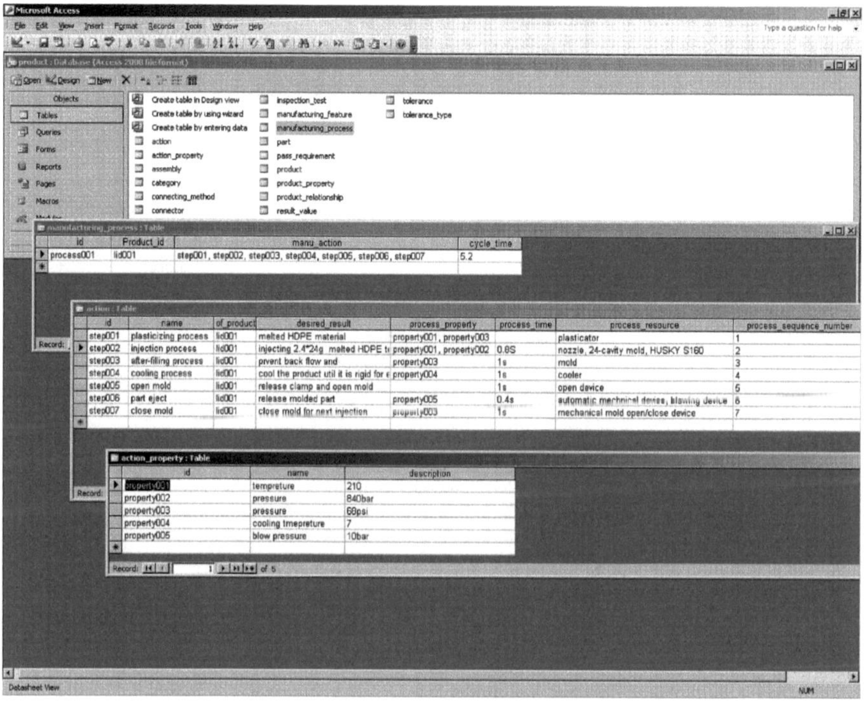

Figure 9.12 Product data for manufacturing process of lid product

Figure 9.13 shows an user interface for inputting product assembly information. There are five fields, which are structured based on the corresponding entities in the assembly_information_schema. For example, the first input field named "assembly basic information" is generated by mapping the structure of the assembly_product. Its input "ID", "name", "sub_part_list", "sub_assembly_list" and "connector ID" refer to the five attributes of assembly_product: id, name, sub_parts, sub_assemblies, and connection_information.

Figure 9.14 shows the detailed search results of nine instances of the product clamp. There are nine components of the clamp product. For example, the flange nut part product data are presented as: "part009", "flange nut", " "[6], ("subassembly", "subassembly001"), 3 and 003. They are sequentially matched to the attributes of the part, which are id, name, description, super_item, lever_in_assembling_hierachy, and connection_information.

The product data can be retrieved from the PDMS. For example, consider the "part009" shown in Figure 9.14; the "super_item" is "subassembly" and the "super_item_id" is "subasseblyO02". The detailed information of "subassembly002" can be retrieved by querying in the subassembly table. The "part009" is listed in the "sub_parts" column. This example presents the relationship between subassembly

[6] " " means that no data are input for that column.

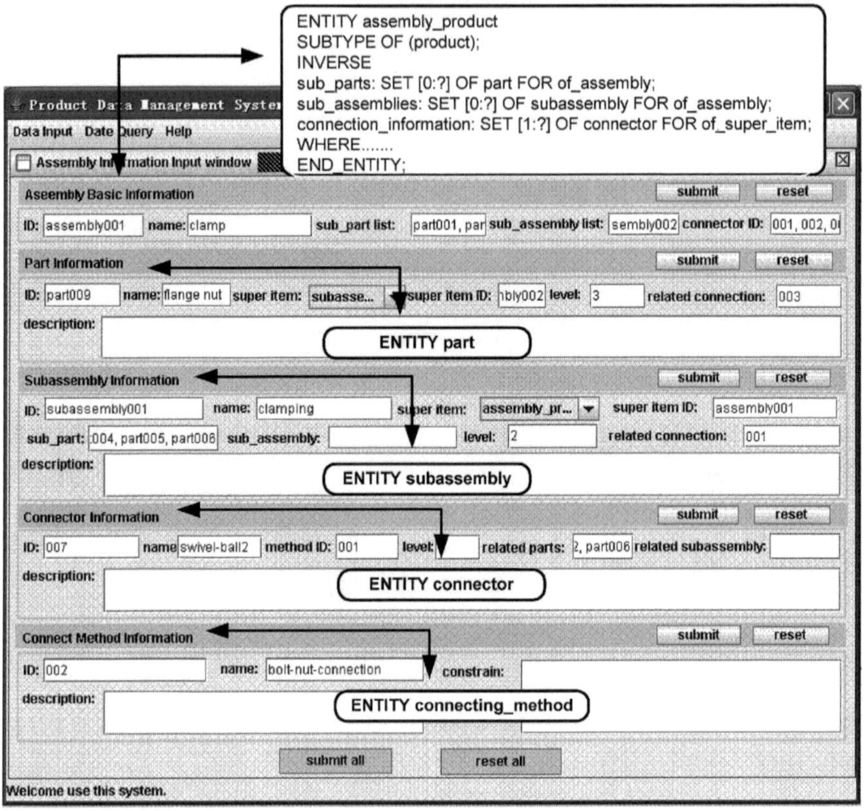

Figure 9.13 Assembly information input

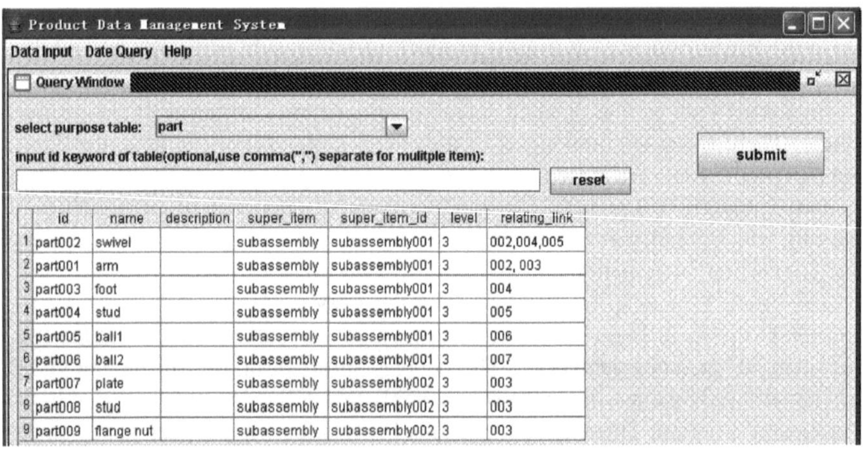

	id	name	description	super_item	super_item_id	level	relating_link
1	part002	swivel		subassembly	subassembly001	3	002,004,005
2	part001	arm		subassembly	subassembly001	3	002, 003
3	part003	foot		subassembly	subassembly001	3	004
4	part004	stud		subassembly	subassembly001	3	005
5	part005	ball1		subassembly	subassembly001	3	006
6	part006	ball2		subassembly	subassembly001	3	007
7	part007	plate		subassembly	subassembly002	3	003
8	part008	stud		subassembly	subassembly002	3	003
9	part009	flange nut		subassembly	subassembly002	3	003

Figure 9.14 "Part" data querying results

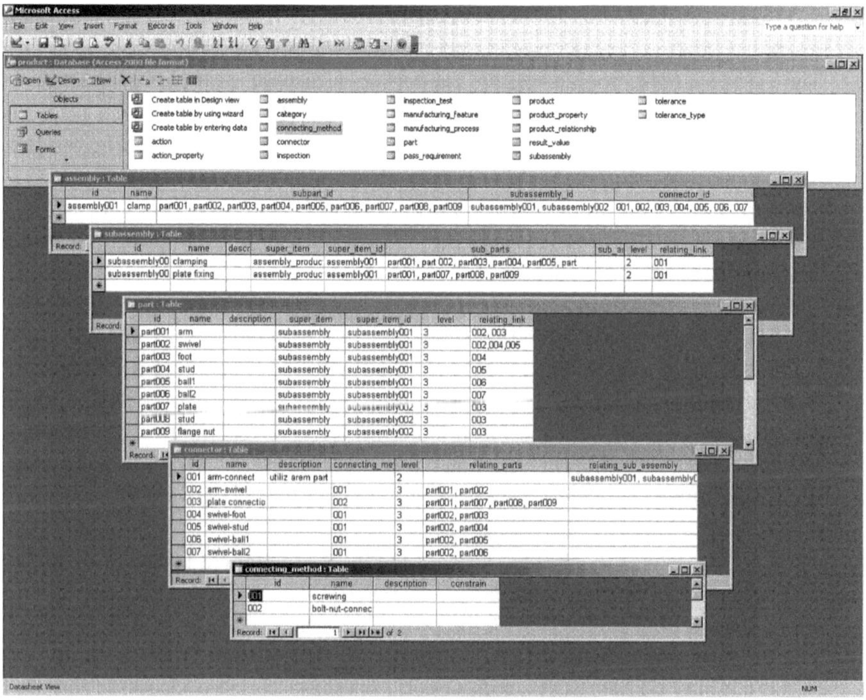

Figure 9.15 Assembly data for the clamp product stored in the PDMS

and part. The super_item attribute of the part can be valued by an instance of subassembly; the sub_parts attribute of subassembly are defined by a set of instances of the part.

Figure 9.15 shows a summary of all the assembly data for the clamp product, which are stored in the five tables (assembly, subassembly, part, connector, and connecting_method) of the product database in the prototype PDMS.

9.8 Conclusion and Future Work

This chapter presents two case studies to illustrate how the STEP-enabled GPMF can be used to support efficient information sharing and exchange for integrated PD. The focus of the study is to develop a generic modelling methodology for modelling products of different types. This is achieved through definition of the schemas and the proposed modelling methodologies. A prototype PDMS system has been developed to demonstrate how the GPMF works. From the case studies, the following conclusions are drawn:

1. The proposed STEP-enabled GPMF is compatible with modelling products of various types. The sample products used in the case studies are chosen from

different manufacturing applications. Through the GPMF, the product data are modelled into the product model and stored in the proper formats.

2. The GPMF is able to support the modelling of a wide range of product data, especially product manufacturing data. In the case studies, five aspects of product data are modelled through the proposed GPMF. The corresponding product models provide a comprehensive view of the product.
3. The case studies show that all the parts of the GPMF are associated with a core – the EDM. The entire product modelling process is tightly dependent on the data structure defined in this data model.
4. The EDM is flexible in implementation. It has four modules. Each module of the EDM, even each EXPRESS schema, can be considered as an individual EXPRESS data model. They can be applied with the EDM data exchange and sharing methods to model the corresponding product data. In three case studies, the product general information module, the product geometric data module, the inspection_information_schema, the process_planning_shcema, and the assembly_infomration_shcema are used individually to support modelling product data.
5. Three EDM data exchange and sharing methods are applied in the case studies with related software environments. Any one of these methods can be chosen to support modelling products with the proper format.
6. The prototype system PDMS is developed to demonstrate how the proposed STEP-enabled GPMF works with general database systems for modelling and managing product data. This is based on the third level of EDM data exchange and sharing method.
7. The GPMF provides a well-established mechanism to support the integration of manufacturing systems through the proposed product modelling methodologies. However, more work needs to be carried out before it is applied in an actual integrated manufacturing environment to support the integration of PD systems in a modern manufacturing environment. The future work in the area is enormous and not limited to the following three areas.
8. The first area is to further develop the prototype PDMS system. This includes the input/output interfaces for the integration of various computer aided systems, such as CAD, CAPP and CAM. The system needs to provide a standard interface to transfer product data with proper format to the end user.
9. The second research area is to further validate the proposed GPMF by applying it in modelling products of other types. Our research work in modelling sheet metal and injection moulding products has shown that this is a complicated process (Tu and Xie 2001, Xie and Xu 2006). Future work in this area requires great effort to define new schemas as STEP itself is still at the development stage.
10. The third area of work is to explore the possibility of extending the proposed GPMF to support Web/Internet-based manufacturing. Integrating web technologies into the modelling framework can significantly improve information exchange effectiveness through the internet/intranet. This can enable the proposed GPMF to be used in a distributed PD environment to support

Web-based PD activities. The possible implementation method involves utilising mapping between XML and EXPRESS, which is defined in STEP Part 28 (ISO 2003).

References

Ai, Q.S., Xie, S.Q., Zhou, Z. D., Liu, Q., Tao, L., Yang, W.Z., 2010, STEP-compliant Knowledge-base in support of Customized Product Development for SMEs. Advanced Materials Research, Vols. 97–101, 3571–3574.

Cai, C. T., Li, Y. Y., Dai, Y. H. and Liu, X. Y., 2002, Design method of application protocol of the machine parts based on STEP, Jisuanji Jicheng Zhizao Xitong/Computer Integrated Manufacturing Systems, CIMS, 8, 892–895 (in Chinese).

Chin, K. S., Zhao, Y. and Mok, C. K., 2002, STEP-based multview integrated product modelling for concurrent engineering, International Journal of Advanced Manufacturing Technology, 20, 896–906.

Gu, P. H. and Chan, K., 1995, Product modelling using STEP, Computer-Aided Design, 27, 163–179.

ISO, 1994, Industrial automation systems and integration: Product data representation and exchange: Part 203: Application protocol: Configuration controlled 3D designs of mechanical parts and assemblies, Reference number: ISO 10303-203:1994(E), First edition, Switzerland.

ISO, 1998a, Industrial automation systems and integration: Product data representation and exchange: Part 45: Integrated generic resource: Materails, Reference number: ISO 10303-45:1998(E), Second edition, Switzerland.

ISO, 1998b, Industrial automation systems and integration: Product data representation and exchange: Part 49: Integrated generic resources: Process structure and properties, Reference number: ISO 10303-11:1998(E), First edition, Switzerland.

ISO, 2000, Industrial automation systems and integration: Product data representation and exchange: Part 41: Integrated generic resource: Fundamentals of product description and support, Reference number: ISO 10303-41:2000(E), Second edition, Switzerland.

ISO, 2002, Industrial automation systems and integration: Product data representation and exchange: Part 204: Application protocol: Mechanical design using boundary representation, Reference number: ISO 10303-204:2002(E), First Edition, Switzerland.

ISO, 2003, Industrial automation systems and integration: Product data representation and exchange: Part 28: Implementation methods: XML representations of EXPRESS schemas and data, Reference number: ISO 10303-281:2003(E), First edition, Switzerland.

ISO, 2004, Industrial automation systems and integration: Product data representation and exchange: Part 118: Application protocol: Ship structures, Reference number: ISO 10303-118, First Edition, Switzerland.

ISO, 2005, Industrial automation systems and integration: Product data representation and exchange: Part 224: Application protocol: Mechanical product definition for process planning using machining features, Reference number: ISO/DIS 10303-224, Third Edition, Switzerland.

Jasnoch, U. and Haas, S., 1996, Collaborative environment based on distributed object oriented databases, Computers in Industry, 29, 51–61.

Krause, F. L., Kimura, F., Kjellberg, T., Lu, S. C. Y., Van derWolf, A. C. H., Ating, L., ElMaraghy, H. A., Eversheim, W., Iwata, K., Suh, N. P., Tipnis, V. A. and Weck, M., 1993, Product modeling, CIRP Annals: Manufacturing Technology, 42, 695–706.

Lamit, L. G., 2004, PRO/ENGINEER® WILDFIRE™, Brooks/Cole, a division of Thomson Learning™ Inc.

Li, H. L., Han, J. H., Dong, J. X. and Wang, Y., 1996, Feature-based, parametric modelling system for CAD/CAPP/CAM integrated system, Industrial Technology, 1996. (ICIT '96), Proceedings of The IEEE International Conference on 2–6 December, Shanghai, China.

Meng, M. C., Yang, L. and Bai, L. K., 1997, Feature modeling system based on STEP, Jishanji Jicheng Zhizao Xitong/Computer Integrated Manufaturing System, CIMS, 3, 34–38.

Ming, X. G., Mak, K. L. and Yan, J. Q., 1998, A PDES/STEP-based information model for computer aided process planning, Robotics and Computer Integrated Manufacturing, 14, 347–361.

Murphy, G., 1950, Similitude in Engineering, Ronald Press Company, New York.

Shah, J. J., and Mathew, A., 1991, Experimental investigation of the STEP form-feature information model, Computer-Aided Design, 23, 282–296.

Shaharoun, A. M., Razak, J. A. and Alam, M. R., 1998, STEP-based geometric representation as part of product data model of a plastics part, Journal of Materials Processing Technology, 76, 115–119.

Song, Y. Y., Cheng, Y., Cai, F. Z., Xiao, Y. B. and Tang, D., 1999, Study on knowledge-based integrated design for manufacture of mechanical parts, Qinghua Daxue Xuebao/Journal of Tsinghua University, 39, 21–24 (in Chinese).

Tang, D., Zheng, L., Li, Z. and Chin, K. S., 2001, STEP-based product modelling for concurrent stamped part and die development, Computers in Industry, 46, 75–94.

Tu, Y. L. and Xie, S. Q., 2001, An information-modeling framework for sheet metal parts intelligent and concurrent design and manufacturing, International Journal of Advanced Manufacturing Technology, 18, 873–883.

Usher, J. M., 1996, STEP-based object-oriented product model for process planning, Computers and Industrial Engineering, 31, 185–188.

Xie, S. Q. and Xu, X., 2006, A STEP-compliant process planning system for sheet metal parts, International Journal of Computer Integrated Manufacturing, 19, 627–638.

Zha, X. F. and Du, H., 2002, A PDES/STEP-based Model and System for Concurrent Integrated Design and Assembly Planning, Computer-Aided Design, 34, 1087–1110.

Zhao, W. and Ma, W., 1999, Feature modeling for aeroengine blades according to STEP, Beijing Hangkong Hangtian Daxue Xuebao/ Journal of Beijin University of Aeronautic and Astronautics, 25, 535–538 (in Chinese).

Chapter 10
Information Framework for Rapid OKP Product Development

Abstract Information integration is an important issue in supporting integrated and concurrent PD. This chapter explores the definition and the structure of an information framework for rapid development of sheet metal parts. This framework aims to build an information bridge to fill the gap among sheet metal part design, process planning and manufacturing systems. It is based on the principles of zero thickness and zero bend radius, which are used to abstract the geometry entities of sheet metal parts in order to facilitate part modelling and information modelling. In this chapter, a tree-based step-structure information modelling methodology for sheet metal parts is proposed and a case study is given.

10.1 Background and Motivation

Today's manufacturing industry relies strongly on computer technology to support activities through a product's life cycle. Effective and efficient information sharing and exchange among computer systems have been criticial issues. An information modelling methodology that describes information flow unambiguously throughout the PD life cycle is an enabling technology that facilitates the development of a large-scale and networked, computer environment that can support RPD.

For the development of sheet metal parts, traditional product design and manufacturing processes are carried out in a series of stages, especially for bent sheet metal products. With this traditional design and manufacture method, a sheet metal product is frequently designed without systematic consideration of downstream PD requirements, such as process planning, manufacturability, production scheduling and manufacturing optimisation (Shao *et al.* 2009). Also, the feedback from these downstream processes to the product designer can only be made after the product is designed or possibly after it is manufactured. This often results in an expensive and time-consuming rework (Verlinden *et al.* 2008). Consequently, it affects the quality, cost and delivery time of the product.

To fill the gap between the design and other downstream PD processes, efficient and effective information interchange among different software systems, different departments and different people in the company is critical for concurrent design and manufacturing of a sheet metal product. Correct and effective information exchange throughout the PD life cycle can shorten the PD lead time, improve the production efficiency, achieve high quality, and, at the same time, cut down the production cost.

In order to achieve intensive information interchange on sheet metal design and manufacturing processes and intelligent decision support in the earlier design stage, an integrated PD system where different software packages and people can share information simultaneously has been presented in Chapter 4. To support the development of this system, an information framework that allows the use of information in a shared environment is proposed in this chapter.

The information framework is used to build information models for information sharing among different software packages so that these software packages work concurrently in an integrated environment. Also, the framework can be used for building design and manufacturing knowledge bases, which can be used as a reference for intelligent decision support. This work has been used to build a WWW-based intelligent design and manufacturing knowledge base and has been tested in a sheet metal company.

10.2 The Role of the Information Framework

Owing to the numerous and complicated interactions among the tasks of concurrent sheet metal PD processes, an Internet-based integrated sheet metal PD system was proposed in Chapter 4. The platform contains several modules, including an unfolding module, an Internet-based data integration environment, knowledge bases, data communication tools among different modules (Li *et al.* 2007), a CAPP module (Xie and Zhang 2008), a CAD module, a CAM module, a cost estimation module (Verlinden *et al.* 2008), a computer simulation platform (Lu *et al.* 2010, Cia *et al.* 2009), GUIs, *etc.*

In this system, the relationships between the information framework module and other modules are shown in Figure 10.1. Figure 10.1 shows that the information framework provides feasible solutions to build automatic links between data for product design, process planning, unfolding software package, logistics management (*i.e.*, supply, partner and subcontractor chains management), CAM and costing. Product data models can be built to manage information flow among these application software packages. The fundamental problem with data sharing among these software tools is that the output of one tool is often required as the input of another tool, yet each system uses proprietary data formats that are not easily transferable to other systems. It is difficult, therefore, to move data from one step of the design process to another step without performing data translation. Past solutions have been focused on building translators between individual tools. For N different software tools as shown in Figure 10.2, this approach needs $o(N^2)$ data translators. This is obviously not a practical solution to the problem of data exchange. For example, extensive real-time data exchange between different systems via the Internet will cost

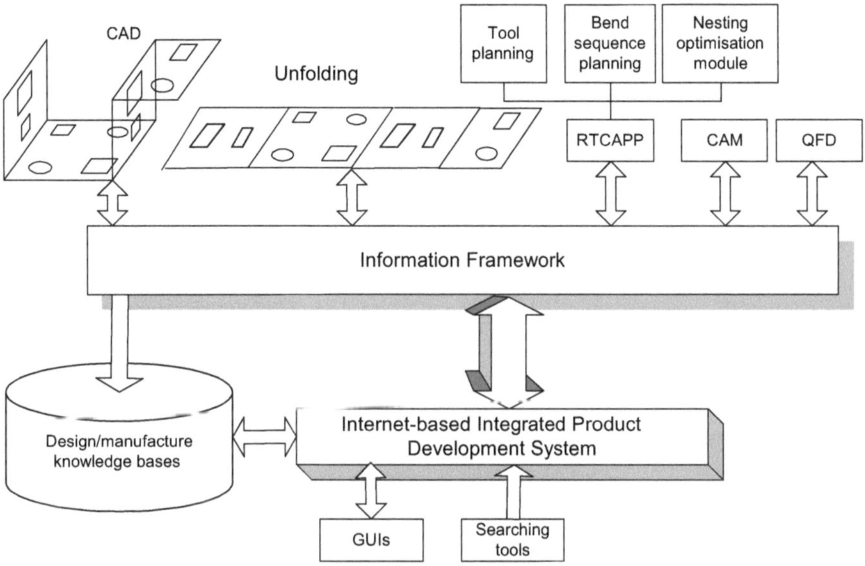

Figure 10.1 Role of the information framework in the integrated PD system

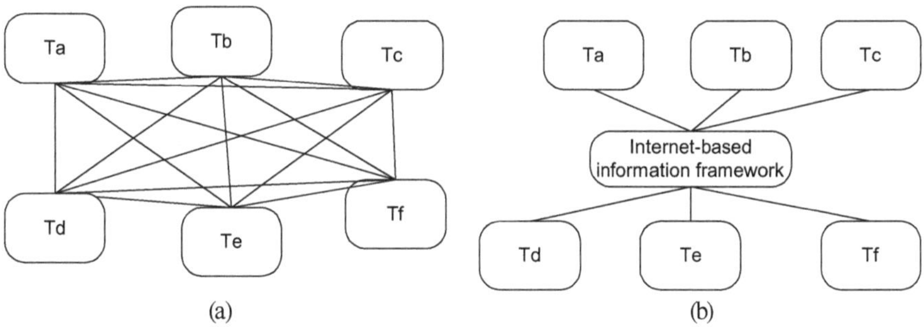

Figure 10.2 Data translators versus integrated data environment

more time and thus greatly influence the speed of Internet communication. Furthermore, it is important to the history of the design process to provide more than just a translator to move data from one tool to another. It is also important to record different versions of design files, as well as the relationships that exist between the data in such files. In some cases, some tools may require combined input from several different tools. Hence, a solution as shown in Figure 10.2b would be ideal to resolve the issues, where N different design and analysis tools only require O(N) translators, and information communication occurs through some common, shared data source. Such a solution depends heavily on standards for data exchange combined with the use of database technology for managing exchange files and data relationships.

This chapter presents the results of the development of an information framework. The information framework can be used to solve the problem of information modelling and real-time exchange between different software tools in an Internet-based concurrent and integrated PD system. This chapter discusses how to build an information framework to support information /data modelling in the Internet environment.

10.3 The Zero Thickness and Zero Bend Radius Principle

This section discusses a generic handling methodology for bent sheet metal parts, as it plays an important role in product design and modelling processes and has been used in this research.

This research considers a generic handling methodology for its parametric shape of bent sheet metal parts. The details of the generic handling principles are:

1. Zero thickness principle: supposing that the thickness of bent sheet metal parts is zero; designers do not need to consider the thickness of the parts.
2. Zero bend radius principle: supposing that the bend radius is zero; designers not need to consider the changes of parametric shape in the bend process.

The above principles are based on the assumption that the length of the middle layer of the sheet metal parts is the same before and after the bend process. The middle layer will change when the following situations change, *e.g.*, product materials, bend method, and mould structure. A database was built for selecting the position parameter of the middle property layer (Hua 1989). As shown in Figure 10.3, the length of the sheet metal part after unfolding can be calculated as follows (Hua 1989, Chang 1991, Liu 1998):

$$L = L_1 + L_2 + \frac{|\alpha| \cdot \pi}{180°}(r + k \cdot t)$$

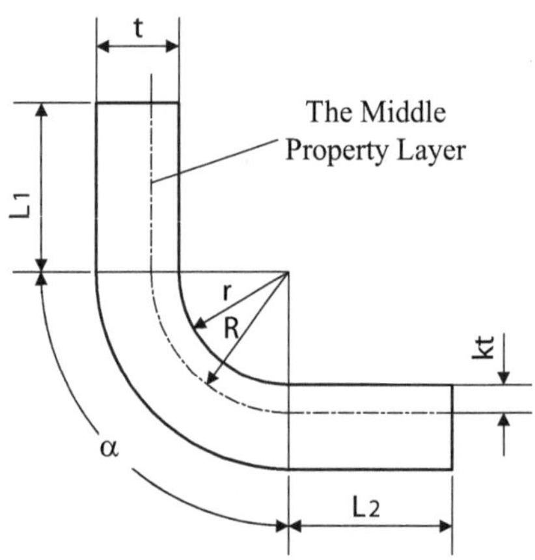

Figure 10.3 Position of the middle property layer

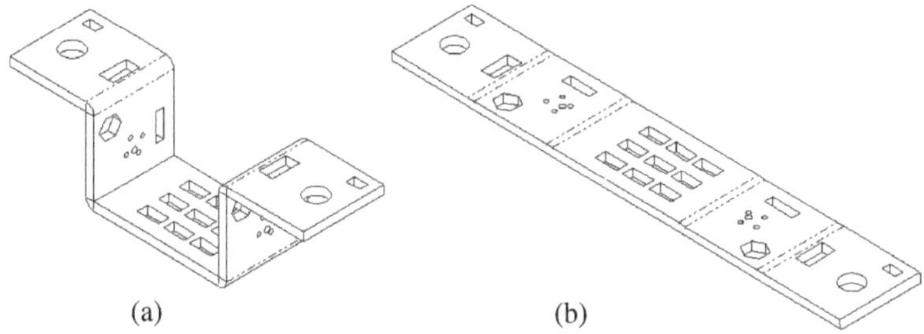

Figure 10.4 Generic handling method for bent sheet metal part: (a) example sheet metal part; (b) unfolded example sheet metal part

where L_1, L_2 are the length of two linear sides;
α is the bend angle (degrees);
k is the position coefficient of the middle property layer;
r is the bend radius of the inner surface (mm);
t is the thickness of the sheet metal part (mm).

With these two principles, the bend parametric shape of sheet metals can be considered as a group of continuous zero-thickness planes connected to each other. For example, Figure 10.4a can be expressed as Figure 10.4b without consideration of the thickness of the sheet metal parts. The software will consider the part thickness and bend radius automatically.

The advantages of using the zero thickness and zero bend radius principle to handle the parametric shape of bent sheet metal parts are as follows:

1. Simplicity: a complicated 3D product model can be simplified into a 2D model. This means complicated hand calculation can be avoided, which makes the information modelling process easier.
2. At the design stage, one does not need to consider the plastic deformation of the products during the bend process (the software will automatically consider it).
3. One does not need to consider part thickness at the modelling process stage and 3D entities can be transformed to related 2D plane entities.

The method of using the zero-thickness and zero bend radius principles to handle the parametric shape is for the convenience of the product designer. This generic handling method can transfer 3D shape feature entities into 2D entities while relationships among these entities can be modelled using an object-oriented scheme structure.

10.4 Information Integration Framework

The information integration framework aims to build an information bridge to fill the gap between sheet metal part design, process planning, simulation and manufacturing systems. It is an important issue in supporting integrated and concurrent PD.

This information integration framework consists of a tree-based information modelling methodology to represent information objects and relationships. These objects represent various items of information of a sheet metal product, such as parametric shape, topology and the design specifications (*i.e.*, performance, materials, tolerance, surface and quality requirements), and manufacturing information (such as cost, manufacturability, process specifications, operational data). The relationships among different information objects (*i.e.*, behaviour, state, *etc.*) can also be modelled and can be used to build relational knowledge bases to support concurrent product design and manufacturing. The information framework can also be used to build information models for information sharing among different software packages so that these software packages can work concurrently in an integrated environment.

10.4.1 The Step Structure Information Framework

This research proposes an information framework, termed a step structure information framework as shown in Figure 10.5. The framework contains four top-down information layers, which include a parametric layer, a feature layer, a parts layer and a knowledge layer. The parametric layer contains the geometric data of the shape feature of the sheet metal parts and tool features. The feature layer contains all the feature information which includes not only the feature information (*i.e.*, attributes) but also relationships with other feature-level information objects and objects defined by users. The part layer contains all the part information including feature information and relationships among different part-level information objects. The knowledge layer contains not only the part information, but also "knowledge related" information objects and an inference engine. The knowledge in the knowledge layer is extracted from part-level knowledge and feature-level knowledge, formed by information objects and relationships among them. The knowledge

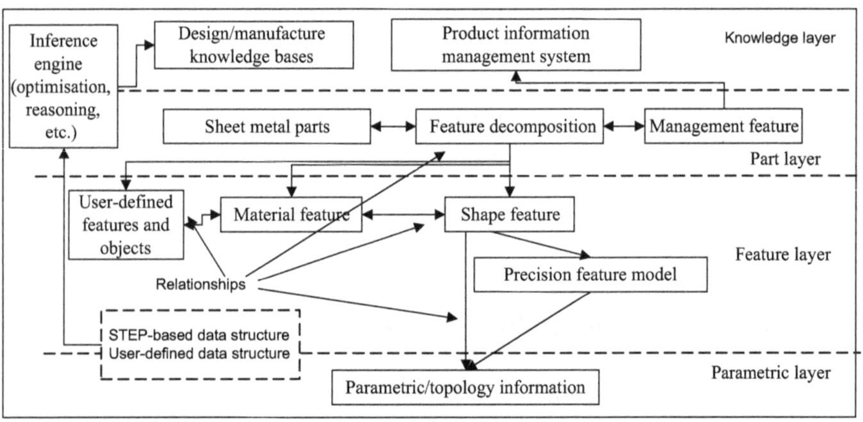

Figure 10.5 Information integration framework for sheet metal parts

in the knowledge layer can be directly used to support intelligent concurrent design and manufacturing. The management feature is used to manage all the information of a certain part, which can be saved and used as part of a company database. Application objects defined by users according to the requirements of the project can be put in the feature and part layer. After a new object is defined, its relationships are created by either the users or automatically by the optimisation algorithm and existing knowledge. This object with its relationships can be regarded as new knowledge, which can be used in new product design and manufacturing processes.

In addition to geometry, the step structure framework that has been developed within this project includes representation of function and behaviour, physical decompositions, functional decompositions and the relationships between the physical and functional domain. In this step structure information framework, the relationships are used to represent the relationship between sets of objects, including a physical decomposition, a functional decomposition and other kinds of relationships. The overall sheet metal part relationships among shape feature representation is comprised not only of the collection of objects that represent physical entities, but also of other objects, relationships, interconnections between them as well as various attributes and their values.

10.4.2 Information Relationship Representation of Sheet Metal Parts

From the previous discussion, we know that one of the challenges in developing a sheet metal information framework is to delineate and represent the complex relationships among product entities and diverse types of sheet metal definition data. In conventional CAD/CAM systems, sheet metal parts are represented and stored as a complete geometric and topological 3D solid. Such a representation is quite suitable for display and for performing geometric computation-intensive tasks such as engineering analyses and simulation. However, it is not appropriate for tasks that require decision-making based on high-level information about geometric entities and their relationships. Therefore it is necessary to define sheet metal geometric entities and their relationships so as to provide high-level information required in these applications.

Tree-based feature models are able to represent the relationships among data and can model sheet metal parts with the following results: 1. the representation of sheet metal geometric features, their properties, relationships and functions; 2. a wide definition of data and entity relations; and 3. representation of semantics for different levels of applications. Knowledge can also be modelled using this tree structure as it can be considered an "efficient" combination of different information objects by an inference engine.

To represent the feature model and its relationships to the sheet metal part, the tree-based information models should not only represent the shape features, but also the relationships among different features. Usually, shape features contain both main

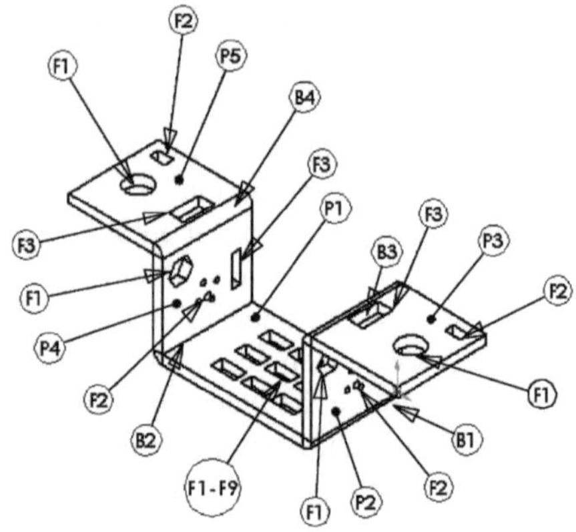

Figure 10.6 An example for building feature relationships graph for a sheet metal part

features and auxiliary features. According to the relationships among main features, and between main features and auxiliary features, it is possible to build a relationship graph among different features. The shape feature relationship graph can be used to represent the shape modelling process. Figure 10.6 is an example of a sheet metal part that will be used to show how to build a shape feature relationship graph. There are five planes contained in this part named P_1, P_2, ..., P_5, respectively. There are several features (circle, rectangle, *etc.*) in each plane and there are four bend features named B_1, B_2, B_3, B_4 in this part. For this typical sheet metal part, the information object of each feature can be built according to the principles below.

The information modelling principle is as follows: The model starts from a reference plane, and then joins other planes together based on this reference plane. When defining a bend feature, a reference plane feature that connects with this bend feature is chosen. Theoretically, from the information modelling point of view, the reference plane can be any plane in the part. But, usually one of the planes that will be a fixture plane during the manufacturing process will be selected as the reference plane. When defining a plane feature, besides the reference plane, it is necessary to choose another bend feature which shows how the plane that contains this plane feature connects with its father plane. When defining an auxiliary feature, it is necessary to choose a plane feature from the same plane, then define this auxiliary feature according to the working coordinate of this plane. Figure 10.7 shows the tree-based relationship graph among different features for this example part when P_1 is assumed to be the reference plane, the relationships among different features (A_i, B_i) and planes (P_i) are clearly shown in Figure 10.7.

Based on the principles above, two branches of the tree structure can be used to build any product model, which makes it easy to achieve different application modules, *e.g.*, nesting optimisation, path planning and CAM module. The details of the two-branch tree-based information modelling methodology are as follows. The reference plane is regarded as the starting point of the tree. When defining a bend feature, as shown in Figure 10.7 it will be inserted into the left point of the tree

Figure 10.7 Tree structure of shape features

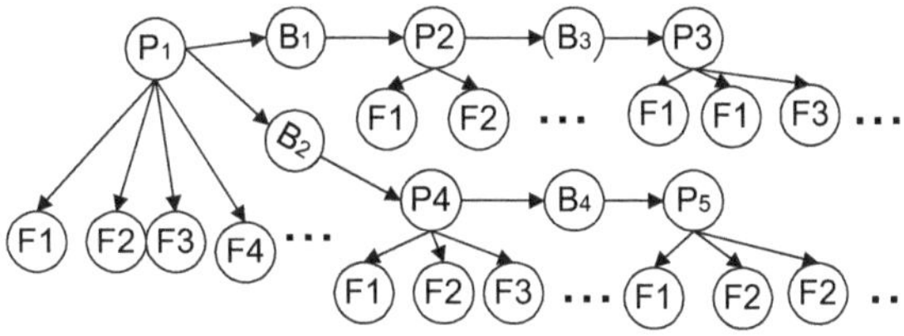

or the left point of the left point of the tree, which contains the father feature that connects with this bend feature. This structure shows that the bend feature has a "son–father" relationship with its "father" feature; when defining a plane feature, it is being inserted into the right point of the bend feature that connects this plane with its father plane; when defining an auxiliary feature, it is being inserted into the right point of the tree or the right point of the right point of the tree, which contains a plane feature, which means the auxiliary features have "brother" relationships with each other. Based on this modelling methodology, the example part can be represented as a two-branch tree structure, which is shown in Figure 10.8.

This two-branch tree-based data structure can also be used to describe the relationships among features, information objects and other user-defined objects. This tree structure can be extended as sheet metal parts become more complicated. However, building information models that clearly show the relationship among features in different planes is difficult but also very important for the information framework. It should be mentioned that the relationships here contain static relationships and dynamic relationships. Static relationships mean that the relationships are not changeable (*i.e.*, predefined material, tool–tool feature). Dynamic relationships mean that the relationships can be changed (*i.e.*, a new punch tool object for a feature, an information object changes its connection to another object). Dynamic relationships are usually used to represent renewable knowledge and relationships among information objects. In an actual application process, a data structure that shows the relationships among different features may change as the information model goes

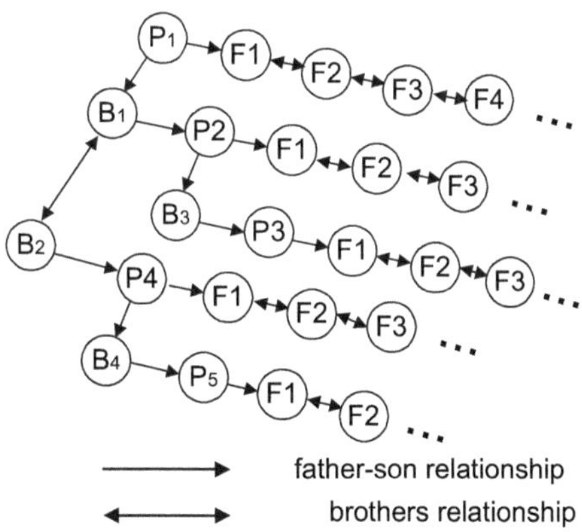

Figure 10.8 The two-branches structure of shape-features

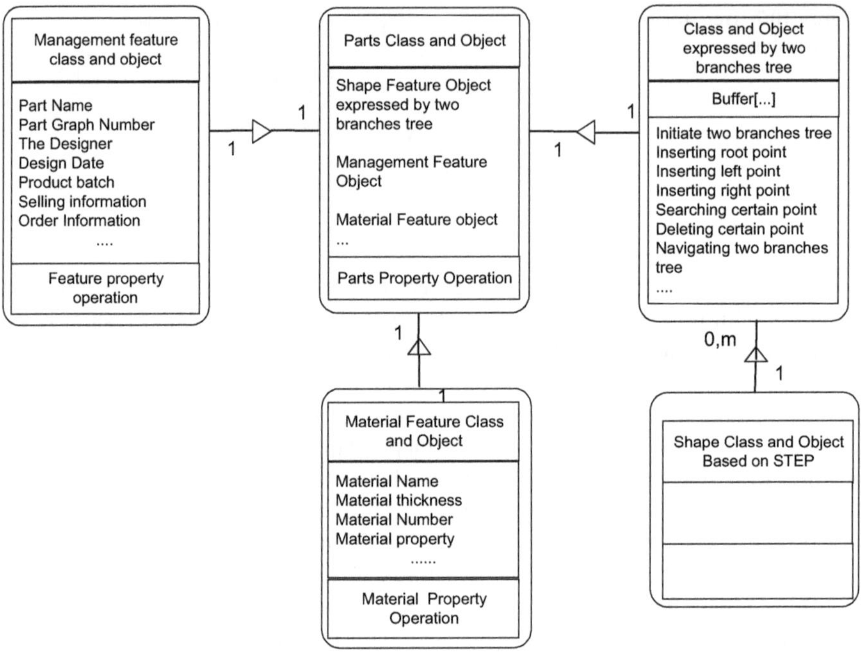

Figure 10.9 Common information model for sheet metal parts

through the application software, *i.e.*, the relationships among different features in a CAPP information model will change according to the automatic arrangements of the shortest path optimisation algorithm. Evidently, the tree structure can be extended as sheet metal parts become more complicated and can also be used to represent distributed dynamic relationships.

In order to facilitate sheet metal product modelling and to build distributed information management systems, a common information model is built in this research. Figure 10.9 shows the structure of a common information model. It depicts a management object with its relationships to other downstream information objects, based on the information integration framework. The contents of information objects are defined by users based on the actual application (*i.e.*, cost, tool sequence, lead time) and can be filled by software and users through the integrated system.

10.4.3 Modelling Languages and Methodology

The modelling language used to achieve the step structure information framework should represent sheet metal parts as sets of objects and relationships of objects. It should represent physical entities such as assemblies, subassemblies, and components, as well as non-physical concepts such as function and behaviour in the Internet environment. In recent years, quite a few information modelling languages have been developed or are under development, *e.g.*, the Integrated Computer Aided Manufacturing (ICAM) Definition Language 1 Extended (IDEF1X) (Appleton D. company), the EXPRESS language (ISO 10303-11:1994(E)) and the Unified Modelling Language (UML)(UML Website). These information-modelling languages provide various ways of formally representing the information framework.

The modelling languages that have been used for this research include Visual C++, VBSCRIPT, JAVASCRIPT and EXPRESS. Visual C++ is an object-oriented (O-O) language, which supports class definition and object operations. It is used for developing most of the sheet metal application modules and tools. VBSCRIPT and JAVASCRIPT are used for knowledge sharing and client–server communication. EXPRESS is based on programming languages and the O-O paradigm, and can be used for object definition and specification of constraints on the objects defined. A number of languages have contributed to EXPRESS, *i.e.*, C++, SQL, *etc.* These languages make it easy to achieve all the application modules of the integrated system.

Figure 10.9 shows a management object and its relationships with other information objects that can be represented by EXPRESS or other O-O languages. It includes a two-branch tree object for shape features, management feature object, material feature object, *etc.* The contents of an information object can be defined based on the actual application (*i.e.*, cost, tool sequence, lead time, *etc.*). The shape feature objects of the part are stored in a two-branch tree. The material feature object, the precision feature object and other objects are connected with the shape feature object. The relationship between precision feature object and shape feature object can be shown by the relationship between these two objects. The information object in Figure 10.9 can be extended as the two-branch tree connects with more features and information objects.

10.5 Case Study

This section discusses how to build an information model for a typical sheet metal part

Figure 10.10 shows a simple sheet metal part, which includes two plane features P_1, P_2 and one bend feature B_1. There are two inner rectangular features F_1 and F_2 in plane P_1. Assume P_1 to be the reference plane, and that feature F_1 is positioned by the reference plane P_1. The feature F_2 is positioned by the reference plane P_1 and the feature F_1's position. The shapes and dimensional positions of these features have been shown on Figure 10.10.

Figure 10.11 shows the step structure information model of this part and the details of information models being built for different features of this part. The O-O expression of this part is shown in Figure 10.11a, which includes the relationships among different information objects, a management object and connection

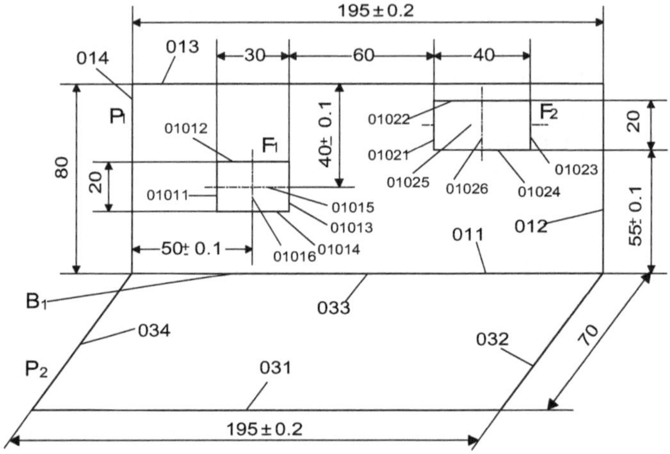

Figure 10.10 Example sheet metal part for object-oriented information model

to other information objects. The relationships among different features, the relationship between feature and related information objects, *etc.*, can be shown using EXPRESS-G. Other information models that represent other features are discussed below.

The O-O expression of the plane feature is shown on Figure 10.11b, which includes information, such as the restraint value, the length and width of the plane feature. The O-O expression of the rectangular feature is shown in Figure 10.11c, and includes information such as the restraint value, the length and width of the rectangle. The O-O expression of bend feature B_1 is shown in Figure 10.11d, which includes information such as bend radius and angle. The O-O expressions of two plane features P_1 and P_2 are similar; the O-O expressions of two rectangular features F_1 and F_2 are similar. The feature number of plane feature P_1, bend feature B_1 and plane feature P_2 are 01, 02 and 03, respectively; the feature number of rectangle inner hole features F_1 and F_2 are 0101 and 0102. The whole information model of this part, which includes all these sub-information models, is shown in Figure 10.11a. As we can see from this structure, the information model can be extended and the tree-based data structure can be divided into different layers when more classes or objects are connected. The detailed contents of the information model may also change as the application changes, but the step information structure of this part would remain unchanged.

The description above indicates that if we add a new feature into this part, a new information model will be created. This information model and its relationships with other features will be automatically added as they go through design and manufacturing systems. The data value of each object can also be changed by the program automatically. For example, the value that means manufacturing sequences of the feature in a certain feature object will change as the information object goes through the shortest path optimisation algorithm. The cutting tool object has some relation-

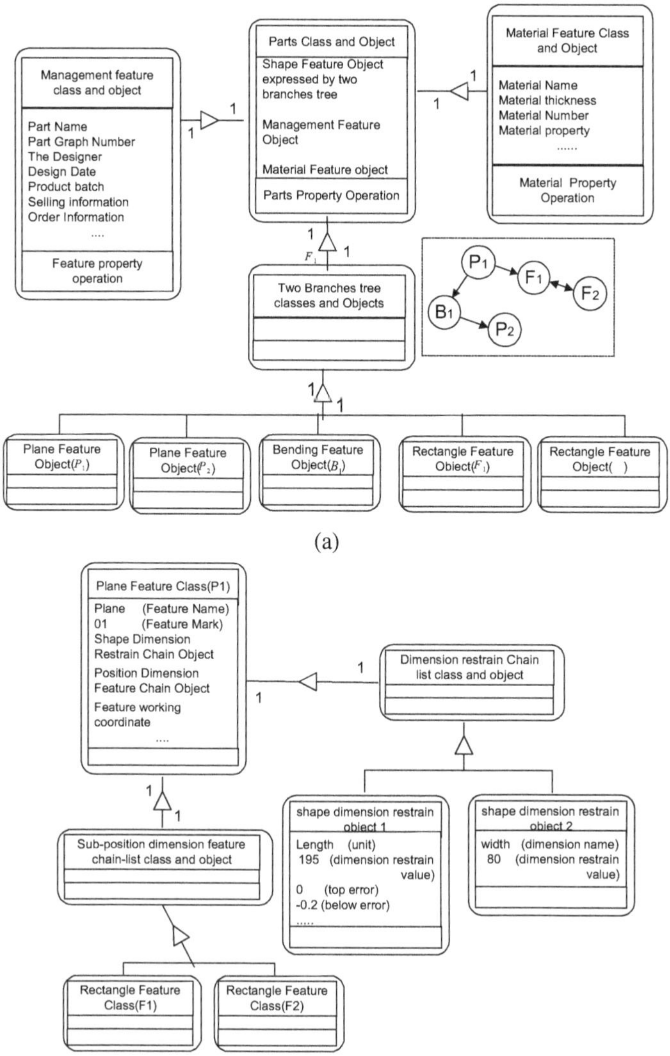

ship with these features. The tool planning module will assign each feature a relationship with a selected tool that will be used to manufacture this feature according to the optimisation results. The object view of the tools contains all the associated information elements (attributes). The information elements are, for example, the tool ID, tool name, description, *etc.* Furthermore, there is also usage information, which will maintain the information related to the level of inventory, tool life, tool location, *etc.* As we can see, the information schemas and relationships between features, tools, *etc.*, can be implemented in the form of a company database.

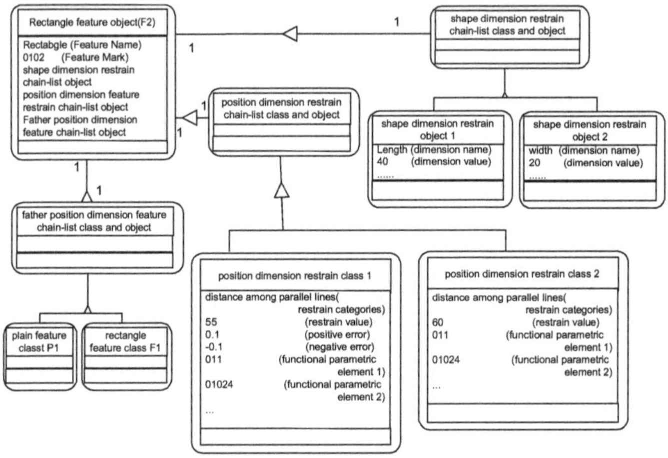

(c)

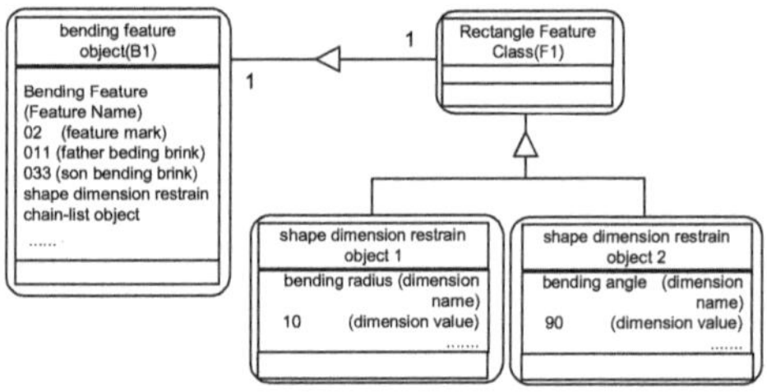

(d)

Figure 10.11 Object-oriented information expression of the sheet metal part: (a) overall model; (b) modelling a plane feature; (c) modelling a rectangular feature; (d) modelling a bend feature

10.6 Conclusion

In order to support the rapid development of sheet metal parts/products, an information framework that aims to provide an information modelling methodology to support sheet metal parts intelligent concurrent design and manufacturing has been proposed and explained. This information framework divides information into four layers; a parametric layer, a feature layer, a parts layer and a knowledge layer. It is an important step towards the development of information models of different parts that are responsive to the needs of emerging manufacturing and engineering technologies. This framework can also be used to achieve intelligent decision support in the earlier design stage through correctly defining the structure of the company's knowledge bases. This includes the representation, capture, sharing and reuse of corporate design and manufacturing knowledge. The models extend well beyond representing solid geometry, and can, in principle, represent other important engineering information, such as properties, attributes and functions, and can also be extended to the modelling process for other products.

References

Cai, Z.Y., Wang, S.H., Xu, X.D., Li, M.Z., 2009, Numerical simulation for the multi-point stretch forming process of sheet metal. Journal of Materials Processing Technology, Volume 209, Issue 1, Pages 396–407.

Chang, W. W., 1991, The Technology Handbook of Sheet Metal Parts Unfolding (Shengyang: The Publishing House of East-North Technology University).

Li, M.Z., Cai, Z.Y., Liu, C.G., 2007, Flexible manufacturing of sheet metal parts based on digitized-die. Robotics and Computer-Integrated Manufacturing, Volume 23, Issue 1, Pages 107–115.

Liu, J. Q., 2000, Design & Manufacturing of parts with complicated shapes, Ph.D. thesis, Huazhong University of Science and Technology, P.R. China.

Lu, X.F., Li, H.F., Zhang, Y.R., Zhou, Y., 2010, Simulation and experiment on workability for cold pressure forming of sheet metal part with step cross-section. Chinese Journal of Mechanical Engineering, Volume 23, Issue 2, Pages 161–165.

Shao, W., Guo, J.J., Zhou, A.W., 2009, A Framework for Measurement and Analysis of Sheet Metal Parts. Proceedings of the 2009 Second International Conference on Intelligent Computation Technology and Automation, Volume 2, Pages 363–366, IEEE Computer Society, Washington, DC, USA.

Verlinden, B., Duflou, J.R., Collin, P., and Cattrysse, D., 2008, Cost estimation for sheet metal parts using multiple regression and artificial neural networks: A case study. International Journal of Production Economics, Volume 111, Issue 2, Pages 484–492.

Xie, S.Q. and Zhang, W.J., 2008, Optimal sheet metal process planning: an agent-based approach. International Journal of Mechatronics and Manufacturing Systems, Volume 1, Issue 1, Pages 131–153.

Zhou K.H., 1989, The Calculation Methodology of the unfolded parts. National Industrial Publishing house, Beijing.

Part IV
Product Development Methods, Algorithms and Tools

Part IV of the book introduces the new methods, algorithms and tools we have developed to support rapid OKP product development. These methods, algorithms and tools have been developed for typical applications in the OKP field and have been implemented in practice. Our findings are that these methods, algorithms and tools are capable of being used in the development of other OKP products. In this book, detailed examples are given so that they can be used as references for readers to develop their own OKP systems.

Chapter 11 presents a cost estimation and optimisation system, which is a part or subsystem of an integrated PD system. The PD cost under consideration in this chapter is the sum of the costs of product design, production and logistics, which is the main cost of a product. Some other possible costs associated with a PD cycle, *e.g.* overhead costs, are normally proportional to this main PD cost and in practice they are estimated by multiplying a constant percentage to the PD cost. Unlike some earlier research which focused on the cost estimate in a single manufacturing company, this cost estimation system aims at the PD cost estimate and optimal control under the complexities of a global manufacturing environment. The proposed cost estimate and optimal control system has been implemented in a sheet metal manufacturing company.

Chapter 12 reports on research work that aims at developing an integrated data structure to support RPD in an Internet environment. The emphasis is placed on integrated data management and reuse of past product development experience to support a company's aim to shorten its PD cycle. The integrated global data structure model was modelled using EXPRESS from STEP with consideration of real-time data communication within the Internet environment. In terms of this data structure, a design/manufacturing knowledge base was developed as a major part of the WWW-based PD system. The basic principles and concepts of the knowledge base and the WWW-based knowledge management system are presented. An industrial implementation is also reported.

Because efficient management of product information that covers the whole life cycle is critical to the enhancement of corporate competitiveness, Chapter 13 explores the design and development of a WWW-based PD information management system for a cross-nation manufacturing corporation in New Zealand. Since product data are often managed in a distributed computing environment, CORBA is employed to ensure interoperability among distributed information objects.

Internet-based "Design for X (DFX)" systems have been recognised as efficient tools for implementing concurrent engineering and playing a key role in RPD. Internet-based DFX or IDFX systems can be applied by manufacturing companies to rapidly produce high quality products with low costs and thus higher profits. However, the implementation of IDFX systems is not an easy task. Chapter 14 presents two typical applications of IDFX, *i.e.* Internet-based design for manufacture (IDFM) and Internet-based design for cost (IDFC) systems, for rapid and economical tool-/mould-making.

Process planning is very important in the OKP industry as global competition to reduce product cost intensifies. An optimal process planner can maximise the utilisation of costly raw material resources, improve machining efficiency, and hence reduce product cost. Chapter 15 explores algorithms for solving the path planning issue. This work uses sheet metal products as examples and investigates the problems using genetic algorithms. The proposed genetic algorithm approach uses a genetic encoding scheme and a genetic reproduction strategy to reach an optimum solution. The effectiveness of the genetic algorithm path planning approach is compared with the "ant colony" algorithm.

Chapter 11
Cost Estimation and Optimisation Framework for Rapid Product Development

Abstract The ultimate goal of mass customisation is to achieve economies of both scope and scale. This goal implies a conflict between customisation and economy of scale (or mass production) in the traditional manufacturing paradigm. However, recent developments in computer and Internet communication technologies, along with concurrent engineering, as well as modular design methodology provide concepts, methods and technology infrastructure for realising mass customisation. One of the findings from numerous research efforts on mass customisation is the use of e-commerce technologies to manage a product development chain that links customers, suppliers and manufacturers together to approach concurrently customised products in a short time and at the low cost level of mass production, which is the very definition of mass customisation. To ensure the success of mass customisation in a product development (PD) chain, a rapid, automatic yet accurate cost estimate and control system is needed. This chapter presents a novel cost index structure, together with two novel cost estimate methods, namely the generative cost estimate method and the variant cost estimate method, used for the development of a semi-automatic or fully automatic computer aided cost estimate and control system in mass customisation. Finally, an industrial case is reported to illustrate the principles and feasibility of the proposed data structure, methods and system framework.

11.1 Introduction

In the early stages of a PD process, an estimation of the development product life cycle cost is the primary piece of information that will be needed to determine the profitability of the product. The more reliable the cost estimate process is, the more likely the right decision will be made. This has become an important issue for RPD. In this chapter, an Internet-based cost estimation and optimisation system, a part or subsystem of the IRPD system, will be reported. The PD cost under consideration in this chapter is the sum of the costs of product design, production and logistics, which are the main costs of a product. Some other possible costs associated with a PD cycle, *e.g.*, overhead costs, are normally proportional to this main PD cost and in practice are estimated by multiplying the PD cost by a constant percentage. Unlike earlier research that focused on the cost estimate in a single manufacturing company, this Internet-based cost estimation system aims at estimating the PD cost estimate and optimal control under the complexities of a global manufacturing environment. The proposed "cost estimate and optimal control" system has been implemented in a sheet metal manufacturing company. This industrial implementation will also be reported in this chapter to demonstrate the principles of the system.

In today's competitive global market, a short PD cycle time and high quality product with reasonable cost are becoming more and more critical for small or medium sized manufacturing companies to survive. In particular, the PD cost normally determines the market share, profitability and return on investment of small or medium sized manufacturing companies. To control PD cost at early PD stages is, in turn, a problem of how to accurately estimate the possible cost associated with a PD cycle. Estimates of PD cost are closely related to the selection of a mix of production development processes, which can fully meet the customer's requirements with a lowest sum of costs of these selected processes. Hence, it is necessary to develop new methods and tools for small or medium sized manufacturing companies to rationally select proper PD processes, accurately estimate and optimally control the PD cost.

Cost estimation and reduction have always been of great concern in a PD cycle (Romero Rojo *et al.* 2009). In recent years, research aiming to improve a manufacturing company's PD ability has been focused on two aspects: how to shorten the PD cycle time; and how to reduce the PD cost. In fact, these two aspects are interrelated. A shorter product cycle time will normally result in a lower PD cost. On the other hand, optimally controlling the PD cost usually leads to a shorter PD cycle time (Castagne *et al.* 2008). Hence, research projects studying either PD cycle time or PD cost aim to achieve the same goal, *i.e.*, rapid and economical PD, from two different angles.

To estimate and optimally control PD cost in today's global competitive manufacturing market, several issues need to be researched. First, PD in a global manufacturing environment has been proved to have the advantages of agility, often providing quick and cheap solutions for some parts or even the whole PD from the partners or subcontractors, and yet effectively applying state-of-the-art technology to address the customer's requirement in a global competitive environment. However, a well-established PD management system (*e.g.*, cost estimate and optimal control system) in a global manufacturing environment is still under development. Although a relatively large number of papers and reports have addressed the cost estimate and control problems in various PD processes (such as design, manufacturing and logistics chain management), these methodologies were normally developed based on cases in individual companies rather than in a global manufacturing environment. In a global manufacturing environment, a manufacturing company (normally called the master company) may carry out a PD cycle through collaboration among its partners or subcontractors. These partners and subcontractors may use different PD management systems and computer aided engineering software systems. To effectively communicate, share the necessary information, integrate the effort and optimally control the PD cost between the master company and its partners/subcontractors are the problems which need to be addressed by this research, which aims to develop a PD cost estimate and optimal control system in a global manufacturing environment.

Second, no matter whether it is a global manufacturing environment or an individual manufacturing company, PD cost estimate and control is an interdependent and correlated problem. It is influenced and dynamically determined by a number of preconditions. The cost of product design, for instance, will be strongly influenced by the quality of the product definition. Correctly understanding the customer's requirements and hence having a clear product definition will normally result in a quick and successful product design with a lower product design cost, which is mainly determined by the total engineer hours used (Tu *et al.* 2002). Likewise, the cost of manufacturing a part is normally affected by the design of the part. The final form of the part design, which includes the geometry, surface, tolerance and property specifications of the part, either directly or indirectly influences the selection of manufacturing processes and hence the cost of manufacturing the part. The bad experiences gained from manufacturing industries, particularly from sheet metal manufacturing companies, show that decisions made at the design stage without PD costing considerations normally result in a wrong PD cycle/process chosen. A wrong PD cycle will let the company suffer a longer PD cycle time due to reworks, a poorer product design and manufacturing process that is often indicated by lower customer satisfaction, and hence higher PD cost.

Finally, I should point out that most developed or existing cost estimate and control methods and systems as reported in the literature have problems which cannot be directly adopted in the IRPD system. The detailed discussions on the shortcomings or limits of these developed methods and systems are given in Section 11.2. These methods and systems need to be modified or further developed so that they can be applied to address the complexity and uncertainty of a PD process in a global manufacturing environment.

The PD cost estimate and control is no doubt an effective tool to monitor a PD cycle. The cost estimate results can be used as important reference parameters or indicators to check the feasibility and profitability of developing a new product. It is well known from experience that rework downstream in a PD process, *e.g.*, rework in manufacturing, is more costly than corrections made in earlier PD stages, *e.g.*, in design. PD cost estimate and control is obviously a planning and control activity which has taken place at earlier PD stages. Accurately estimating and optimally controlling the cost of PD is crucial to shorten PD cycle time and reduce PD cost. To develop useful methods, algorithms and a system framework to support this planning and control activity at early PD stages is very important for OKP manufacturing companies in order to improve their position in the market, particularly for small or medium sized OKP manufacturing companies.

This chapter starts with a review of related research work and discussions of limits and shortcomings of these developed methods, algorithms and systems as reported in the literature for PD cost estimate and control. Then, following this literature review, an Internet-based system for PD cost estimate and control in a global manufacturing environment is proposed. This system was developed as a decision support tool for product developers to make decisions or selections among possible design alternatives, manufacturing processes and different ways of managing a product logistic chain under considerations of the PD cost and cycle time. To estimate and control the cost of product manufacture and logistics, based on product structure, a "cost index structure" is proposed. To address the communication and data transfer problems in a global manufacturing environment, the STEP has been adopted in the cost index structure to model product data related to costing and other relevant information through a PD life cycle. To demonstrate the principles of the proposed system, data structure, methods and algorithms for PD cost estimate and control in a global manufacturing environment, a case study for developing a sheet metal parts/products is reported.

11.2 Literature Review

The ability to rapidly and economically develop a product has attracted much attention from industrial practitioners and academic scholars. Quite a few research papers have been found which address the problems of rapid and economic PD from the point of view of PD cost estimation and control, such as design for cost (DFC) and economical manufacturing (Ou-Yang and Lin 1997), neural-network-based costing and control (Bode 2000), cost estimation based on information management (Hubert *et al.* 1999), concurrent costing (Ou-Yang *et al.* 1997, Bernard *et al.* 2007), and activity-based costing (Koltai *et al.* 2000). However, all these papers addressed problems in an individual manufacturing company, and did not take into consideration the dynamic relationships among the manufacturing partners and the complexity of cost estimation and control in a global manufacturing environment.

As indicated by Becker and Prischmann (1993), the accuracy or reliability of a PD cost estimate increases as PD progresses. A PD cost estimate at an early PD stage is normally very rough and unreliable. Sometimes, an overestimated PD cost of a product at an early PD stage may mislead a company into giving up a product that may have had great potential in the market. On the other hand, an underestimated PD cost may trap the company into an extensive investment, and end with a product that has only little or no market share. Hence to correctly estimate the PD cost of a product at an early PD stage is very important for rapid and economical PD.

To support decision-making in the early PD stages, much research effort has been made in cost estimation at the early PD stages. Typical approaches include parametric-oriented costing, similarity-based costing, activity-based costing, detailed cost estimation, life cycle costing, learning curve costing and neural-network-based costing (Smunt 1999, Tamas *et al.* 2000, Bode *et al.* 1998, 2000). Monden and Lee (1993) reported a cost estimate system called Kaizen Costing. This system was used for budgetary control and cost reduction of PD. Horvath *et al.* (1994) proposed a standard costing system to meet cost limits (or standards) set by the management. Kawada and Johnson (1993) presented an accurate cost estimation system to support strategic management accounting. Cawthorne-Nugent *et al.* (1989) developed a computer aided costing system which employed knowledge base technology and the cost models associated with part features. Marx *et al.* (1998) further reported another application of knowledge base technology for PD cost estimation. At early PD stages, there is not enough information and clear functional relationships between the product attributes and costs as well as the ambiguity of prices or costs from suppliers. Some artificial intelligence (AI) techniques and fuzzy logic techniques, such as learning curves (Becker and Prischmann 1993, Nadeau *et al.* 2010), neural networks (Tu and Jiang 1997, Ju and Xi 2008), and self-learn methodologies (Zhang and Fuh 1998) have been applied to PD cost estimation.

Manufacturing cost is always an important factor influencing early decision-making on product design, process planning, product manufacture and final market share of the product. It is a major part of the overall PD cost as it is the cost of making the product. Usually, to determine the manufacturing cost of a product, a lot of detailed data will be needed, such as selected manufacturing processes and the sequence of the processes, processing times of the processes, hourly rates for labour forces and machines, raw material cost, subcontractor cost, transportation cost, logistic chain (*i.e.*, supply and delivery chain) management cost, overhead cost, and equipment depreciation and maintenance cost. Traditionally, manufacturing cost is estimated by adding up all these costs. The computer is often used as a computational tool to speed this summation. Typical examples applying this traditional manufacturing cost estimation method can be found in Casey (1987), Goldberg (1987) and Ostwald (1984). Winbourne and Toolsie (1991) proposed a manufacturing cost estimation system that can be used at an early product design stage to estimate the manufacturing cost of the product under the uncertainty and ambiguity of the manufacturing process plan. A major drawback of this approach is that the proposed system requires some initial estimates of cost-related product design and manufacturing parameters, *e.g.*, surface treatment, machine setup or tear-down times, *etc.*

Like the aforementioned PD cost estimate research work, knowledge base technology has also been used in manufacturing cost estimation through storing the cost partners indexed by the cost-related product design and manufacturing parameters (Geiger and Dilts 1997). However, all these approaches did not indicate how to correctly estimate the values of these cost-related product design and manufacturing parameters. Wang and Bourne (1995) presented a manufacturing cost estimating system, called "totally integrated manufacturing cost estimating system" (TIMCES), which integrated the computer aided design (CAD) system, process planning system and cost estimate system. Inputs to this system include the selected manufacturing processes, the sequences of the processes, and cost data for standard components and manufacturing processes. It is obvious that the TIMCES is another application of the traditional manufacturing cost estimate method. It needs a large amount of complicated input data. Kamarthi *et al.* (1993) proposed a manufacturing cost estimate system for sheet metal parts/products manufacturing. This system in fact used the cost as an important parameter to select manufacturing processes and available machine tools with lower costs at every process planning stage. The system only considered the standard process sequence and 2D shape of sheet metal parts/products.

In summary, all the aforementioned cost estimate methods or systems can be considered as the application of one or more of the following three basic cost estimate methods:

1. The function-based costing method: this method estimates PD cost through decomposing a product into generic or standard functional modules and quotes the costs for these generic or standard functional modules according to previous experience or the market price of these functional modules.
2. The similarity-based costing method: this method predicts PD cost by searching for cost information on similar products made in the past.
3. The expenditure-based costing method: this method estimates PD cost by listing all the possible expenditures associated with the PD cycle and quotes the costs for these expenditures based on reference prices stored in a costing knowledge base. The key point in applying this method is to properly decompose a PD process into basic expenditures to be easily quoted according to standard market prices or previous experience. The activity-based cost estimate methods or system are typical applications of this cost estimate method.

These basic cost estimation methods are also used alone or in combination in the PD cost estimate system to estimate the cost of a manufacturing process, a component, or a consumed manufacturing resource. However, the major drawbacks preventing direct adoption of these cost estimate systems (as described in the literature) are:

1. All the research reported in the literature was focused on problems in an individual manufacturing company rather than in a global manufacturing environment.
2. Nearly all the research work reported assumed that the processes through a PD cycle were carried out sequentially rather than concurrently, which is more common in today's small or medium sized manufacturing companies in order to rapidly and economically develop a product to meet customer needs.
3. The proposed cost estimate systems are not able to dynamically and flexibly cope with some problems peculiar to the global manufacturing environment. These include relatively frequent changes of product design and process plan due to customer requirement changes or the involvement of a partner or subcontractor; a quicker price or cost change in a global market; higher uncertainty and ambiguity of product design and manufacturing processes at early PD stages for a product to be developed in a global manufacturing environment; complicated cost estimation and optimal control through searching an available manufacturing resource in a global manufacturing environment.

In this chapter, a PD cost estimation system is proposed. It is a part of a rapid and economic PD system in a global manufacturing environment. This cost estimate system communicates with the other subsystems or modules in the PD system through computer intra/Internets. It was developed as a decision support system to help the selection of alternative design proposals, manufacturing processes, manufacturing resources and partners/subcontractors from a cost optimisation point of view. As stated earlier in this chapter, to control PD cost is in fact to speed up a PD cycle or shorten the PD cycle time. It interactively communicates and works with all the other subsystems and models of the PD system to incrementally and concurrently develop a product feature by feature. To test the feasibility of the proposed cost estimate system, it has been implemented in a sheet metal manufacturing company to develop a sheet metal part. This industrial implementation is also reported in later sections of this chapter.

11.3 Overall System Structure of the Rapid and Economic Product Development

Figure 11.1 shows the overall structure of the rapid and economic PD system, which was developed for small or medium sized manufacturing companies to quickly respond to and meet the customer's needs. The system includes five subsystems, *i.e.*, product design, manufacturing process planning and optimisation system, operations management system, cost estimate and optimal control system, and logistics management system. These five subsystems can communicate with each other via

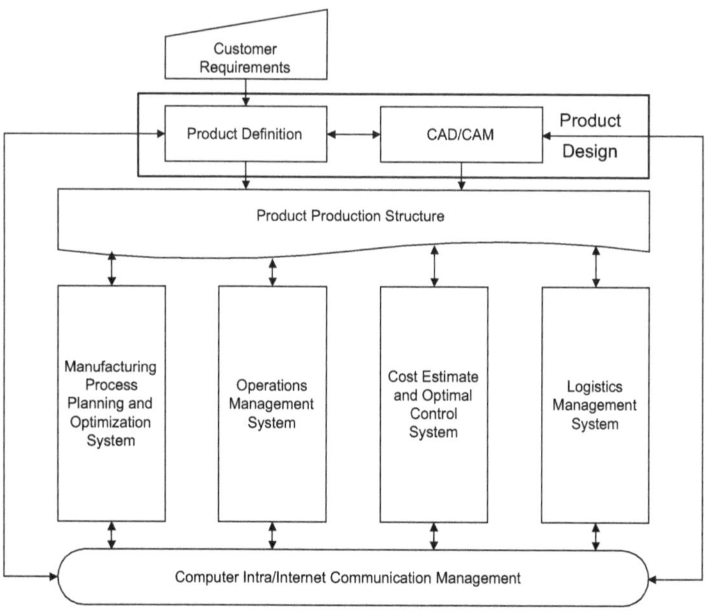

Figure 11.1 The overall structure of the rapid and economic product development system

computer intra/internets. At early PD stages, these five subsystems work together to generate an electronic file called the "product production structure" as shown in Figure 11.1. A product production structure is an integrated product data model, which can store all the necessary data through a product life cycle hierarchically in a common object. In this common object, the data is recorded in EXPRESS from STEP. Hence, all kinds of computer aided engineering and management software packages should be able to operate this common object. This data model provides feasible solutions to build automatic links between the data for product design, process planning, logistics management (*i.e.*, supply, partner and subcontractor chain management), production scheduling and costing. For a detailed description of this product model, please refer to Chapter 11. In later PD stages, *i.e.*, product prototype development and final production stages, these sub-systems, as shown in Figure 11.1, adapt to changes to the product production structure to cope with possible changes required by the customer, suppliers, partners, subcontractors and the company itself. During the production stage, these subsystems work as optimal planning and control systems to control the production to stick to what was designed and planned in the product production structure.

In the PD system as shown in Figure 11.1, the "product design" subsystem converts customer's requirements into a product design, *i.e.*, an engineering drawing plus product property and manufacturing requirement specifications. However, this design subsystem does not work like a traditional product design system, which designs a product in one go. It follows a concurrent engineering principle called "prototype-based incremental PD" (Tu *et al.* 2000a) to incrementally design a product feature by feature and concurrently search or plan all the necessary data for making the designed feature. This product design system consists of a computer aided customer interface which was developed through applying the QFD method to convert the customer's requirements into the technical attributes of a product as addressed in Chapter 3. Based on the technical attributes, the search model searches for a similar product or similar products made in the past. The similar product(s) are used as the prototype(s) of the product to be developed to meet customer's needs. The product designer can design products through modifying the features of the prototype(s) or adding new features to the prototype(s), according to customer's requirements, in an incremental or feature-by-feature manner.

After a feature is modified or designed by product designers, the "manufacturing process planning and optimisation system" searches for suitable manufacturing processes from a GT (group technology) code index manufacturing process knowledge base, or plan the processes through computer simulation technology or shop floor test if the processes cannot be found in the knowledge base (Tu *et al.* 2000c). It also uses a manufacturing process optimisation model to optimise the manufacturing processes, *e.g.*, tool selection and shortest path planning for a CNC cutting process. Through the "logistics management system", it can search for alternative processes from the company's partners and subcontractors. As reported in Chapter 9, this process planning and optimisation system plans alternative or optional manufacturing processes to create the modified or designed feature of the product. Selection among these alternative or optional processes is determined by the estimated costs of these processes, which will be provided by the "cost estimate and optimal control system" as shown in Figure 11.1. Finally, the selected manufacturing processes will be added into the product production structure as part of the product data.

The "operations management system" provides information on the production schedule, manufacturing resource availability or plan, and inventory of raw materials or/and components, which is relevant to a manufacturing process as planned by the manufacturing process planning and optimisation system. It should be pointed out that the operations management system manages the necessary manufacturing resources and materials not only inside a manufacturing company but also from the company's suppliers, partners and subcontractors via the logistics management system. The data managed and provided by this sub-system is in a global manufacturing environment. These global operations management data will be added into the product production structure as a part of the product data together with the designed feature and planned processes.

The logistics management system consists of a database called "logistics knowledge base" to record all the necessary data of the company's suppliers, partners, and subcontractors, such as their addresses, products (used as parts or components) or services (used as possible outsourced processes), prices, delivery times, quality/reliability/reputation (measured by "excellent", "good", "average", and "poor"), manufacturability (*e.g.*, special equipment, unique experience or process, special surface treatment, *etc.*), and transportation means and costs. This logistics knowledge base can be accessed and retrieved via intra/internets (Tu and Xie 2000). In addition to this knowledge base, the logistics management system also consists of an "enquiry and quotation" process model, a web page management model and an automatic e-mail sending, receiving and processing system. The enquiry and quotation process model can be used to formulate an enquiry on the company's web page for purchasing a material, component, or a manufacturing process from possible suppliers, partners or subcontractors globally. It links directly with the logistics knowledge base. Through opening part of the logistics knowledge base to the public, the possible suppliers, partners or subcontractors can directly "write" their responses into the logistics knowledge base, which can be further sorted and processed by the enquiry and quotation process model to select the best supplier, partner or subcontractor. The web page management model is used to manage the company's web page. The automatic e-mail sending, receiving and processing model manages and controls the company's communications with its suppliers, partners, and subcontractors. Normally, these e-mails are in a standard format, which looks like a data table but with hyperlinks. They are automatically processed by the model and the necessary data is extracted from these reply e-mails, such as company's names and detailed addresses, costs/prices of the materials or processes, delivery times, transportation means, sample products or past works (in forms of electronic photos or engineering drawings) to show their ability to provide a wanted component or manufacturing process, *etc.* These logistics data are added into the product production structure together with the feature, manufacturing processes, and operations management data.

The "cost estimate and optimal control system" consists of a cost index structure, a knowledge base and some cost optimisation models. It will be further discussed in the following sections.

The computer aided PD system as shown in Figure 11.1 can be operated by a product designer or developer to design a product, to plan how to make the product and to quote the cost of the product concurrently or incrementally in feature by feature manner. Since all the subsystems described above communicate with each other through intra/internets, several engineers at different places in a company or the world who collaborate in the development of a customised product can also use this PD system. We believe this system is much more efficient and advanced than today's commercial CAD/CAM systems which are widely used to support product design in the manufacturing industry.

Before discussing the cost estimate and optimal control system, I would like to define PD cost:

$$\text{PD cost} = C_{dd} + C_p \tag{11.1}$$

where C_{dd} is product definition and design cost and C_p is product production cost.

Product definition and design cost is determined by the time consumed by the product design engineers. To directly control these times does not make any sense since the main goal of product definition and design is to design a product with high quality. A high quality product means the product will provide maximum satisfaction for the customer. Hence a linear programming model was developed to maximise the customer's satisfaction under the constraints of customer's requirements, technical constraints, resource constraints and a given product definition and design budget (measured in dollars), which is normally decided by the company's management according to the company's marketing strategy and past experience (Tu *et al.* 2002). This means the product definition and design cost will not be estimated and controlled by the cost estimate and optimal control system. The costing subsystem in the PD system as shown in Figure 11.1 estimates and controls the total product manufacture and logistics cost.

The product production cost is the sum of the costs of all the manufacturing processes. The cost of a manufacturing process includes:

1. Test and possible rework fees due to the difficulty of the process.
2. The total consumption of the necessary manufacturing resources to carry out the process, including the consumptions of workforce, machines, and tools.
3. The total logistics cost for purchasing and/or transporting raw material, component or outsourced manufacturing processes, and relevant logistics management cost (*e.g.*, communication cost and paper/data management cost). If a manufacturing process is outsourced or subcontracted, it only includes this logistics cost. If a raw material or component is within the company, the logistics cost for this raw material or component equals the price of the material or component plus inventory and material handling cost in the company. In short, to simplify the cost estimate and control process in a global environment, material handling and purchasing cost are considered to be part of the logistics cost.

11.4 Cost Index Structure

To estimate and control the manufacture cost and logistics cost of a PD, a cost index structure was developed, and is illustrated in Figure 11.2. In accordance with the product production structure, which is a hierarchical common object data model, the cost index structure adopts the same hierarchy as the product production structure.

As shown in Figure 11.2, a manufacturing process is associated with a cost index. A cost index includes three generic data classes, viz. adding, resource and logistics, which are associated with the possible three costs of a manufacturing process as discussed at the end of Section 11.3. "Adding" means adding cost to a manufacturing process. It is used to pay for the possible test and rework due to the difficulty of the process. The difficulty of a process is determined by the design of a product, part, component or feature, *e.g.*, geometry, surface finishing or treatment requirements, special tolerance and machining requirements, *etc.* "Resource" means resource cost or consumption of manufacturing resources as discussed in Section 11.3. "Logistics" means logistics cost as discussed in Section 11.3. Likewise in the product production structure, these cost data classes are written in EXPRESS from STEP so that they can be accessed and retrieved by different computer software systems.

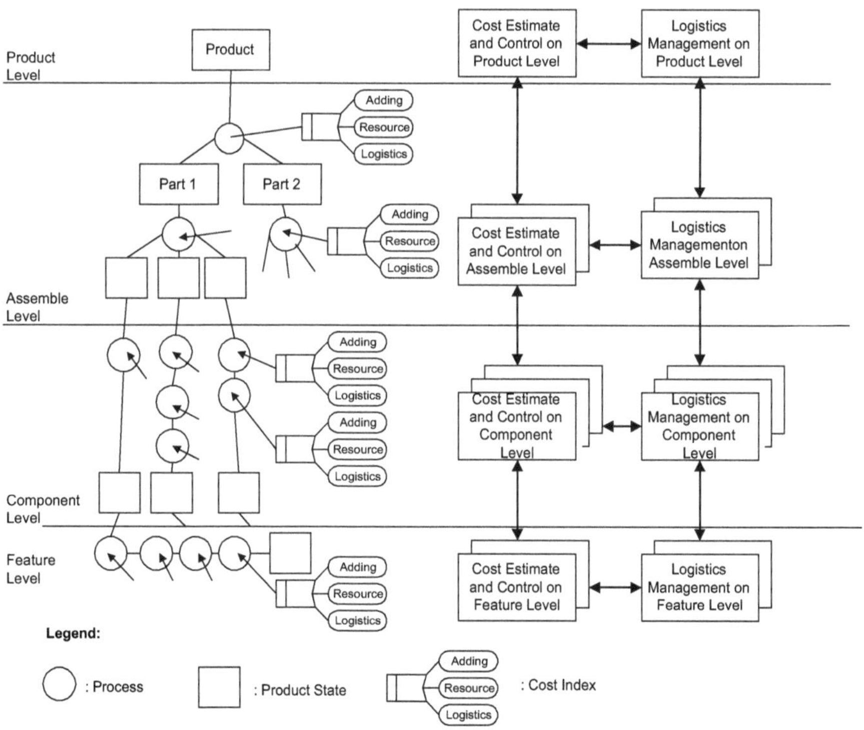

Figure 11.2 Cost index structure

The idea to build the cost index structure in a hierarchy is to keep in line with the overall product production structure. This is also a natural or logical way to estimate, control and record the cost of a product production, from bottom up (feature level) to the top product level. It is obvious that the cost of a feature, which is added to a component, is a part of the cost of the component. Likewise, the costs of components that are used to assemble a part are a major part of the part cost, and the product cost is determined by the costs of the parts that are finally assembled into the product. Hence, the cost estimate and optimal control system estimates the cost of the production of a product from the bottom level up to the top level of the hierarchy, as shown in Figure 11.2. The models, *e.g.*, cost estimate and control models on the component level, will try to estimate or search for the cost data (*i.e.*, adding, resource, and logistics) for the generic cost index associated with each manufacturing process by using the cost data from the cost modules at the lower level (*e.g.*, cost estimate and control models at feature level), the costing knowledge base, and the logistics management system. As shown in Figure 11.2, the logistics management system also includes a number of logistics models which are hierarchically organised from the feature level at the bottom up to the product level at the top.

Using the cost index hierarchy as shown in Figure 11.2, a product production cost can be decomposed from level to level into detailed expenditures or indexes.

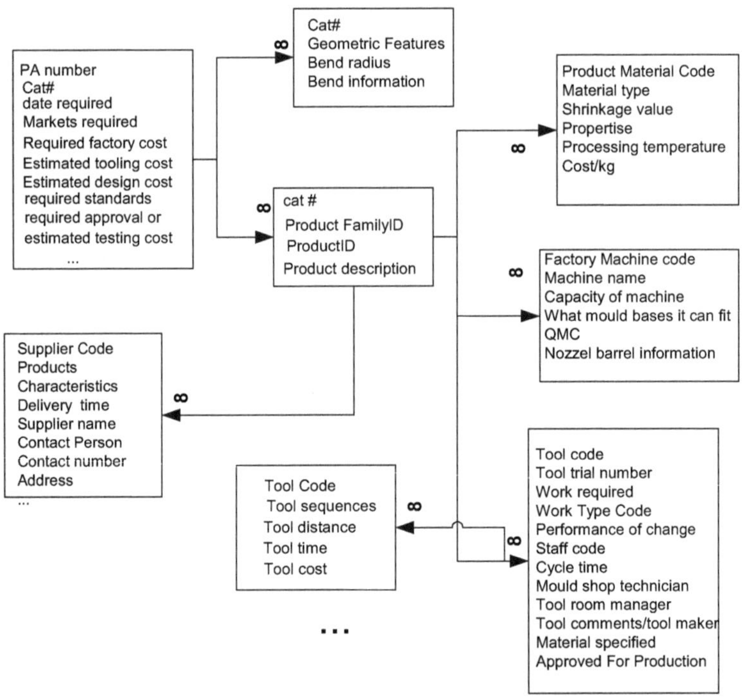

Figure 11.3 Costing knowledge base for sheet metal manufacturing

In fact most of these expenditures on lower levels of the cost index structure can be directly estimated using the price of a standard component, part, or the cost of a common manufacturing process. To minimise PD cost design engineers always try to design the product using as many of these standard components, parts and common manufacturing processes as possible. Also in practice, the costs of quite a few manufacturing processes normally on a higher level of the cost index structure can be derived using linear combinations of these standard prices or costs of common processes. Hence, a costing knowledge base was developed as an important part of the cost estimate and optimal control system to structurally record these standard prices and costs of common processes. However, it is difficult to propose a general data structure to build such a costing database for all kinds of manufacturing businesses. The data structure of a costing knowledge base for a particular manufacturing business (*e.g.*, sheet metal manufacturing) is related to the special manufacturing requirements and practices in that type of manufacturing. Figure 11.3 shows a costing knowledge base that was developed for sheet metal manufacturing. It is obvious that this costing knowledge base does not fit, say, mould/tool manufacturing.

However, the cost index structure as proposed in Figure 11.2 is generic, and can be applied to all kinds of manufacturing. To be in line with this cost index hierarchy, the costing knowledge base needs to be developed to store costing data in a similar hierarchical structure, *i.e.*, the data tables in the database are linked and indexed from a table, which records product data information in top down data tables that record the features data information at the bottom.

11.5 Cost Estimation and Optimisation

Using the cost index structure, the cost estimation and optimisation of the cost estimate and optimal control system can be described in the following two stages.

Stage 1. *Cost estimate*. In this stage, the system needs to estimate the various costs of the cost indexes associated with the processes (see Figure 11.2) from the feature level at the bottom up to the product level at the top. The inputs to the system at this stage include manufacturing processes (including all the possible alternatives) planned by the manufacturing process planning and optimisation system (see Figure 11.1), manufacturing resources (including all the possible alternatives) and their cost rates and inventory records from the operations management system (see Figure 11.1), quotations or prices from all possible suppliers, partners and subcontractors, which are searched and provided by the logistics management system (see Figure 11.1), and the costs or prices of these standard components and manufacturing processes from the costing knowledge base. With these input data, the cost estimate and optimal control system will estimate costs or instances for these generic cost indexes associated with the planned processes and their alternatives using two cost estimate methods, viz. the "generative cost estimate method" and the "variant and knowledge based cost estimate method". These two methods are described in the following subsections. The cost estimate is an incremental or evolutional process, which estimates the cost indexes feature by feature in line with the overall incremental product design, and production planning process as implied by the concept of prototype-based incremental PD (Tu *et al.* 2000c). It is not developed to estimate various costs of making a product in one go after a product has been completely designed and planned for production. This incrementally designs a product, plans the manufacturing processes, schedules manufacturing resources and inventory, manages logistics and estimates costs feature by feature from the bottom level of a product production structure up to the top level, and can provide the overall PD system with strong flexibility to cope with the changes and uncertainties in early product design stages. In particular, this incremental way of developing a product and estimating the cost can provide in-time feedbacks or references to support engineers' decision-making on the selection of alternative designs, manufacturing processes, manufacturing resources and alternative suppliers, partners and subcontractors. The shortages of this incremental method are the lack of global optimisation, and frequent and complicated communication management between these subsystems as shown in Figure 11.1.

Stage 2. *Cost optimisation*. Cost optimisation means selection among alternative designs, manufacturing processes, manufacturing resources, inventories, suppliers, partners, and subcontractors, or combinations of these alternatives with lowest cost or total costs. Manufacturing process optimisation, *e.g.*, to find the shortest path for CNC cutting, and operations optimisation, *e.g.*, an optimal production schedule with the shortest makespan, are not the tasks of cost optimisation. They are optimising tasks of other subsystems, as shown in Figure 11.1, *e.g.*, the manufacturing process planning and optimisation system for sheet metal manufacturing includes a shortest CNC path planning model as addressed in Chapter 9, and the operations management for sheet metal manufacturing includes an optimal production scheduling model (Tu 1997). It is obvious that these optimisations are carried out before the cost optimisation, as they will help cost reduction. For cost optimisation, an optimal algorithm was developed for the selection of alternative operating routines and supplier chains using dynamic programming (Tu *et al.* 1997). More generally, a commercial computer simulation package, *e.g.*, ProModel (by ProModel Corp., Orem, UT), Quest and VNC (by Deneb Inc., Dayton, OH), is used to develop simulation models to solve cost optimisation problems. If a manufacturing process or a logistic chain consists of random variables or stochastic processes, Markov chains (Zijm 1984) may be used to model the process or the logistic chain and hence a cost optimisation heuristics or algorithm could be developed. This is now under research.

It should be mentioned that cost optimisation may be in conflict with other optimisations. An operational routine suggested by the cost estimate and optimal control system from the lowest cost point of view may be in conflict with a production schedule suggested by the operations management system from the minimum makespan point of view. This in fact leads to a problem of decision-making with multiple objectives, lowest cost vs minimum makespan. The problem of decision-making with multiple objectives can be solved using rule-based or heuristics/algorithm-based methods, such as the analytic hierarchy process (AHP), pareto optimality and trade off curves, and goal programming (Winston 1994).

11.5.1 Generative Cost Estimate Method

The generative cost estimate method is a pump line method. With the cost index structure as shown in Figure 11.2 and the input data from all the other systems as well as the price and cost data from the costing knowledge base, the cost estimate and optimal control system can calculate the data or instances of the cost index associated with a process from the bottom level up to the top level of the cost index structure. The cost index of punching a hole in sheet metal (a process at the feature level), for example, can be determined by calculating adding cost, resource cost and logistics cost. As mentioned before, the adding cost is decided according to the difficulty of the process, *i.e.*, punching a hole in sheet metal for this example, and it is the sum of the costs of the possible tests and reworks. The resource cost for this example will be calculated by multiplying the standard cost rates of the manufacturing resources used (*e.g.*, punch machine, tool and machine operator) by the times of these manufacturing resources used. The standard cost rates were recorded in the costing knowledge base and the times used can be provided by the manufacturing process planning and optimisation system. The logistics cost is determined according to whether this process needs to take a material or component into consideration and how and where this material or component can be provided. To avoid the duplicated cost data in these cost indexes, only the first process, which is amongst a group of sequential processes to be carried out on a blank or a sheet metal, will take the logistics cost of the blank or sheet metal into account. All the other succeeding processes will not take the logistics cost of this blank or sheet metal into account in their cost indexes under the logistics, or a zero (0) will be assigned to the logistics. However, if a succeeding process needs to add a new material or component, under the logistics of its cost index, the logistics cost of this new material or component will be estimated and recorded. As mentioned before, the logistics cost is the sum of the costs of the purchased item, transportation (material handling plus inventory cost if the item is in the company), and communication. Normally, a supplier will quote the transportation cost as a part of the quotation for providing an item. Hence the logistics management system can provide the costs of the purchased item and relevant transportations by enquiring for a quotation from suppliers. The communication cost for sourcing items depends on the difficulty of sourcing this item. According to previous experience, the management of a company can set up some guidelines or controlled budgets for sourcing various items. These controlled budgets are recorded in the costing knowledge base and can be used by the cost estimate and optimal control system to quote or estimate the communication cost as a part of the logistics cost of getting a required item. If a process is outsourced, *e.g.*, a subcontracted process, it has only the logistics cost in its cost index and the other two terms will be zero.

When all the cost indexes have fixed data or instances, the production cost of a component, part or product can be calculated by the cost estimate and optimal control system through adding up all the cost data in the cost indexes associated with these processes which are under and linked (directly or indirectly) to the component, part or product in the cost index structure.

11.5.2 Variant and Knowledge-based Cost Estimate Method

As implied by the discussions in the previous section, the generative cost estimate method can only be used to estimate the costs of standard components and processes. For non-standard or "new" components and processes that are required by the product design to address particular customer requirements, the variant and knowledge based cost estimate method needs to be employed. This method estimates the cost of a component or process through searching and comparing the similarities between the developing product and the products that were made in the past. As shown in Figure 11.1, all the products previously made would have product production structures that are integrated product data models. According to the discussions in this chapter, it is obvious that these product data models include cost indexes for all the manufacturing processes. In fact, after a product is made in the company, its final version of product data model (or product production structure) is recorded in the product database, which is the main part of the design/manufacturing knowledge base of the PD system, as shown in Figure 11.1. By searching this design/manufacturing knowledge base, the cost estimate and optimal control system may find the cost index of a process, which is in a previously-made product production structure. This is similar to the process used in the development of the product. The cost estimate and optimal control system search for a similar process by following a GT (group technology) coding system (Tu *et al.* 2000c) and the product production structures are stored and retrieved in the design/manufacturing knowledge base according to the same GT coding system. Similar processes in the design/manufacturing knowledge base have the same process codes. If a process in the new product is a combination of several primary processes, the cost estimate and optimal control system will search for the cost indexes of the these primary processes and then determine the cost index for this combined process through adding up the cost indexes of these primary processes. A reference number rather than a GT code will be assigned to this combined process that is stored in the design/knowledge base as a part of the product production structure. It will not be re-used or searched by the cost estimate and optimal control system or the manufacturing process planning and optimisation system.

11.5.3 Considerations of Cost Estimate

Based on the experience gained from this research and industrial implementations, by using the two cost estimate methods as discussed in Sections 11.5.1 and 11.5.2, the cost estimate and optimal control system can estimate cost indexes for most planned manufacturing processes. However, there are still some processes that cannot be estimated automatically using either method. In this circumstance, the system shows these processes on the screen and asks the product developers to determine cost indexes for these processes. For a combined process, if the system cannot find the cost indexes for some of the primary processes from the design/manufacturing knowledge base, the system shows these primary processes on the screen, waits for a manual costing for these primary processes, and then decides the cost index for the combined process.

Another problem for cost estimating is that the cost of a component or outsourced process is within a fuzzy cost domain, *i.e.*, from a lowest to a highest cost. In this circumstance, a neural-network-based algorithm plus some evaluation rules can be used to infer a most likely cost of the component or the outsourced process. Tu and Jiang (1997) discussed this algorithm and the evaluation rules.

The third problem in cost estimation and optimisation is a conflict between cost optimisation and the reliability of suppliers. In practice, the cheapest cost normally implies an unreliable supplier. The management of a company can solve this problem through choosing proper quality and marketing policies, *e.g.*, only accept quotations from reliable suppliers, partners or subcontractors.

11.6 Implementation of the Cost Estimate and Optimal Control System

The proposed cost estimate and optimal control system has been implemented in a sheet metal manufacturing company. The main principles of the cost estimation and optimisation system are discussed using the example of development of a sheet metal part as shown in Figure 11.4.

11.6.1 Example Scenario and System Description

As shown in Figure 11.4, the example is a sheet metal part which is part of a sheet metal product. To simplify the discussions here, I show only this sheet metal part and

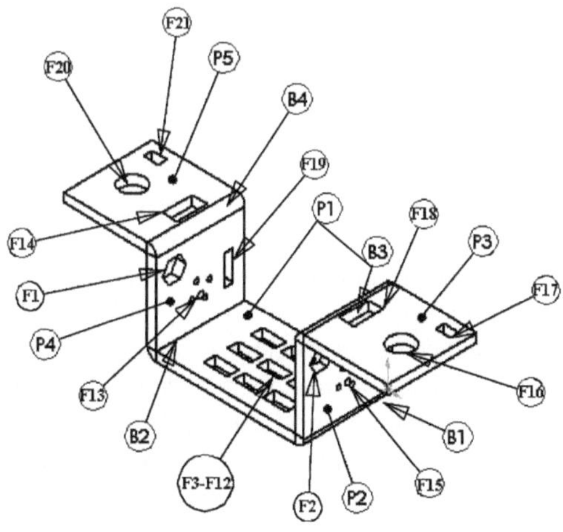

Figure 11.4 Example part

not the whole sheet metal product. However, the readers can imagine that through a number of assembly processes this part will be finally assembled into the product.

This sheet metal part can be made by using CNC sheet metal punching machines or a CNC laser-cutting machine to punch or cut these holes, rectangles, and polygons on a flat metal sheet, and it is then formed by a bending machine into the final shape. The punching or cutting processes for making these holes, rectangles and polygons are indicated by F1 through F20, as shown in Figure 11.4. The part starts as a flat sheet that includes only one face, as indicated by P1 in Figure 11.4. After four bending processes, as indicated by B1 through B4 as shown in Figure 11.4, the final part will have five different faces, *i.e.*, P2 through P5 as shown in Figure 11.4, in addition to P1.

The overall PD system as its structure illustrated in Figure 11.1 includes the following subsystems:

1. Computer aided product definition system: this system was coded using Visual C in association with Microsoft Excel. By using the QFD functions, the system converts the customer's requirements into the technical attributes of the product, as addressed in Chapter 4.
2. Computer aided product design system: Pro/ENGINEER is used as the computer aided product design system. However, a prototype search model and an interface model were coded using C and integrated with the Pro/ENGINEER system through its application program interface (API). The prototype search model is able to search for prototypes, *i.e.*, products made in the past, which have the most similar technical attributes to the product to be developed. These prototypes are modified into the customer-wanted product according to the concept of prototype-based incremental PD. The interface model is used by Pro/ENGINEER to read and write the product production structure.
3. Manufacturing process planning and optimisation system: this system was coded in C++. To meet the special requirements of sheet metal manufacturing, this system includes a nesting model and a CNC tool path-planning model in addition to the process-planning model. The nesting model can be used to automatically plan a cutting layout so that the trim loss can be minimised. The CNC tool path model was developed to optimally plan the CNC cutting or punching paths so that the total cutting time can be minimised. A screen printout of this system is shown in Figure 11.5 (Xie *et al.* 2001a).
   Figure 11.5 shows a tool path planning process to be automatically carried out by the system. The parts as shown in Figure 11.5 have been nested by the system, *i.e.*, properly placed on a given size of sheet so that the material waste or trim-loss is minimised. The part as shown in Figure 6.4 does not need the nesting optimisation but it needs the tool path planning so that these features could be punched or cut efficiently.
4. Operations management system: this was coded in C++ and used to schedule shop floor operations and control the inventory through the communication with the company's MRP (manufacturing resource planning) or ERP (enterprise resource planning) system.

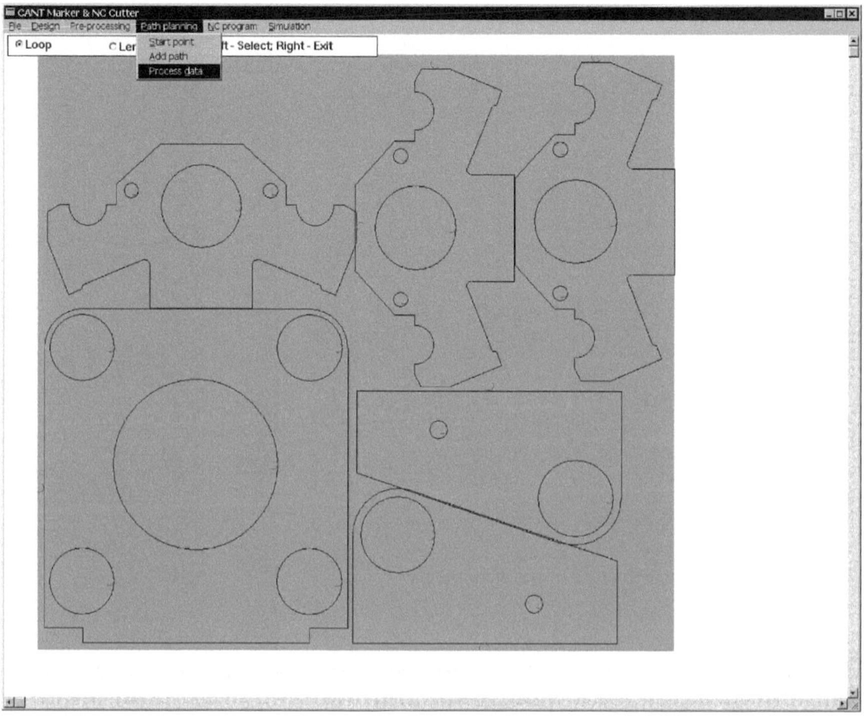

Figure 11.5 A screen printout from the manufacturing process planning and optimisation system for sheet metal parts

5. Logistics management system: this was coded in Java. The HTML and XML languages as well as common gateway interface (CGI) were used to write the web pages and to manage communication through computer intra/internets (Tu *et al.* 2000a).
6. Cost estimate and optimal control system: this was code in C++ based on the cost estimate and optimisation methods and algorithms as discussed or mentioned in Section 11.5.

All these subsystems can communicate with each other through computer intra/internets.

11.6.2 Cost Estimate and Optimisation for Making the Example Part

First, based on the generic cost index structure as shown in Figure 11.2, the cost index structure for the case part can be drawn and illustrated in Figure 11.6. As shown in Figure 11.6, the part can be made by sequentially carrying out 24 punching or

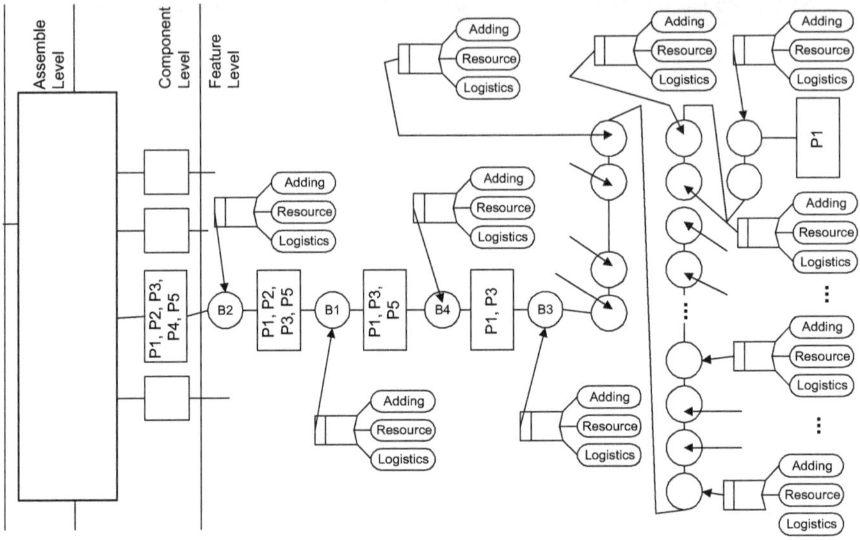

Figure 11.6 Cost index structure of the example part

laser cutting and sheet metal forming processes. Some product states have been inserted into the manufacturing process chain. "P1" means the blank sheet which has one face P1. After the folding process B3, the semi-finished part will have two faces, P1 and P3. After the other forming processes, the part will have its final shape that includes 5 faces, P1 through P5. The manufacturing process planning and optimisation system determines the sequence of the processes as shown in Figure 11.6.

The first process, *i.e.*, F1 as shown in Figure 11.6, consists of the logistics cost in its cost index. The rest of the punching or cutting processes will not include a logistics cost if they can be carried out in one machine and do not need to be transported to other machines. The first folding process, B3, includes a logistics cost in its cost index. This is the cost for transporting the semi-finished part from the punching or laser cutting machine to a folding machine. The logistics cost of the rest of the folding processes will be zero if these folding processes can be completed in one folding machine.

In short, by using the methods (including the manual cost estimate) as discussed in Section 11.5, the cost indexes for all the manufacturing processes can be decided. After these cost indexes have been estimated, the cost estimate and control system carries out the cost optimisation if it is necessary. As mentioned in Section 11.5, the cost estimation is in fact a selection among alternative designs, manufacturing processes, or operation routines based on cost optimal (or cheapest cost) criterion.

First, let us assume that the 20 machining features (*i.e.*, F1 through F20 as shown in Figure 11.4) can be carried out in two alternative ways, *i.e.*, 1. using a CNC punch machine to punch all 20 features, and 2. using a laser cutting machine to cut the two polygons (F1 and F2) first and then using a punch machine to punch the rest of the holes and rectangles. The first alternative has the advantage of completing all processes in one machine, which saves the material handling cost moving between two machines, but will need a special punch tool to punch the two polygons. The second alternative can let the company avoid having a special punch tool but it implies a material handling cost or logistics cost moving between the laser cutting machine and the punch machine, particularly if this laser cutting machine is an outsourced facility, or a subcontracted process. If one ignores other technical factors and uses only the cost factor as a reference to make a selection between these two alternatives, this can be done by using the cost estimate and optimal control system in the following way:

1. the cost index structure as shown in Figure 11.6 is modified as in in Figure 6.7, which includes alternative processes;
2. the cost estimate and optimal control system estimates costs and decides the cost indexes for all processes;
3. calculate the total costs for making the part by two process alternatives according to the calculation process as described in Section 11.5.1. Hence a decision can be made according to these two possible costs.

Second, if the part shown in Figure 11.6 can be designed in another form, *i.e.*, an alternative design, the cost index structure as shown in Figure 11.6 can be modified to include alternative designs (or parts) at the component level. Then these alternative designs are decomposed into processes at the feature level. After this modification of the cost index structure, which can be simply imagined and is omitted from this chapter, the cost estimate and optimal control system estimates the cost indexes for all the processes of the two different designs, and calculates the costs of the two al-

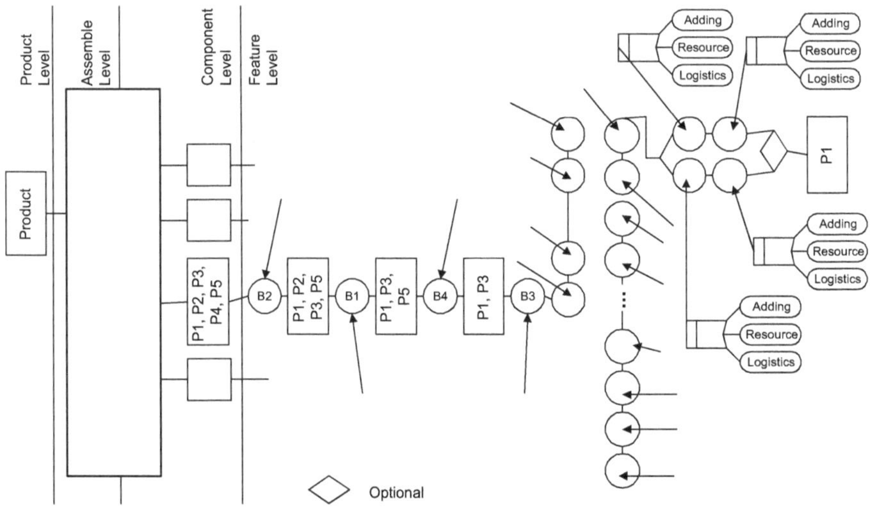

Figure 11.7 Cost index structure which includes alternative processes

ternative designs, just the same as it does to make a suggestion to select alternative processes, as mentioned in the previous paragraph.

Third, if a part can be made using alternative machines and alternative operational routines, as shown in Figure 11.8, the dynamic programming cost optimisation model in the cost estimate and optimal control system can be used to select a routine to make the part that has the minimum total cost. Before the cost estimate and optimal control system can carry out the dynamic programming calculations, it needs to treat these processes on alternative machines as alternative processes, record them in the cost index structure, and then estimate the cost indexes for all these alternative processes. Based on the cost indexes of these alternative processes, the system can calculate the cost for a pair of machines, which equals the machining cost on the next machine plus the material handling cost (or logistics cost) for moving between these two machines. After all the costs of machines in the network, as shown in Figure 11.8, are calculated by the system, the system uses the dynamic programming technique to find a shortest path from the start point to the finish point or an operating routine with minimum total cost.

This dynamic programming optimisation method can also be applied to select the cheapest logistics chain among a number of alternative chains if the machines as shown in Figure 11.8 are thought of as suppliers, partners or sub-contractors. For a more complicated case on a manufacturing shop floor or a global manufacturing environment, computer simulation technology, *e.g.*, ProModel and Quest, can be employed for the cost estimation and optimisation. Tu *et al.* (2000b) reported an application of computer simulation technology to estimate and optimally control the cost of injection mould-/tool-making on a manufacturing shop floor.

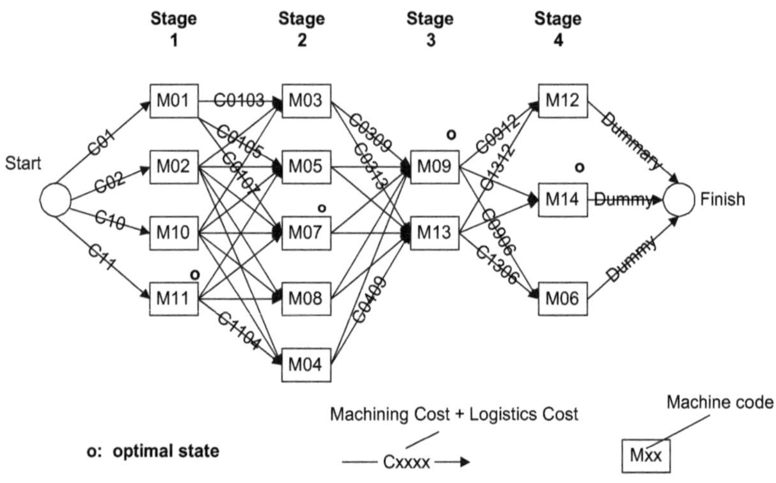

Figure 11.8 Dynamic programming network for making a part through alternative operation routines

11.7 Conclusion

The ability to correctly and rapidly estimate and optimally control PD cost is always a strength pursued by manufacturing companies to improve their place in a competitive global market. Numerous research efforts have been made to automate cost estimation and optimisation in a PD cycle although the research and development task is tedious and enormous sometimes. This chapter presented a feasible computer aided solution to automate the cost estimation and optimisation of PD in a global manufacturing environment. By following the concept of prototype-based incremental PD, it is able to provide in-time cost estimates and optimisation results to the product developers as important decision-making references in a concurrent approach to PD. This concurrent approach enables product developers to design the product, plan the processes, schedule operations on the shop floor, manage logistics, and estimate and optimally control the cost feature-by-feature. This incremental way of developing a product can help product developers to clarify the PD life cycle data at an early PD stage and hence avoid rework, which often occur in the traditionally sequential approach to PD. Avoiding rework in a PD cycle, particularly for a customised product, will reduce the PD cycle time and cost significantly.

The proposed cost index structure is generic and can be applied to all kinds of PD. It is modelled in EXPRESS from STEP as part of the product production structure. Hence it can be assessed by all kinds of computer aided engineering or management software systems. It provides links or indexes to associate various costs to product features through the manufacturing processes that are used to create these features. Using this method of recording the various costs of PD, cost analysis and optimisation in a PD cycle can be easily and automatically done by the cost estimate and optimal control system as presented in this chapter. This cost index structure, as part of the product production structure stored in the design/manufacturing knowledge base, can be further used as knowledge and references for future PD. It is very important for a manufacturing company to record its knowledge and past experience so that they can be reused to improve the company's practices in PD.

The cost estimate methods and optimisation algorithms are effective. They can easily be understood and implemented in industry. The cost estimate and optimal control system as well as other subsystems of the PD system run in an intranet/Internet communication environment. This makes these systems or subsystems run distributed in a computer network. It also improves the system's interdependence and reliability, and at the same time reduces the complexity of the design and development of such systems. To use STEP along with computer intranet/Internet communication technology to develop these systems or subsystems makes these systems compatible with commercial computer aided engineering or management software systems, such as Pro/ENGINEER, SmartGroup's PDM (product data management) system, *etc.*

References

Becker, J. and Prischmann, M., 1993, Supporting the design process with neural network complex application of cooperating neural networks and its implementation, Journal of Information Science and Technology, 3, 79–95.

Bernard, A. , Perry, N., Delplace, J. C., 2007, Concurrent cost engineering for decisional and operational process enhancement in a foundry, International Journal of Production Economics, 109(1–2), 2–11.

Bode, J., 1998, Neutral networks for cost estimation, Journal of Cost Engineering, 40, 25–30.

Bode, J., 2000, Neutral networks for cost estimation: Simulations and pilot applications, International Journal of Product Research, 38(6), 1231–1254.

Castagne, S., Curran, R., Rothwell, A., Price, M., Benard, E., Raghunathan, S., 2008, A generic tool for cost estimating in aircraft design, Res Eng. Design, 18(4), 149–162, DOI: 10.1007/s00163-007-0042-x.

Casey, J., 1987, Digitizing speeds cost estimating, American Machinist and Automated Manufacturing, 131(1), 71–73.

Cawthorne-Nugent, M., Vieira J. D. L. and Watson, P. A., 1989, An intelligent knowledge-based system for cost estimating in the make-to-order environment, Computer-Aided Engineering, 6(4), 121–127.

Goldberg, J., 1987, Rating cost-estimating software, Manufacturing Engineering, 98(2), 37–41.

Geiger, T. S. and Dilts, D. M., 1997, Automated design to cost: integrating costing into design decision, Journal of Computer Aided Design, 29(6/7), 423–438.

Ju, B. and Xi, L.F., 2008, A product cost estimation for the early design of sedans using neural networks. International Journal of Automotive Technology and Management, 8(3), 331–349.

Horvath, P., Niemand, S. and Wolbold, M., 1994, Target costing: state of the art review. Manufacturing Competitive Frontiers, 18(3), 30–40.

Hubert, K., Brinke, E. T., Lutters, E. and Streppel, T., 1999, Integrated Cost Estimation Based on Information Management Univ. Ternte, Lab. Design, Production, and Management, Twente, The Netherlands, Available at: http://opm.wb.utwente.nl/staff/erik/phd/wgp99.pdf.

Kamarthi, S. V., Cohen, P. H. and Demetter, E. C., 1993, CMES: Cost minimisation and estimation system for sheet metal parts, International Journal of Flexible Automation Integrated Manufacturing, 1(1), 69–79.

Kawada, M. and Johnson, D. F., 1993, Strategic management accounting – why and how, Manage. Accounting, 75(2), 32–38, 1993.

Koltai, T., Lozano, S., Huerrero, F. and Onieva, L., 2000, A flexible costing system for flexible manufacturing systems using activity based costing, International Journal of Product Research, 38(7), 1615–1630.

Marx, W. J., Mavris, D. N. and Schrage, D. P., 1998, Knowledge based system integrated with numerical analysis tool for aircraft life cycle design, Artificial Intelligence for Engineering, Design, Analysis and Manufacturing, 12(3), 211–229, 1998.

Monden, Y. and Lee J.Y., How Japanese auto makers reduce costs, Manag. Acc., 75(2), 22–26, 1993.

Nadeau, M. C., Karb, A., Rotha, R., Kirchain, R., 2010, A dynamic process-based cost modeling approach to understand learning effects in manufacturing. International Journal of Production Economics, Volume 128, pp. 223–234.

Ostwald, P., 1984, Cost Estimating, 2nd edn. Englewood Cliffs, NJ: Prentice Hall.

Ou-Yang, C. and Lin, T. S., 1997, Developing an integrated framework for feature-based early manufacturing cost estimation, The International Journal of Advanced Manufacturing Technology, 13(9), 618–629.

Romero Rojo, F., Roy, J. R. and Shehab, E., 2009, A Methodology for Variability Reduction in Manufacturing Cost Estimating in the Automotive Industry based on Design Features, Proceedings of the 19th CIRP Design Conference – Competitive Design, Cranfield University, 30–31 March, p. 197.

Smunt, T. L., 1999, Log-linear and non-log-linear learning curve models for production research and cost estimation, International Journal of Product Research, 37(17), 3901–3911.

Tamas, K., Lozano, S., Guerrero, F. and Onieva, L., 2000, A flexible costing system for flexible manufacturing systems using activity based costing, International Journal of Product Research, 38(7), 1615–1630.

Tu, Y. L., 1997, Production planning and control in a virtual OKP company, Computer in Industry, 34, 271–283.

Tu, Y. L. and Jiang, Z. B., 1997, A neural network based approach for the quotation of a product in Oneof-a-Kind Production, in Proc. 1st Int.Conf. Engineering Design and Automation, Bangkok, Thailand, March 18–21, 1997, pp. 307–310.

Tu, Y. L., Kam, J. J. and Xie, S. Q., 2002, Rapid one-of-a-kind PD in a global and virtual manufacturing environment, presented at the International Conference of Advanced Production Management Systems (APMS 2002), Eindhoven, The Netherlands, Sept. 8–13.

Tu, Y. L., Chu, X. L. and Yang, W. Y., 2000a, Computer aided process planning in virtual one-of-a-kind production, Computer in Industry, 41, 99–110.

Tu, Y.L., Chu, X.L. and Yang, W. Y., 2000b, Computer Aided Process Planning in Agile One-of-a-Kind Production, Computer in Industry, 41, 99–110.

Tu, Y. L., Kam, J. J., Fung, R. Y. K and Tang, J. F., 2002, Resource deployment in earlier PD stages, in Proc. 4th Int. Symp. Tools and Methods of Competitive Engineering (TMCE 2002),Wuhan, China, April 22–26, pp. 509–518.

Tu, Y. L., Xie, S.Q. and Fung, R. Y. K., 2000, Rapid PD in a virtual production environment, in Proc. IFAC Symp. Manufacturing, Modeling, Management and Control (MIM 2000), Patras, Greece, July 12–14, pp. 63–68.

Tu, Y. L. and Xie, S. Q., 2000, A WWW based Integrated Product Development Information Management System, IFAC, MIMO2000, Greek, July.

Wang, C. H. and Bourne, D. A., 1995, Using features and their constraints to aid process planning of sheet metal parts, IEEE International Conference of Robotics and Automation, pp. 1020–1026, 21–27 May 1995, Nagoya, Japan.

Winbourne, J. P. and Toolsie, G. M., 1991, Computer-aided tool cost estimating (CATE), in Proceeding 1991 ASME International Computers in Engineering Conference and Expo., Computer Engineering, 1, 617–621.

Xie, S. Q., Tu, Y. L., Aitchison, D., Dunlop, R. and Zhou, Z. D., 2001, A WWW based Product Development Platform for Intelligent and Concurrent Sheet Metal Products Design and Manufacturing. International Journal of Product Research, 39(6), 3829–3852.

Winston, W. L., 1994, Operations Research: Applications and Algorithms, 3rd edn. Belmont, CA: Duxbury, 772–818.

Zhang, Y. F. and Fuh, J. Y. H., 1998, Neural network approach for early cost estimation of packaging products, Computers & Industrial Engineering, 34(2), 433–450.

Zijm, H., 1984, The Optimality Equations in Multichain Denumerable State Markov Decision Processes with the Average Cost Criterion: The Unbounded Cost Case, C. Q. M. Note 22. Eindhoven, The Netherlands: Philips B.V., Center for Quantitative Methods.

Chapter 12
A Global Data Structure for Supporting Rapid Product Development

Abstract This chapter reports the research work that aims to develop an integrated data structure to support rapid product development (RPD) in the Internet environment. The emphasis is placed on integrated data management and the reuse of past PD experience to support a company's aim to shorten its PD cycle. The integrated global data structure model was modelled using EXPRESS from STEP with the consideration of real-time data communication within the Internet environment. In terms of this data structure, a design/manufacturing knowledge base was developed as a major part of the WWW (World Wide Web)-based PD system. The basic principles and concepts of the knowledge base and the WWW-based knowledge management system are presented in this chapter. An industrial implementation is also reported.

12.1 Introduction

Today an increasing number of manufacturing companies have realised that the ability to quickly develop a customised product in an economic and efficient way is critical for them to survive in the keenly competitive international market. It has been widely conceived that the capability of RPD is one of the key issues that need to be considered to enhance the competitiveness of a company.

In order to quickly and successfully deliver new OKP products to the market, the development of innovative products needs to be accelerated. This may require the support of new technologies (*i.e.*, information technologies, Internet technology), in particular for companies whose business partners are distributed throughout the world. In past decades, the computer and computer communication technologies have been widely applied to shorten the PD cycle, *e.g.*, the use of powerful CAD technologies, information technology, optimisation technology, virtual reality (VR) and virtual manufacturing technologies. Asynchronous transfer mode (ATM) networks and fast Ethernet enable quick and reliable exchange of relevant data and thus have changed the PD process tremendously. The Internet provides a new "platform" to access relevant information from all over the world in near real-time. Communication and cooperation among the members of a PD team located in different areas of the world, for instance, can be realised by computer-supported cooperative work (CSCW) tools via the World Wide Web.

However, as addressed in Chapter 2, how these technologies can be integrated to support RPD is still a problem. One of the reasons for not being able to solve this problem is that there is no common integrated product data structure that can be combined with these new technologies and used in different stages of PD processes. Hence this chapter attacks this problem by presenting a global integrated data structure for supporting WWW-based RPD. This integrated data structure has been used to build a WWW-based knowledge base system to record and manage the data and information flow through a PD cycle in a leading New Zealand manufacturing company. The industrial test results show that the system can be used to support the PD process and effectively shorten the PD cycle.

12.2 Overview of System Structure

The overall structure of the WWW-based rapid product (industrial switches or control units) development system is shown in Figure 12.1. This system has been developed for a leading New Zealand manufacturing company to support and speed up its PD cycle. In this system, the integrated data structure is used to build an environment to support the overall PD process, *i.e.*, marketing, product design, tool making, and manufacturing. The integrated data structure can be used to build STEP-based product databases, and design and manufacturing knowledge bases. The WWW-based information management system that contains these data, knowledge and data management software tools can be used to support the PD process, *i.e.*, help product designers to make decision at early stages of the PD process. Also, other application

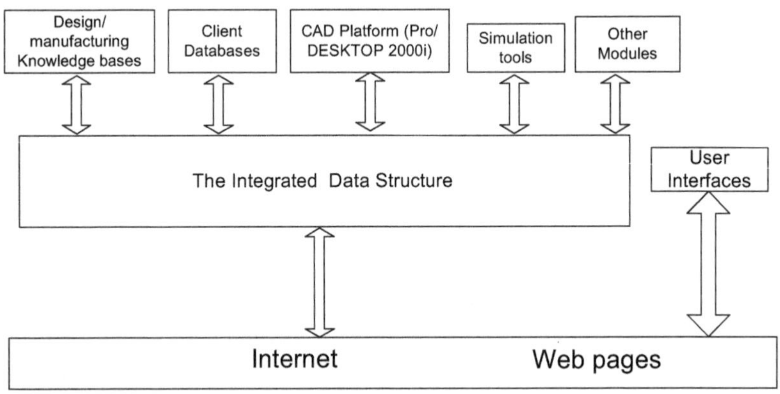

Figure 12.1 Structure of the WWW-based rapid mould PD system

models (*i.e.* cost and simulation) can be achieved based on the integrated data structure. The methodologies that are used to achieve the system and the implementation of these application modules have been reported by Tu *et al.* (2000). The application program interfaces (APIs) of various computer aided engineering software systems (*e.g.* ProE) are secondary development tools that can be used to develop the new application models (*i.e.* product fuzzy search model or supply chain management model). These models can be added into the open structured system. The information changes in different application models can be directly transferred to each other. As can be seen, the integrated data structure can be used to bridge the information gap among commercial software packages, user developed application software, databases and knowledge bases in the WWW environment. The structure of the integrated data environment is described in the next section.

Figure 12.2 shows the home page of the RPD information management system. It can be seen that the system consists of seven computerised models that can be accessed through the inter/intra-net.

The Project Appraisal model is an on-line project management system. Each of the PDs will be treated as a project. The Project Appraisal model records the managerial data about a project, such as the progress of the project (indicated by the achievement of milestones), product design engineer, tool design engineer, product manager, sales estimates, estimated project costs, and comments and endorsements of the departmental managers. This data will be real-time updated and can be accessed via the company's Internet/intranet.

The Design/Manufacturing Knowledge Base was developed according to the data structure to be presented in detail in the following sections of the chapter. This knowledge base can be retrieved through the inter/intra-net. It physically consists of several subdatabases that are allocated to different departments, such as prod-

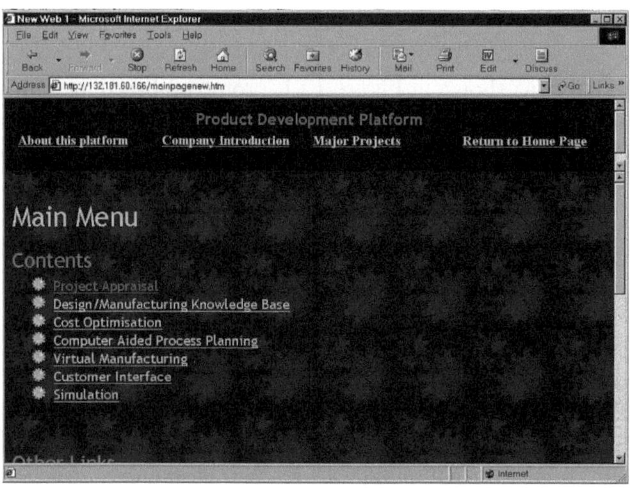

Figure 12.2 The home page of the RPD information management system

uct design, tool design, tool making, shop floor, and inventory control. It records all the necessary information about products manufactured in the past and products under development, *i.e.* ongoing projects. In fact, this knowledge base and its WWW-based management system functions as a communication platform to facilitate the company's integrated data management and concurrent engineering approach at early product design stages.

The Cost Optimisation model was developed using the dynamic programming technique and computer simulation technology (Tu *et al.* 2000). This model estimates the cost of a product at an early design stage by optimally selecting possible manufacturing processes, subcontractors, production facilities, and shop floor scheduling through virtually manufacturing the product on a simulation platform, history data searching and seeking outside sourced services or subcontracting. This cost model was reported in Tu *et al.* (2000).

The Virtual Manufacturing consists of the computer simulation models of the common CNC milling centres in the companies for product (mould/die) machining and a shop floor simulation platform. These simulation models were developed using the VNC developed by Deneb Inc. (Dayton, OH) for CNC machine cutting simulation, and ProModel developed by ProModel Corp. (Orem, UT) for shop floor and production system simulation.

The Customer Interface is an open interface for customers to get involved with PD through the Internet platform. Through this interface, the customers can view the development progress of their products and give comments on their products. This is the only model that can be accessed by customers.

The Simulation includes two packages, VNC and ProModel. Through the Simulation, the staff in a company can access these two simulation packages to further develop simulation models.

The Computer Aided Process Planning model was developed based on a novel concept, called "prototype based incremental process planning". This concept was presented by Tu *et al.* (2000), and is part of the prototype-based incremental PD method.

With the system as shown in Figure 12.1, prototype-based incremental PD follows a specified pattern. First, the design targets (or primary features) that are identified by the user interface are prioritised according to their influence on customer satisfaction, manufacturability, quality, and cost. Correspondingly, different endeavour rates will be assigned to the design targets (or primary features). According to this prioritised list of primary features, the system will automatically search for an existing product that is the closest to the required new product from the design/manufacturing knowledge base.

The existing products are products that were successfully produced by the company in the past. The similarity between the existing product and the new product is measured by the sum of "similarity" rates of the common features between the existing product and the new product. The existing product with the highest sum of endeavour rates is taken as a prototype for the new product.

After a prototype is selected, a relevant process plan for producing this existing product is also found since they are stored in an extendable common object of the data model developed in terms of EXPRESS in STEP (see Section 12.4). Based on the prototype(s) and relevant process plan(s), an incremental process planning method as developed by Tu *et al.* (2000) can be applied so that the product design and process planning are carried out concurrently in a feature-by-feature stepwise manner. The incremental process planning method requires the insertion of a sub-process plan for creating this new feature needs to be inserted into the process plan if a new feature is added to the prototype. If an existing feature of the prototype is modified, the process plan needs to be modified accordingly.

The subprocess plan for creating a feature can be searched from the existing products recorded in the design/manufacturing knowledge base or developed by the design/process-planning engineer by using his/her experience, a shop floor test, or a manufacturing process simulation package, *e.g.* Virtual NC. If a feature cannot be created by any means, either the designer should avoid this feature or feedback should be given to the customer.

After a prototype or several prototypes are developed as described above, a new product will be designed, planned, and virtually manufactured on this system mainly by computer simulations and the data recorded in the design/manufacturing database. Finally, the output from this system is obtained as a fully recorded product data model.

12.3 STEP-based Integrated Data Structure

As mentioned in the previous section, the product design/manufacturing knowledge base is created as an on-line data library to support all kinds of activities through a PD cycle in a desktop computer environment. It was developed according to a STEP-based integrated data structure. Figure 12.3 shows the framework of this data structure.

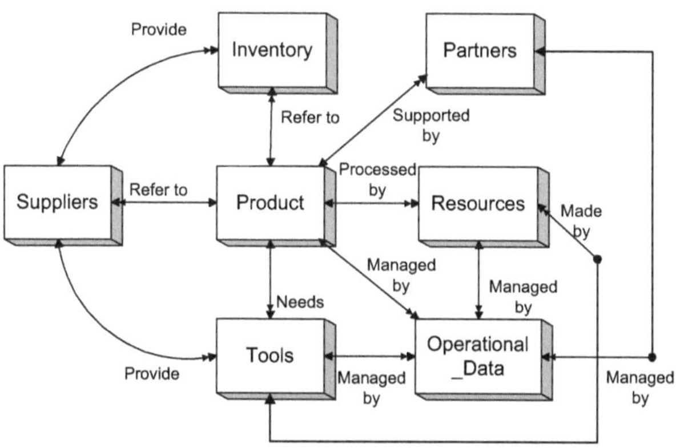

Figure 12.3 Data structure of the design/manufacturing knowledge base

In Figure 12.3, "Product" means all the design and manufacturing data about a product. These data are physically stored in computers in different departments and managed by the system according to an extendable common object data model. An extendable common object data model for an injection mould, for instance, is modelled using EXPRESS-G in Figure 12.4.

As shown in Figure 12.4, the product model is built as a tree structure. It can be envisaged that this type of tree structure product data model can be extended to meet the requirement of information modelling of different products (Tu *et al.* 2000). The integrated mould making data model as shown in Figure 12.4 already contains most commonly used components for a product manufacturing process. The product model in Figure 12.4 includes several entities on different levels. The top level is "Product", which includes a number of sub-classes called "Components" and a subclass called "Assembly". All the necessary data regarding the components of the product will be recorded under the "Components" subclasses. Likewise, all the necessary data about how to assemble the product will be recorded in the "Assembly" subclass. A "Component" sub-class includes several "Primary_Features" entities, several "Materials" entities, and several "Secondary_Features" entities. Accordingly these entities record the primary features, secondary features, and materials of a component.

The primary features are those design and manufacturing features by which the main shape, performance, function, manufacturability and cost of a component are determined. For an injection mould, a cavity will be a primary feature. The secondary features are normally for supporting functions and the integrity of the component (or product) or even decoration of the component (or product), *e.g.*, a screw hole, a slot, a pocket for parts assembly, the surface treatment of a mould, *etc.*

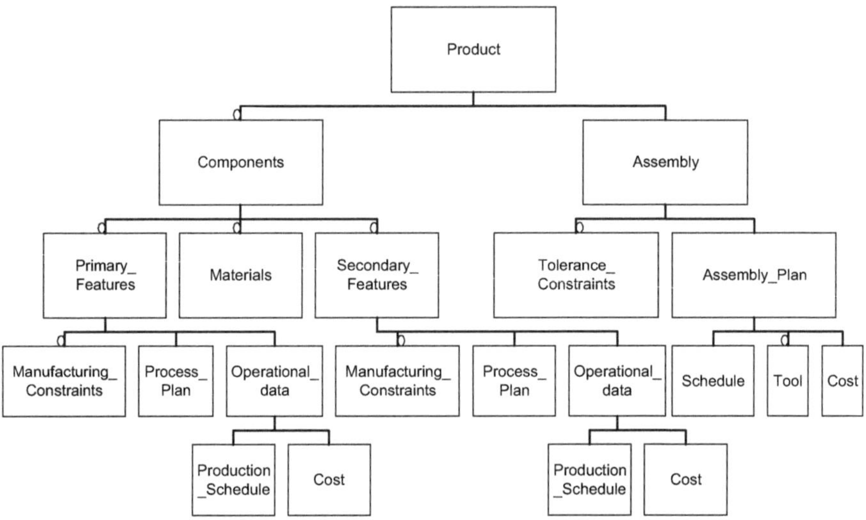

Figure 12.4 Integrated product data model in EXPRESS-G for an injection mould

In Figure 12.4, either a “Primary_feature” entity or a “Secondary_feature” entity includes three types of sub-data-classes, viz. “Manufacturing_Constraints”, “Process_Plan”, and “Operational_Data”. “Operational_Data” further includes “Schedule” and “Cost” subdata classes. Examples using this data structure to model products can be found in Tu *et al.* (2001) for an injection mould.

“Suppliers”, as shown in Figure 12.3 includes all information about the company’s suppliers, *i.e.* contact information, product specifications, reference prices, delivering lead times, reputation assessment, and quality assessment. “Inventory” records the company’s present inventory level, which is part of the company’s MRP system. “Tools” records the company’s tooling information. “Resources” keeps records of the company’s equipment and employees. “Partners” records detailed information about the company’s sub-contractors. It contains important data on evaluation of the subcontractors’ manufacturability, reliability, and quality of their service or products. “Operational_Data” contains present and previous production planning and scheduling data records. This is part of the company’s production planning and scheduling system.

12.4 The WWW-based Knowledge Base System

Since efficient management of information flow in PD is critical to reduce the PD cycle time, new approaches and methodologies to achieve a WWW-based information management system have been continually put forward. CORBA (common object request broker architecture), for example, is employed to ensure interoperability among distributed objects (Heck *et al.* 2009, Gruber *et al.* 1992). STEP and KQML (knowledge query and manipulation language) are used to achieve openness of the information system.

Figure 12.5 shows the structure of the WWW-based information system that has been developed for a manufacturing company in New Zealand. This system comprises of the WWW-based software tools for information management and sharing among different departments, the integrated data environment and interfaces, the CORBA-based distributed environment, customer interfaces and several knowledge bases. The WWW-based database publishing tool and object communication tool have been described in Tu and Xie (2000).

To manage and share the distributed data and software systems/packages (or objects) in the company, CORBA technology has been employed in the project to build a computer communication network or information sharing platform in the company (see Figure 12.5). In fact, in recent decades, CORBA technology and the distributed component object model (DCOM) have been developed and applied for building such distributed object communication and management networks. Many successful cases can be found in literature, *e.g.* (Mowbray 1995).

The customer interface was developed using the QFD method (Tu and Xie 2000). It first provides inter/intra-network access for customers to input their requirements in the system. According to the customer’s requirements, the product designers can

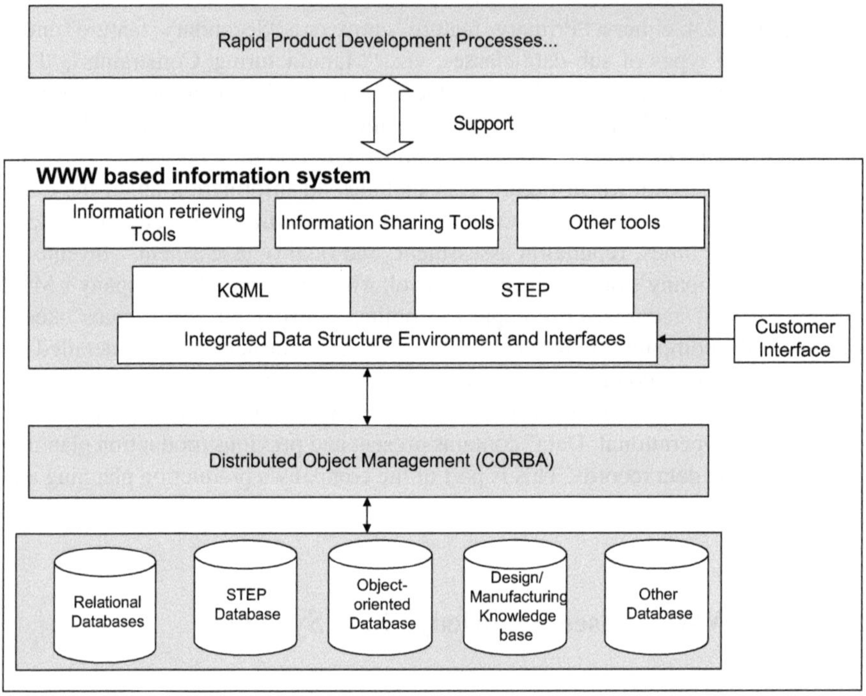

Figure 12.5 Structure of the WWW-based information management system

use the QFD models built into the system and the product design and manufacturing knowledge bases, which were developed according to the company's previous experiences, to identify the technical features (or attributes) of the product to address the customer's needs. A linear programming model has also been developed to help the PD planner allocate resources (mainly staff times) under the constraints of a fixed lead time and a given PD budget to obtain various technical attributes of the product so that the maximum overall customer product satisfaction can be achieved. The system also provides the supporting functions for RPD, such as incremental process planning, manufacturing process simulation, shop floor operation simulation, cost estimation, subcontract planning, and supplier selection.

12.4.1 Knowledge Base Structure

In terms of the STEP-based integrated data structure as discussed in Section 12.4, a general PD knowledge base building framework has been developed to create and manage the knowledge element, knowledge relationships and knowledge objects. The knowledge base building framework contains four information layers, which, from top to bottom, are the knowledge layer, parts layer, feature layer and element layer. The element layer contains relevant data or information about the shape feature of a product. The feature layer contains all the feature information including the feature attributes and relationships between the information objects in the feature layer. The information objects are defined by different production departments for a PD, such as feature geometric objects, manufacturing process objects, resource objects, and operation objects. The part layer contains all the part information including part feature information and relationships between different information objects in the part layer. The knowledge layer not only contains the part information, but also "knowledge related" information objects and a reference engine. Here, the knowledge is extracted from the part layer and the feature layer, and is organised by information objects and relationships between information objects. Therefore, the knowledge includes part-layer knowledge and feature-layer knowledge. A common information structure that contains the relationship of the information objects for a PD in a manufacturing company has been addressed in Section 12.4. The knowledge in the knowledge layer can be directly used to support intelligent concurrent design and manufacturing. Here we add a management feature to the part features to manage all the information objects of a certain part. The management feature can be saved in a knowledge base and used as a part of knowledge data to manufacture the part. For different applications, the users can define their own application objects in the part layer and feature layer through a graphical user interface (GUI) according to the requirements of the PD. After defining the application objects, the relationships between the objects can be created by the users or automatically by the optimisation algorithm and existing knowledge. These objects with their defined relationships are the new knowledge, which will be added into the knowledge bases for future use.

12.4.2 Industrial Implementation

As mentioned in this chapter, the system described has been implemented in a leading New Zealand electrical appliance manufacturing company. The company produces various industrial switches, control systems, and electrical appliances. In terms of the data structures and general system framework, the product design/manufacturing knowledge bases for RPD and the STEP-based product databases have been built to support the company's rapid industrial switches development, which includes PD and mould/tool making.

In order to share the data and knowledge among the people in the company's branches, which are located worldwide, the system has been built on a WWW- based network environment. Hence, the WWW technologies and Internet software development tools, such as JAVA, CORBA, Visual InterDev, XML and KQML have been used to develop this system. In the following section, a case will be used to illustrate how the system works.

According to the customer's requirements for a particular type of electrical switch, the search model in the system can find a prototype, *i.e.* a product produced

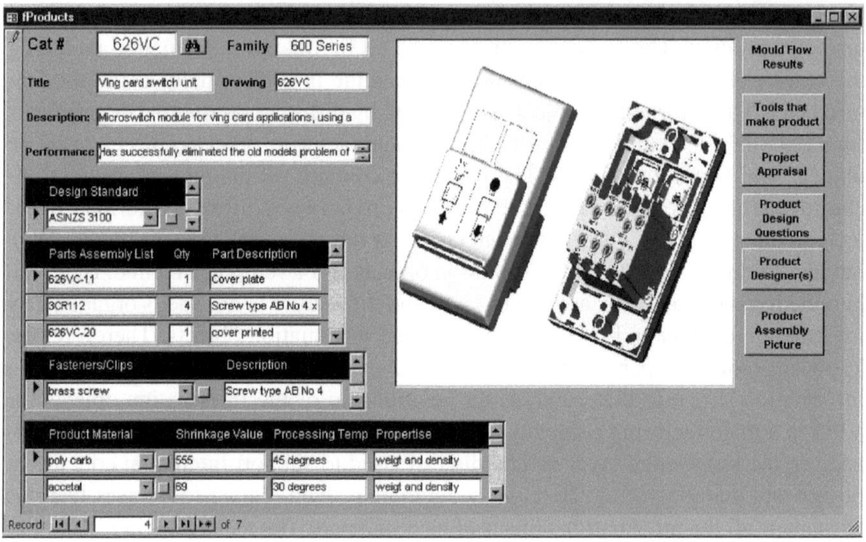

Figure 12.6 The product interface

by the company in the past that has the maximum sum of endeavour rates. For this case, the prototype found by the system is the 626VC industrial switch in the 600 series. The output from the system is shown in Figure 12.6.

In fact, Figure 12.6 shows a dynamic user interface on the web page. From this interface, a user can quickly search the necessary information for producing the 626VC. The user, for instance, can quickly get the CAD drawing of the product by double clicking "626VC" in the Drawing box (Figure 12.6), the assembly illustration by clicking the "Product Assembly Picture" button on the right hand side of the page, designer's information by clicking the "Product Designer(s)" button, comments and discussions made at the earlier product design stages by clicking the "Product Design Questions" button, project appraisal information as mentioned in Section 12.1 by clicking the "Project Appraisal" button, and toolmaking information by clicking the "Tools that make product". The toolmaking information is shown in Figure 12.7.

Figure 12.7 shows the user interface for toolmaking information. From Figure 12.7 it is obvious that the product 626VC was made using the injection mould M-5112. From this tool interface, a tool designer can get the relevant information for making a tool, such as cavities and parameters of the tool, other products made using the same tool, factory moulding machine data, manufacturing information (*e.g.*, CNC machine, operator, NC code, *etc.*), tool components, tool materials (including the information about the suppliers), tool designers, and tool trials. For example, viewing general information on surface finish requirements on the tool M-5112 is simply performed by clicking the button directly located at the end of the surface finishing field box. The tool surface finishing requirements for mould M-5112 are shown in Figure 12.8.

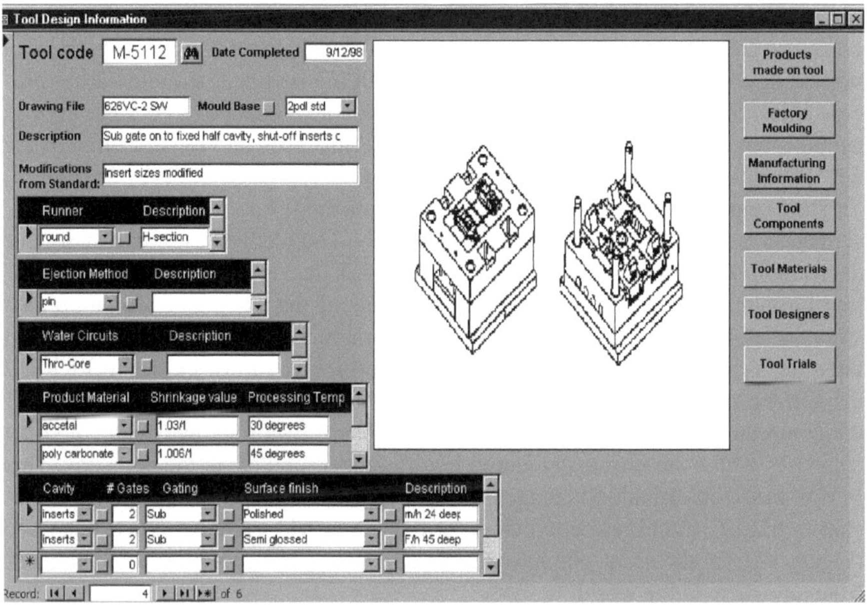

Figure 12.7 The tool interface

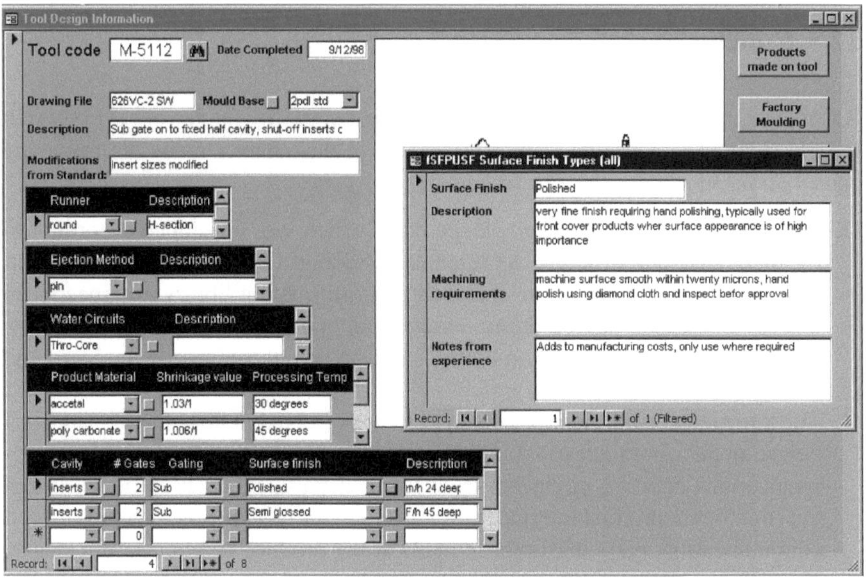

Figure 12.8 General information on surface finish requirements of mould M-5112

In terms of this prototype and the customer's requirements, the product designer(s) and tool designer(s) can quickly develop a new product and the necessary mould/tool through modifying the design and manufacturing information of switch

626VC and mould M-5112. Following the concept of incremental process planning as described in Section 12.3, the new design features for the product or the tool will be developed one by one concurrently with the pertinent manufacturing process plans. Some past experiences gained through developing 626VC were used as useful references to help the designers to avoid mistakes and rework through the new PD cycle. This helps the company to dramatically reduce its PD cycle time.

The computerised models and the general knowledge base building frame as described in Section 12.4 were used to develop the knowledge bases or planning or decision support tools through the PD cycle. All the relevant data and information through a PD cycle were recorded in terms of the extendable common data structure as presented in Figure 12.4. The information models behind the product interface as shown in Figure 12.6, the tool interface as shown in Figure 12.7 and the surface finishing information in Figure 12.8 are the instances or attributes of the common object 626VC that are extracted from several knowledge or product databases which are physically stored in different computers in the company. This WWW-based information management system automatically brings the data relevant to 626VC from these computers (or databases) via a front-end user interface (Figure 12.6) to the users who have the right to access this communication network from anywhere in the world.

12.5 Conclusions

As indicated throughout the chapter, the main goal of this research work is to shorten the PD cycle. The research work as presented in this chapter made the following contributions to achieve this research goal:

1. The WWW-based design knowledge base, which was developed according to the integrated data structure as presented in Section 12.4, can be used by a manufacturing company as an on-line data and information library. Engineers can quickly and easily source the necessary data and information to support new PD. This data and information library can also help the company to record and manage its past experiences.
2. The knowledge base management system as presented in this chapter can function as an inter/intra-net communication platform to facilitate and integrate communications between engineers in different departments. The product design question in the product interface (Figure 12.6), for example, is a record of such communication, particularly those made at the product's earlier design stages. Normally in a manufacturing company, extensive communication involved at the early product design stages often results in misunderstandings and confusion between the departments due to poor management and recording of this communication. These misunderstandings and confusion can easily prolong the PD cycle. By using the WWW-based knowledge base management system as presented in this chapter, this communication can be clearly recorded and easily

accessed and viewed via the company's inter/intra-nets. An engineer in any department can simultaneously view the comments made by the engineers in other departments. This will help the company to save time usually spent in tedious cross-functional department meetings.

3. As shown by the product interface (Figure 12.6) and tool interface (Figure 12.7), the knowledge base as presented in this chapter included a lot of descriptive information (such as product design questions, tool surface finishing requirements, designers' information, manufacturing process and comments, *etc.*) rather than the abstract data given in a traditional database. According to our experience with manufacturing companies, these descriptive comments are very important references to remind designers and manufacturers to avoid making the same mistakes or spending time on a problem that was solved in the past. With these descriptive comments in the knowledge base, the knowledge base can be used as a training tool to help graduate engineers quickly grasp the company's design and manufacturing experience.
4. As described in Section 12.3, the integrated data structure was developed in terms of STEP. Since most computer aided design, and engineering and management software vendors support STEP, the system as presented in this chapter should have good compatibility with other computer aided software systems in advanced manufacturing.

References

Gruber, T. R., Tenebaum, J. M. and Weber, J. C., 1992, Toward a knowledge medium for collaborative product development. In Proceedings of the Second International Conference in Artificial Intelligence in Design, ed. J.S. Gero, Pittsburgh, PA, 22–25 June.

Heck, B. S., Wills, L. M. and Vachtsevanos, G. J., 2009, Software Technology for Implementing Reusable, Distributed Control Systems. Intelligent Systems, Control and Automation: Science and Engineering, Volume 39, Part VI, 267–293, DOI: 10.1007/978-90-481-3018-4_11.

Mowbray, T. J. R., 1995, The Essential CORBA: Systems Integration Using Distributed Objects, Wiley, New York.

Tu, Y. L. and Xie, S. Q., 2000, A WWW based Integrated Product Development Information Management System. IFAC, MIMO2000, July, Greek.

Tu, Y. L., Chu, X. L. and Yang, W. Y., 2000, Computer aided process planning in agile one-of-a-kind production, International Journal of Computers in Industry, 41, pp. 99–110.

Tu, Y. L. and Xie, S. Q. and Zhou, Z. D., 2001, An Information Modelling Framework to Concurrent Product Design and Manufacturing. IAM'2001, March 17–21, 2001, American University in Dubai, U.A.E.

Chapter 13
An Internet-based Product Information Management System

Abstract Efficient management of product information that covers the entire life cycle is critical to the enhancement of corporate competitiveness. This chapter explores the design and development of a World Wide Web (WWW)-based PD information management system for a cross-nation manufacturing corporation that is headed by a holding company in Christchurch, New Zealand. Since product data are often managed in a distributed computing environment, CORBA is employed to ensure interoperability among distributed information objects. The WWW-based information management system discussed in this chapter includes two major components: 1. WWW-based product design and development distributed object oriented databases; 2. A WWW-based integrated system platform. Several submodels are introduced; these include an object oriented database structure, a WWW-based information management system, a WWW database tool, an information access tool, the incremental process planning method and an integrated software platform for the integration of CAD, CAPP, CAM and the WWW-based information management system.

13.1 Introduction

Widespread use of the Internet and WWW has had a significant effect on the method of inter- and intra-communication of OKP manufacturing companies. These efficient Internet and intranet communication tools have helped OKP manufacturing companies to save production costs, shorten the lead time to market, implement globalisation and concurrent engineering, and make companies rethink the issues at the heart of competitive manufacturing. In order to enhance competitive ability, a number of changes are needed for OKP manufacturing companies, especially for those with globally distributed "partners". It has been widely conceived that having the capability of rapid PD is one of the key factors needed to enhance corporate competitiveness (Bernard *et al.* 2009). This may require the support of information systems, especially for companies whose business partners are distributed over the globe. Our example manufacturing corporation, which is managed by a holding company in Christchurch, New Zealand, is such an international manufacturing group which has subbranches in Australia and Malaysia. For this company, integration of various information systems through the WWW is a holistic approach to manage the complexity of its product design, development, manufacture, and distribution, so that it can quickly respond to customers' requirements.

At present, this company is facing increasing demands from customers for product varieties, low cost and short delivery times. However, there are some hurdles that impede the meeting of these customer requirements. Firstly, suitable discussions are needed at different stages of its tool/mould design and making process, so that a tool/mould can actually be developed subject to various manufacturing constraints. These discussions are normally among different engineers in the company's branches located in different countries through documented letters or e-mail. These are time consuming and very inconvenient to manage. Secondly, a number of important tool-mould-making process data are not recorded and these data are lost after projects have been finished. Besides the information lost in tool-mould-making, delays, errors and long product definition procedures are also caused by lack of supporting information. Sometimes, the same product has to be redesigned owing to the loss of historical records.

Hence, the development of a product information management system that is able to record various data through a whole PD cycle and simultaneously provide an integrated platform for information sharing among different partners or departments, is very important for the company to shorten its PD cycle and cut down production costs. A WWW-based information management system for rapid and integrated PD has been developed. It consists of several distributed databases with object-oriented structure to store all the tool-mould-making and PD information. Through the intra/internet, all the information is published on the Web. This provides an easy way of accessing all the necessary information throughout the whole life cycle of a PD process. However, information accessibility does not mean that all the information can be used to directly support the design and manufacturing processes. Sometimes people will get confused when they face so much information and do not know what is useful for them. Therefore, the integration of the available information is a key step for different partners in a PD cycle to effectively share information (Dong and Agogino 1998, Chiu *et al.* 2006). In order to solve this problem, much research effort has been made to develop an integrated information sharing platform in a computer network environment. Some prototypes have been developed, *e.g.*, Dong and Agogino (1998), Xue *et al.* (1999), Chiu *et al.* (2006), Cutkosky *et al.* (1993), Boynton (1993) and Gruber and Russell (1996). However, there exists some limitations to the application of these prototypes, especially in a WWW environment. The first is the requirement for a formal WWW-based distributed database structure with suitable product data models for a manufacturing company, *i.e.*, a decision must be made on what information should be shared and how to represent and record the information from a PD and relevant tool-mould-making processes. In practice, the company production manager usually decides what information is needed, when it is needed and how it will be used depending on the context of the current problem (Boynton 1993). Also, the data structure of the product information management system may vary with the structure and culture of the company. Considering these questions in advance, collecting all the critical pieces of information for the decision, and finally integrating the information and decisions into a product model is usually a hard job (Gruber and Russell 1996).

The critical issues in managing and sharing information for a cross-nation corporation are:

1. How to build the databases/knowledge bases with WWW publishing and real-time information accessing mechanisms in a user-friendly operating environment.
2. How to develop a system, which can capture in real time the related data and knowledge, and use and share the data and knowledge to support PD processes.
3. How to develop a platform to integrate the information management system with existing software packages that are employed at different stages of the PD process. Boynton (1993) suggested that the critical issue for developing a design information system to support company-wide design practice is not the product modelling aspect but the information dissemination aspect.

This chapter discusses the problems mentioned above through the presentation of a WWW-based integrated PD information management system. This system can directly support product design and manufacturing processes in a WWW environment.

13.2 The System Framework

The overall structure of the WWW-based information management system is shown in Figure13.1.

The system was built in two parts, namely the WWW-based information management system and the integrated platform. The first part includes distributed re-

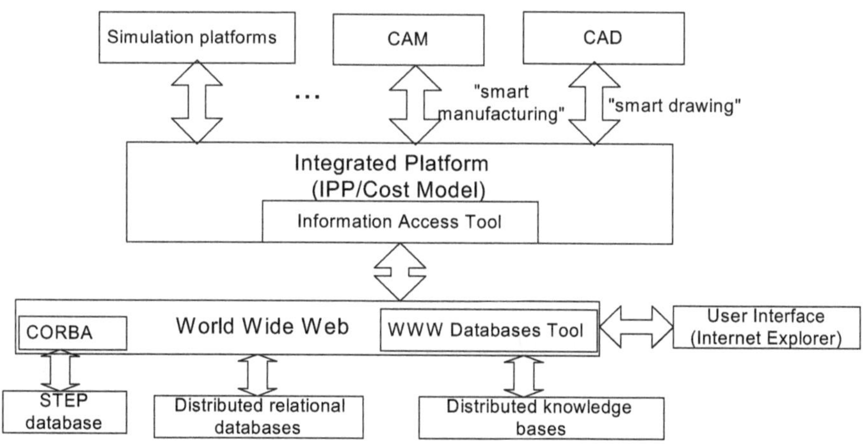

Figure 13.1 Framework of the WWW-based integrated PD information management system

lational databases, STEP databases and knowledge bases, WWW database tools (WDTs), and user interfaces for different departments of the company to manage the product information. The second part includes a collaborative communication tool, an information access tool, incremental process planning (IPP) user interface and a cost optimisation model. This platform is achieved using current WWW development tools (*i.e.*, visual Interdev, JavaScript). The application tools are developed as agents that run in a distributed environment.

The distributed relational databases and knowledge bases record all the data and knowledge gathered from previous PDs in the company's history. The user interfaces were developed and can be viewed through a web browser such as Internet Explorer, which can be used by the different departments to change and manage the databases and knowledge bases on the intra or internet. The integrated platform was developed to integrate CAD, CAM, computer simulation packages, and the WWW-based information management system to support so-called "smart drawing" and "smart manufacturing". The incremental process planning user interface together with the cost optimisation model were developed to select or plan suitable manufacturing processes against design features from a cost optimisation point of view. The multi-objective cost optimisation algorithm was discussed in Chapter 10. The information access tool was designed as a search tool to access all related databases and knowledge bases according to the requirements of the integrated platform. Some major models or tools mentioned above will be further discussed in the chapter.

13.3 The Internet-based Information Management System

Today's design and manufacturing practice in an OKP manufacturing company often involves complicated communication, interaction, and data exchange between individuals inside and outside of the company, such as engineers, suppliers and customers. For example, to design a part, a design engineer may need to access information on process planning to determine the production requirements or the product data management system to find a similar part and revision levels. The challenge is to create related information and knowledge so they can be shared in real time. The development of intra/internet communication technology has provided a feasible solution to company-wide knowledge sharing and real-time communication. Hence, a WWW-based distributed database and knowledge base management system is a fundamental part of the information integration of a manufacturing company.

13.3.1 Product Data Model

When data is added to a database, it becomes a model of that part of reality to which the data refers. As there is an increased need for up-to-date information, an automated database management system (DBMS) was developed based on groups of formalised data modelling rules called product data models. Nowadays a product data model is usually object oriented. The EXPRESS data modelling language in the STEP provides a useful tool to represent various product data, and STEP has set up an international standard architecture for modelling a product. Hence, STEP and the object oriented modelling method were used in this research project to model the products as addressed in Chapter 11. Candadai (1994) also reported another application of STEP to build a product data model. However, STEP is still a developing data modelling architecture and has some limitations for product data modelling, such as instance data or type data (Dong and Agogino 1998). It does not include a mechanism for using classification and inheritance for modelling products in a particular company (Manniso *et al.* 1998). Owing to the limitations of STEP, the data model discussed in this chapter was built by both a STEP modelling method and the traditional relational data modelling technique. For data that STEP cannot describe, related objects are created using an object-oriented method and connected to its STEP object using relationships. One of the advantages of using the object-oriented database (OODB) is that all the data in the object can be extracted by object identity or found by key. The newly created objects can be stored as well. The structure of the OODBS and STEP database will be discussed in the next section. All data in an object can be easily read and written using object oriented programming languages (OOPLs) such as EXPRESS, C++ and Java.

13.3.2 Product Information Management System

13.3.2.1 OODBMS Structure

Figure 13.2 shows the framework of the distributed object oriented information management system for a plastic mould design and manufacture, which was developed for a manufacturing company in New Zealand. It presents a generic structure of an object oriented database management system (OODBMS). As shown in Figure 13.2, this information system includes product information, tool information, manufacturing information, and supplier information. These structures of informa-

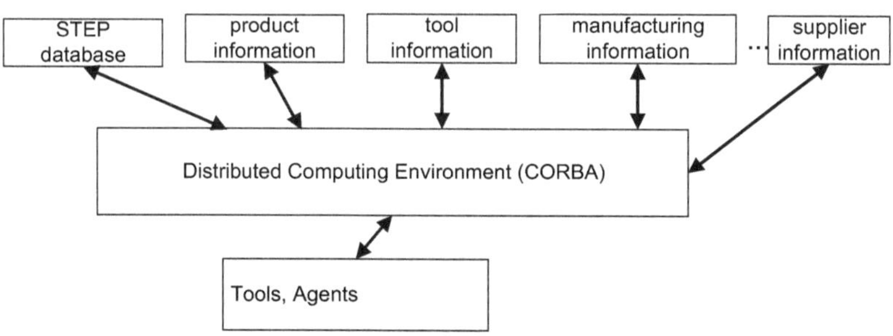

Figure 13.2 Distributed information structure for plastic mould design and manufacturing

tion systems are built according to the requirements and structures of the company, which means the product design database contains all the design specifications, geometric information and other information from the downside stream (*i.e.*, manufacturability, cost). These information systems were built on an object oriented database structure and managed by using CORBA under the distributed computing environment. The detailed information with regards to how CORBA is used to integrate structured databases under multi-operation systems and a multi-language environment are addressed by Fang *et al.* (2000), Yeongho *et al.* (2001) and Mowbray and Zahavi (1995). The distributed information management system eliminates the need to follow the predefined access paths to reach the target data, and makes data access more flexible even under different operation systems. The structure of the information system facilitates uninterrupted queries and is well suited to the manufacturing environment. All the databases in the system were developed using the ORACLE software package. ORACLE supports large size databases with multi-user access which are suitable for a distributed multi-user system. All the data in ORACLE databases can be accessed via ODBC and JDBC interfaces by using object oriented programming languages (OOPLs) such as Visual C++, Borland Delphi and Java. This flexible programmable interfacing ability makes the databases more accessible and easier to develop.

The object hierarchy of an object oriented database developed for a mould/tool design and manufacturing company in Christchurch, which contains detailed contents of the data class created based on the structure of the information in different departments is shown in Figure 13.3. To support various types of data in an industry, different types of objects were developed. The structure of these objects and the relationships between the objects are very important for both the database management system and system integration. For instance, certain product design knowledge in a knowledge base needs to be associated with a product number or other identity number so that it can be correctly and effectively retrieved during production. Figure 13.3 also shows several custom interface objects, such as a product interface, tool interface and manufacture interface. These interface objects were developed for different departments in a company to access the databases using Internet Explorer according to predefined privileges.

13.3.2.2 STEP Database Structure

Product data is one of the most important factors to be considered when deciding the dynamic changes of the PD processes. Under the WWW environment, as different CAD/CAM systems are employed by companies, data sharing or exchange between different systems is very important to achieve RPD. Databases based on STEP can support product data exchange among heterogenous CAD/CAM systems, whereas proprietary file formats are not suitable for data exchange among different systems. STEP is promising in that it is becoming a new emerging standard for the exchange of product data throughout the entire life cycle of products in distributed network environments. A STEP-based information framework to cover

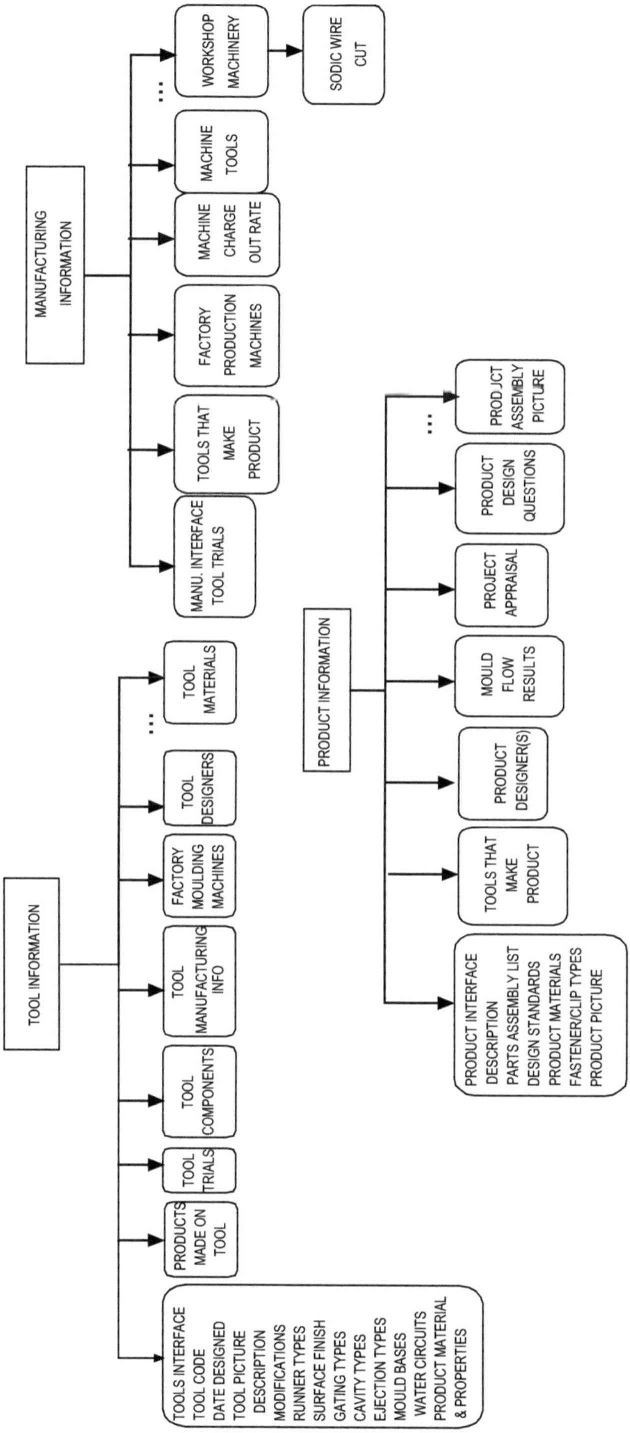

Figure 13.3 Structure of an OODBMS for mould PD

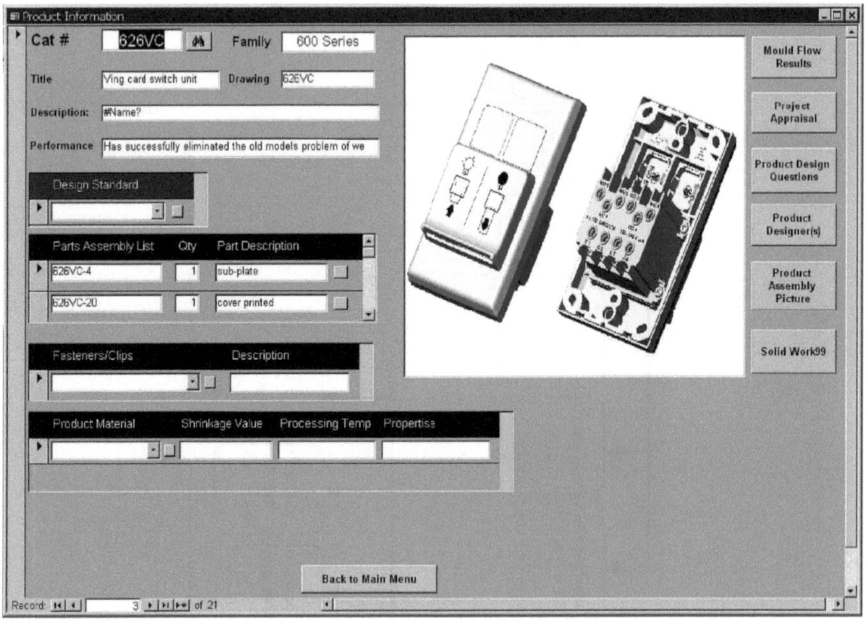

Figure 13.4 A knowledge base example for mould/tool product design/manufacturing

the entire life cycle of sheet metal parts design and manufacturing processes has been presented in Chapter 8. This step-structure information modelling framework is easily extended to other PD processes. A STEP knowledge base based on the information framework above is built for rapid mould/tool PD. Figure 13.4 shows a product information form that is based on the STEP-based information framework structure above. Product designers and manufacturers can use this form to review and modify related information through Internet Explorer after it is published on the WWW.

13.3.3 WWW Database Tool

Although database technology has been in use for a long time, an effective way to make databases accessible through the intra/internet so that data in the databases can be shared on a global scale is still unavailable (Manniso *et al.* 1998). Therefore, it is necessary to develop WDTs that can publish and manage all relational data or knowledge bases over the Web. The WDTs developed by the authors are used to automatically publish and manage databases in a distributed WWW environment.

As shown in Figure 13.5, the WDTs were developed using several computer languages, which include programming languages used on the Web, such as JavaScript, hypertext makeup language (HTML), web authoring language, and CGI. CGI provides a standard protocol for communication between the client browser and the

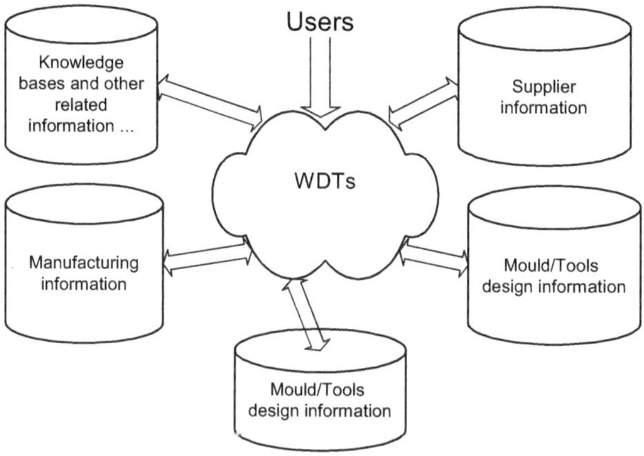

Figure 13.5 WWW database tool developed using multi-computer languages

Web server. It was used to develop the dynamic data linkers between the Web server and the database management system for different departments to access the databases through the intra/Internet. HTML language is used to design the Web pages. Visual C++ was used as a programming platform for developing the WDTs. Based on this platform, the programming entities, such as CGI linkers and HTML web pages, were coded and linked together. WDT proxy and WDT context classes were also developed using C++.

13.4 The Internet-based System Structure

Figure 13.6 shows the integrated platform, which is an intermediate layer between the WWW information management systems and the different software packages that are integrated into the platform as agents. This integrated platform is responsible

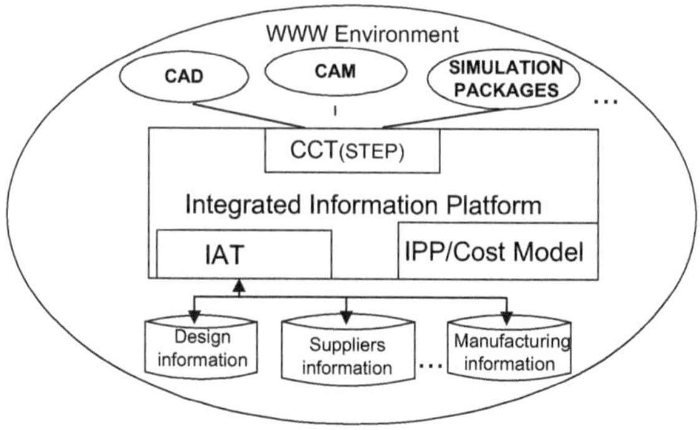

Figure 13.6 Structure of the WWW-based integrated information management system

for production optimisation and real-time information retrieval between the WWW-based information management system and the agents.

This platform consists of several major components, including a collaborative communication tool (CCT), an incremental process planning (IPP) user interface, a production cost optimisation model and an information access tool (IAT). By using this platform, all the software packages, IPP user interface, cost model and databases as shown in Figure 13.6 can be directly accessed through the intra/internet.

In this platform, each agent manager is implemented as a CORBA object and exports specific public methods as an external interface. Other functions can also be achieved in the same way. For example, the detailed implementation method of a WWW data retrieve tool is shown as follows:

1. First define a CORBA object named Retrieve using CORBA IDL, its IDL description is as follows:
   Interface Retrieve {
   Void retrievedata(in string dataFileName, in string schemaName, in string ProductName, out string productspecification, out string dataFile, out string schemaFile, out string ProductContent)
   }
   Users can use function "retrievedata" to get the information needed through product name, dataFilename, *etc.* An interface is provided for users to input all information for data searching. The code is written in C++.
2. Define the interface to the native code in the Java program, create corresponding.h file using JDK tool java.h. Then implement the native function using C++ and compile the C++ code into DLL. When the DLL is loaded, the Java program can invoke the function.

13.4.1 IAT and CCT

Many applications enabled by the WWW, such as virtual university, distance learning, electronic commerce, information gathering and filtering, have a strong need for tools supporting the effective retrieval of information. The goal of designing IAT is to provide a means for users to "on-line search" the data or knowledge according to a particular information enquiry in a PD process, and to "real-time communicate" with the information management systems via the intra/internet. It was designed to enable "real-time control" and to retrieve and search for the data through the intra/internet. The collaborative communication tool (CCT) was designed for dynamically transferring messages, events and data in a WWW environment among the software packages, IPP user interface and cost model.

There still exist some problems in the development of an IAT which can support concurrent multi-users applications, for example, collaborative database accessing, distributed object technology, information retrieval, distributed services and resources management, *etc.* These problems influence the communication speed between the OODBMS and the integrated platform. The search speed of IAT is very important for the system since a lower searching speed will influence the 'dynamic characteristics' of the overall system integration. To accelerate the search speed, the query-oriented search method and the user-oriented search method were combined to develop the search model in this system.

As mentioned earlier in this section, the CCT is a tool to facilitate and manage the real-time communications in a WWW environment between the engineering software packages, the IPP interface and the cost optimisation model. It can dynamically capture data, messages or events and transfer them to the different engineering software packages, the IPP user interface and the cost model in real time through an intra/internet. For example, if a designer adds a feature to a product design by using a CAD package (*e.g.*, Pro/ENGINEER), the CCT can, through the intra/internet, capture this change and transfer it to other software models, *e.g.*, the IPP user interface, CAM package or computer simulation package, for further processing. Feedback can also be sent back to the design platform by the CCT. The CCT was developed using the Pro/ENGINEER application programming toolkit (APT). The API of the Pro/ENGINEER toolkit consists of a library of functions written in the C programming language. These functions can easily be incorporated in a program written in C++. Since the CCT was developed on the Pro/ENGINEER APT platform, it can run on the Pro/ENGINEER operating platform.

13.4.2 Incremental Process Planning (IPP)

Tu *et al.* (2000) presented an incremental product design and process planning strategy for developing customised products or OKP products. This strategy suggests that the company plan the manufacturing process for each new or modified feature, which are then added to the design of a product. Features that cannot be feasibly or economically manufactured should be changed or given up. They also suggested that because of the great variety of OKP products and the usual demand from the customer for lower cost and shorter delivery time, history data and knowledge reuse is very important to an OKP company for successful development of an OKP product.

For the WWW-based PD information management system, the IPP method also plays an important role since the IPP strategy can help a manufacturing company to effectively avoid production rework. History data and knowledge reuse has also been approved as an effective way to cut PD cost and lead time (Tu *et al.* 2000). The IPP user interface was developed for users to plan the manufacturing processes according to the IPP manufacturing strategy. It interacts and communicates with the data or knowledge bases and other models via the intra/internet. The information flows among these models are illustrated in Figure 13.7. The IPP user interface can get information from the WWW database management system, and send the possible manufacturing process plans to the cost optimisation model. After getting the optimisation result from the cost optimisation model, it can save the result into the databases. The IPP user interface can also dynamically communicate with CAD and CAM packages through the CCT.

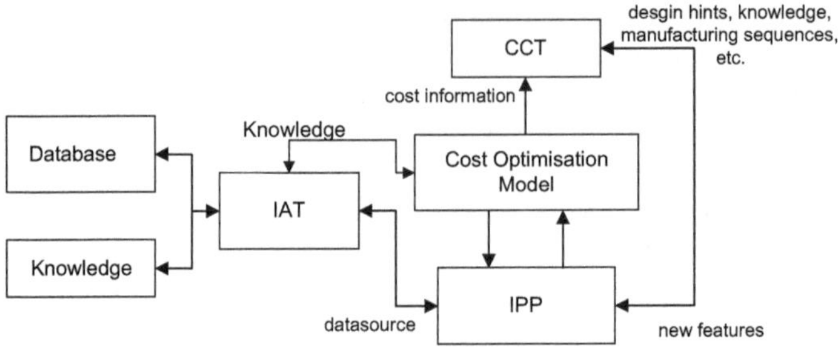

Figure 13.7 Information flows among the different models

13.4.3 Cost Optimisation Model

A generic cost estimation and optimisation framework was reported in Chapter 6. Different from that framework, this cost optimisation model was developed to save PD cost through optimally or rationally scheduling production operations on the shop floor. The inputs to this model are possible manufacturing process plans, which were planned by the process planning engineers using the IPP user interface, and the capacity availability of production resources, which are normally recorded in a company's MRP system. The outputs from the cost optimisation model are a production schedule and a relevant cost estimate. The cost optimisation model was developed using a case-based reasoning method, probabilistic dynamic programming and computer simulation technology. For different products, the cost optimisation function may be different. However, the optimisation algorithms used share common points as the cost optimisation model for sheet metal products cutting has been transferred to a common mathematical model of a NP completeness problem (Xie and Tu 2000). A mathematical model has been built for the cost estimation and optimisation of mould products. The basic cost of a product can be estimated and optimised while considering several possible manufacturing plans. The model can also quickly send the design engineer a cost estimate using case reasoning method to search for a similar solution from the databases via an intra/internet. It can also derive a cost estimate and a production schedule for a designed feature using a probabilistic dynamic programming technique and computer simulation technology to virtually schedule and manufacture the part or product (Tu *et al.* 2000).

13.5 System Implementation

Key challenges that still exist in the development of WWW-based support systems include intelligent search engines, Web accessibility, collaborative and distributed application environment, scalability of Web servers, intelligent agents, server security, the limitations of programming languages like HTML, XML, *etc.* All these challenges have to be considered while developing the WWW-based information management system for rapid and integrated PD. They will influence the functionality, reliability, and the realisation of the integrated system. In this section, we aim to demonstrate the main functions and the feasibility of the system by briefly describing a prototype system, which was developed for a New Zealand manufacturing company to support their mould PD.

Figure 13.8 shows the overall structure of the prototype system, which includes several components. This system can be extended and new functions can be easily added and integrated into the system through agent technology. A CNC machine simulation package called Virtual NC was used to develop the manufacturability test bed and a shop floor simulation package called ProModel was used to develop the virtual manufacturing shop floors for the cost optimisation model. These software packages can be used as real-time simulation of the product manufacturing process. The simulation result is recorded in the WWW-based information system. This information can be retrieved for decision support for similar PD. A Web client called Pro/INTRALINK in Pro/ENGINEER makes it possible to access or manipulate the product data through the Web. Specific interfaces have been designed for people in different departments to activate functions that are provided by the integrated platform, *i.e.*, change/search information or see the manufacturing simulation process. C++ is employed to code all the software models as described in the chapter, such as the IPP user interface, the WDT and the CCT. These tools can be used to locate related information and can achieve intelligent search functions. With the integrated platform that contains these tools, the WWW-based information manage-

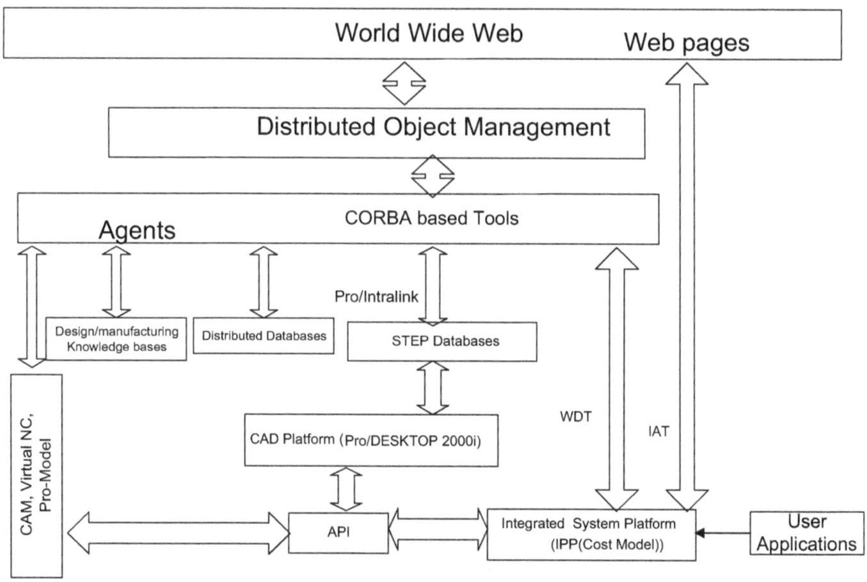

Figure 13.8 WWW-based integrated information management system for mould products design and manufacturing

ment system can be used in different stages of the PD processes. The earlier the design stage in which the system is applied, the earlier the designer can search existing WWW-based database/knowledge bases for existing similar products through the WWW-based interface designed for product designers. This knowledge or information (cost information, lead time, manufacturing ability) can help the designer to make decisions. The integrated platform supported by the WWW-based information management system can also be used as a concurrent PD platform through information integration and sharing at each stage of the PD process. Instead of following the traditional, sequentially arranged PD process, the system can incorporate considerations such as manufacturability, assemblability, serviceability and recyclability into the product design or planning stage. Time consuming rework can be avoided and PD time and costs can be cut.

13.6 Conclusion

It has been well recognised that WWW technology provides an efficient tool and a very promising direction for manufacturing companies to revolutionize their way of managing and integrating information flows in an enterprise. This chapter presented a framework to develop a WWW-based information management system for rapid and integrated PD. A prototype based on this framework has been developed and implemented in a New Zealand manufacturing company to support their mould PD. The structure of the proposed system reflects two aspects: the basic structure of information sharing and the integrated environment for task execution (*i.e.*, optimisation). Our initial experiments during the development of this system with database management system, automatic e-mail transferring system, manufacturing workflow systems and the cost optimisation module have been very encouraging. The manufacturing company involved in this research project has adopted the system in the mould/tool products development process. Problems existing in the company as mentioned in Section 13.1 have been solved. The PD cost has been cut and PD lead time has been shortened.

References

Bernard, A., Taillandier, G., Karunakaran, K. P., 2009, Evolutions of rapid product development with rapid manufacturing: concepts and applications. International Journal of Rapid Manufacturing, 1(1), pp. 3–18.

Boynton, A., 1993, Achieving dynamic stability through information technology, California Management Review, 35(2), pp. 58–77.

Candadai, A., 1994, Information needs in agile manufacturing, Proceedings of Engineering Database, The ASME Database Symposium, Minneapolis, MN, USA, 11–14 Sept. 1994, pp. 101–109.

Chiu, C. M., Hsu, M. H. and Wang, E. T. G., 2006, Understanding knowledge sharing in virtual communities: An integration of social capital and social cognitive theories. Decision Support Systems, 42(3), pp. 1872–1888.

Cutkosky, M., Englemore, R., Fikes, R., Genesereth, M., Gruber, T., Mark, W., Tenebaum, J. and Weber, J., 1993, PACT: an experiment in integrating concurrent engineering systems, Computer, 26(1), pp. 28–37.

Dong, A. and Agogino, A. M., 1998, Managing design information in enterprise-wide CAD using "smart drawings", Computer-Aided Design, 30(6), pp. 425–435.

Fang, N. S., Jiang, H. and Luo, J. Z., 2000, Application of CORBA/SNMP gateway in Enterprise Intranet management, International Conference on Advanced Manufacturing Systems and Manufacturing Automation, Guangzhou, PR China.

Gruber, T. R. and Russell, D. M., 1996, Generative design rationale: beyond the record and replay paradigm, Design Rationale: Uses, Techniques and Concepts, pp. 323–349, Lawrence Erlbaum Associates, ISBN:0-8058-1567-8.

Kim, Y. H., Kang, S.-H., Lee, S.-H. and Yoo, S. B., 2001, A distributed, open, intelligent product data management system, International Journal of Computer Integrated Manufacturing, 14(2), pp. 224–235.

Manniso, T., Peltonen, H., Martio, A. and Sulonen, R., 1998, Modeling generic product structures in STEP, Computer-Aided Design, 30(14), pp. 1111–1118.

Mowbray, T. J. and Zahavi, R., 1995, The Essential CORBA: System Integration Using Distributed Objects, Wiley, New York.

Tu, Y. L., Chu, X. L. and Yang, W. Y., 2000, Computer aided process planning in agile one-of-a-kind production, International Journal of Computers in Industry, 41, pp. 99–110.

Xie, S. Q. and Tu, Y. L., 2000, An integrated CAD/CAPP/CAM system for compound sheet metal cutting and punching, 7th IFAC Symposium on Automated Systems Based on Human Skill, Joint Design of Technology and Organisation, 15–17 June, Aachen, Germany.

Xue, D., Yadav, S. and Norrie, D. H., 1999, Knowledge base and database representation for intelligent concurrent design, Computer-Aided Design, 31, pp. 131–145.

References

Chapter 14
Internet-based "Design for X" for Rapid and Economical Tool-/Mould-making

Abstract Computer Internet communication technology offers tremendous potential for building computer communication and software platforms for the rapid development of OKP products to meet global competition. In the past few years, a variety of Internet-based systems have been developed for the purpose of RPD. Among these systems, Internet-based "Design for X (DFX)" systems have been recognised as an efficient tool for the implementation of concurrent engineering and playing a key role in RPD. Internet-based DFX or IDFX systems can be applied by manufacturing companies to rapidly produce high quality products with low costs and higher profits. However, the implementation of IDFX systems is not an easy task. This is because many new techniques are involved in the application of IDFX, and PD processes are normally sophisticated and vary with PD environments. In this chapter, as illustrated by case studies, two typical applications of IDFX, Internet-based design for manufacture (IDFM) and Internet-based design for cost (IDFC) systems, are proposed for rapid and economical tool-/mould-making. The structure and the key models of the systems are discussed.

14.1 Introduction

DFX has been recognised as one of the most effective approaches to implement concurrent engineering for the goals of RPD (Xie *et al.* 2001). Efficient DFX tools can be used to gather and present facts about products and processes, to clarify and analyse relationships between products and processes, to measure performance, to provide redesign advice on how a product design and manufacture can be improved, *etc.* The advantages of implementing DFX can be summarised as:

1. it provides measurable competitive improvements, which include improved quality, compressed cycle time, reduced life-cycle cost, increased flexibility, improved productivity, and enhanced customer satisfaction;
2. it improves and rationalises decisions on product design, process planning and resource deployment;
3. it has a far-reaching effect on operational efficiency in PD (Huang 1996).

Internet technology is also becoming popular for DFX analysis (Wagner *et al.* 1997) as it can bring tremendous benefits for manufacturing companies, especially those with partners that are globally distributed. Internet-based DFX services can be used to support concurrent (Xie *et al.* 2001), collaborative and economical PD, to reduce the time and cost of product design, assembly and manufacture. Using Internet to deliver DFX functions can help to achieve better communications, closer cooperation, concurrence, transparency, improved customer and supplier involvement, easier project management, team-building in design work, and rationalising and structuring the PD process. For example, Internet-based DFA technology can be used to reduce the cost and time of assembly in a distributed manner by simplifying the product and process by means of reducing the number of parts, combining two or more parts into one, reducing or eliminating adjustments, simplifying assembly operations, designing for parts handling and presentation, selecting fasteners for ease of assembly, minimising parts tangling, and ensuring that products are easily tested. IDFC (internet-based design for cost) systems can help to eliminate those expensive and unnecessary features of a part that lead to increased product manufacturing costs. IDFM (internet-based design for manufacture) systems can help to determine the manufacturability of the possible designs, manufacturing process, machine tools, *etc.* at early design stages. The application of IDFM systems can help eliminate unnecessary features of a part that are often difficult or impossible to manufacture. The impact of Internet-based DFX or IDFX systems can be found throughout the entire PD life cycle, *e.g.*, application of Internet-based DFA to reduce the number of parts will in turn reduce inventory and inventory management efforts.

Although there are many advantages of developing IDFX systems for RPD, the implementation of IDFX systems is not an easy task. Several problems have occurred in recent years when implementing DFX systems. Among these problems, an important question is whether there exists a basic pattern for the development of these DFX tools (Olesen 1992). Another important issue is how traditional DFX systems can be updated to meet the requirements of the global RPD environment. These problems have become major hurdles to developing and implementing IDFX systems.

14.2 System Structure

Figure 14.1 shows the structure of the Internet-based DFX system. The Internet-based DFM system consists of two major components. The first is the Internet-based DFM data/knowledge base that contains mould product DFX information and knowledge. The DFX data/knowledge base is modelled according to a modelling methodology that will be addressed in Section 14.3. The DFX data/knowledge base is organised in appropriate forms for easy access. The other component is the Internet-based prototype system to deliver the DFX functions through the Internet. This prototype system was developed for a mould manufacturing company, and contains three subsystems: client system, server system and DFX knowledge system.

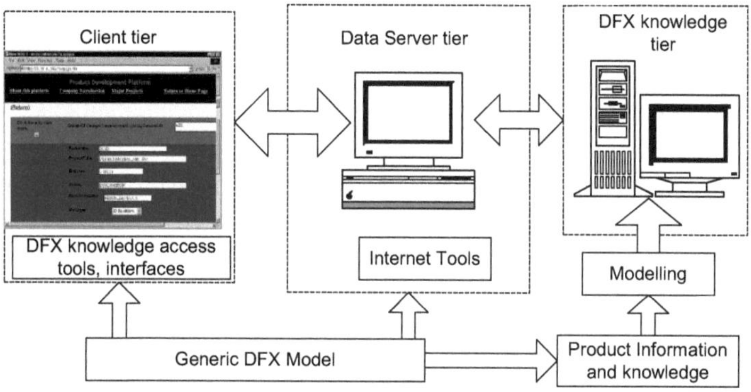

Figure 14.1 Structure of the Internet-based DFX system

The DFX knowledge system is a Microsoft Windows NT server computer that runs a product data management system (Kim *et al.* 2001, Tu *et al.* 2000). The product data management system is stored in a SQL server, which can be accessed from other servers or clients by using designed interfaces and tools. The main function of the product data management system is built based on the DFX principle, which contains all the guidelines of the DFX (mainly DFM and DFC) for plastic injection moulds development (Gregory 1998).

The server system is also a Microsoft Windows NT server computer that runs the Web server software systems such as Internet Information Server (IIS), Blazix Java Web-Server and EJB-Server (http://www.blazix.com/). These servers are used to provide the Internet services. There is a virtual directory in the Web server. All the Web applications of the software systems (*e.g.*, ASP file, HTML file) are stored in the virtual directory and these files can be accessed from either inside or outside the company. In this system, Internet publishing tools have been developed using Visual Basic to automatically publish DFM guidelines on the Web.

The client system is a local computer that runs Internet Explorer or other Web browser software tools. The clients are usually members of the product design and production team. They can use the client system to exchange ideas and discuss the issues in the PD process in real-time. They can also form new DFM data knowledge according to given priorities which can be used as new guidelines for designing and manufacturing similar products.

The IDFX systems also contain several customer interface objects, such as product interface, tool interface, cost interface and manufacture interface. These interface objects were developed for the different functional departments in a company through the use of Internet Explorer to access DFX knowledge databases according to the predefined privileges. Some software tools are developed to support the Internet-based DFM system, *e.g.*, communication tools among these three subsystems, data modelling and searching model, and customer interface. These software tools were addressed by Tu *et al.* (2000).

14.3 DFX Knowledge Base Structure

14.3.1 Knowledge Base Structure

As mentioned in Section 14.2, the DFX data/knowledge base is one of the important components of the IDFX systems. The DFX data/knowledge base has to be built in an appropriate form so that the DFX information and knowledge can be easily managed and reused as guidelines. This section addresses how the DFX knowledge base can be built through creating a distributed DFM knowledge base in an injection mould manufacturing company. Figure 14.2 shows the structure of the DFM data/knowledge base. It represents a generic structure of an object-oriented database management system. As shown in Figure 14.2, this DFM data/knowledge base system contains the product information, tool information, manufacturing information, and supplier information. The structure of DFX knowledge bases is built according to the requirements and the structures of the company, *e.g.*, the product design database contains all the design specifications, geometric information and other information from the downside stream (*i.e.*, manufacturability and cost). These DFX knowledge bases were built using an object-oriented database structure and managed using CORBA (Fang *et al.* 2000) under the distributed computing environment. Detailed information with regards to how CORBA is used to integrate structured databases under multi-operation systems and multi-language environments are addressed by Fang *et al.* (2000) and Kim *et al.* (2001). The DFM data/knowledge base can be used in a distributed environment. It eliminates the need to follow the prede-

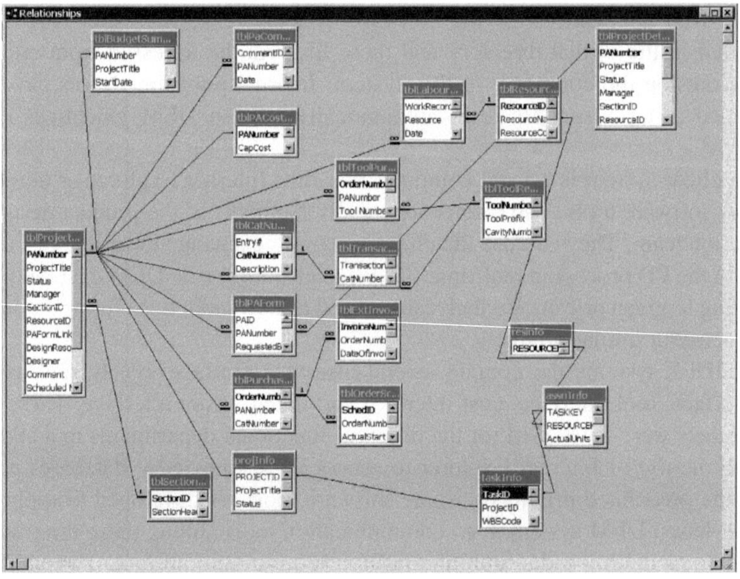

Figure 14.2 DFM knowledge bases structure (shown by relationship)

fined access paths to reach the target data, and makes the data access more flexible even under different operating systems. This data structure facilitates uninterrupted queries and it is well suited to the global manufacturing environment.

In a similar way, other DFX knowledge bases (*e.g.*, DFC knowledge base) are developed for rapid mould development based on an object-oriented and distributed manner. Usually, DFX knowledge bases consist of the detailed knowledge that describes product design knowledge related to activities in other PD stages. To support various types of data in industry, different types of objects were developed based on a generic DFX model, this will be addressed in the following section. This generic DFX model is very important not only for structuring DFX systems but also for integrating the system with other PD systems, *e.g.*, CAD system, product data management system and enterprise resource planning system. For instance, the product design knowledge in a knowledge base usually needs to be associated with a product number or other identity number so that it can be correctly and effectively retrieved in the production.

14.3.2 Generic DFX Model

The need for a basic DFX pattern is essential for the development of IDFX tools. For example, a generic DFX model would speed up the development of specific DFX tools dramatically. It can provide a platform for integrating multiple DFX tools to facilitate the flow of data and decisions between these DFX tools or systems. It also provides a platform for integrating a DFX tool with other decision support systems used in PD such as CAD/CAPP/CAM and CAPM (computer aided production management), and facilitates the flow of data and decisions between the DFX tool and the other systems. Furthermore, a generic DFX model can provide a common basis on which a trade-off can be assessed between competing issues when multiple DFX tools are used. The advantages of establishing a generic DFX model can be further viewed from other prospects, such as speeding up the development of DFX systems and helping to select appropriate DFX tool for a problem (Huang 1996). In this chapter, a DFX data model based on the information framework proposed in Chapter 4 is discussed.

In order to make the DFX (DFM/DFC) systems more compatible with other PD software systems, a generic DFX data modelling methodology is needed for modelling DFX (DFM/DFC) data and knowledge. In the Internet environment, product data is one of the most important factors that need to be considered in the case of dynamic changes of the PD processes. As there are different CAD/CAM systems employed in most manufacturing companies, the data sharing or exchange between different systems is essential to achieve RPD. A DFX data model based on STEP can support product data exchange among heterogenous CAD/CAM systems (Yang and Pei 1999). The proprietary file formats do not meet the need for dynamic data exchange between different systems in a global environment. The combination of STEP and Internet-based data model solutions (*e.g.*, CORBA) provide a solution to solve the product data exchange problems (Zhang *et al.* 2000) and is becoming a new emerging standard for the exchange of product data throughout the whole life cycle of a product in a distributed network environment (Mangesh *et al.* 2000). Xie *et al.* (2002) proposed a WWW-based information management system for rapid and integrated mould product development. Tu *et al.* (2000) presented a STEP-based information-modelling framework to cover the whole life cycle of mould design and manufacturing processes. This information-modelling framework is called step-structure information modelling framework, which contains four top-down information layers, these are the knowledge layer, parts layer, feature layer and parametric layer as discussed in Chapter 12. This framework was first used for information modelling of sheet metal parts and has been extended for modelling mould PD processes. The DFX data model of a mould can be built by following this information framework, which includes the following layers: DFX knowledge layer, DFX product layer, DFX feature layer and DFX parametric layer. The DFX parametric layer contains the geometric data of the shape feature of a product (or mould) in terms of "X" information, *e.g.*, the manufacturing information for a special feature. The DFX feature layer contains all the feature-related DFX information data that include not only the feature information (or attributes) but also relationships between the "X" information objects on the feature level. The DFX part layer contains all the part information data that include down-level DFX feature information and the relationships among part-level DFX information objects. The DFX knowledge layer contains not only the DFX part information, but also DFX knowledge-related information, as well as an inference engine. The knowledge in the knowledge layer is extracted from DFX part-level knowledge and DFX feature-level knowledge, which are formed by information objects and relationships between the objects.

14.4 DFX via the Internet

This section will discuss how two important IDFX systems for rapid and economical development of a plastics injection mould are implemented, *i.e.*, the IDFM system and the IDFC system. The method and process of developing these two IDFX systems are generally applied for the development of other IDFX systems, such as an IDFA system, an IDFR (internet-based design for resource) system.

14.4.1 IDFM for Rapid Mould Development

In order to support rapid mould development, an IDFM system for a mould-manufacturing company in New Zealand has been developed. Figure 14.3 shows some screen printouts from this system. This system can achieve the DFM functions through Internet access. It can be used to develop modular products that are normally assembled by common components, *e.g.*, a plastics injection mould. It helps to prevent unnecessarily complex feature design and manufacturing requirements, such

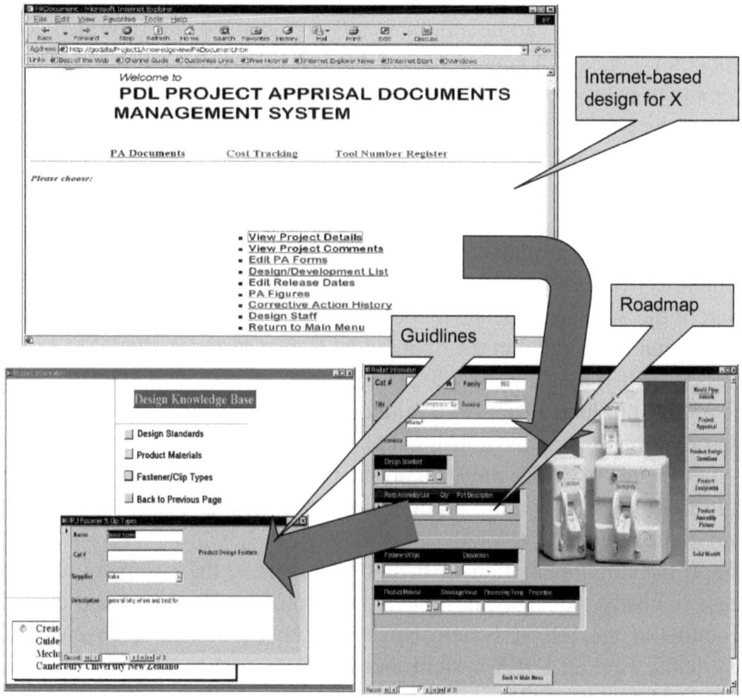

Figure 14.3 DFM via the Internet

as overly smooth surface requirements, a radius that is unnecessarily small, and tolerances that are unnecessarily tight. The objective of reducing the number of parts in the DFM can lead to highly integrated, complicated, and multi-functional parts. On the other hand, the DFM also aims to keep an individual part simple, because extremely complicated parts can hide a high cost that is not initially apparent.

As shown in Figure 14.3, the IDFM system includes the following main functions:

1. An Internet-based DFM data/knowledge base that contains specific guidelines organised in an appropriate form for easy access, which is addressed in Section 14.3.
2. Internet-based tools for data operations (*e.g.*, edit, search and retrieve) of the DFM data/knowledge base and other databases.
3. Various DFM checklists and guidelines, which provide examples of good practice, and remind the designer to check various manufacturing requirements and constraints.

Product designers can easily obtain the knowledge of a product design and manufacture by checking the manufacturing information from the product ID. The Internet-based DFM system is also supported by a computer software system that is able to score a product in terms of ease of assembly, assembly cost and time.

The DFM guidelines can be automatically reorganised and reformed according to the subcomponents of a new mould (or product). This will allow the manufacturing company to meet the dynamic changes in its global market, while reducing the workload of the product manufacture.

Based on the generic DFX model discussed in Section 14.3.2, a product data model is built based on the "common object concept". This data model can be used to model various moulds and can be accessed by using various PD tools (*e.g.*, a CAD/CAM system a product data management system, an Internet-based logistics management system, a computer aided management system, *etc.*). This common object is part of the STEP-based data model that is used to model various products. The information that is included in the data model is also stored in the DFM knowledge bases. As described by Tu *et al.* (2001), a product that is modelled in this system is called a "soft product". The soft product in Internet-based DFX systems provides a detailed roadmap for manufacturing a physical product, *e.g.*, an industrial switch as shown in Figure 14.3. The physical product is called a "hard product" by Tu *et al.* (2001). It is obvious that if a soft product is developed, a hard product can be rapidly produced in the company.

Through the implementation of this IDFM system, the company has reported a decrease in costly redesigns, design errors, manufacturing costs and cycle times. The structure of the IDFM system can easily be extended for developing other IDFX systems, *e.g.*, an IDFA system and an IDFC system. The DFA technology is closely linked to the DFM technology, but is oriented primarily to assemblies, semi-assemblies, and the final product.

From the discussions above, it is obvious that product developers, through IDFM systems (including the related product data management system), are able to access information that is useful for them to improve the design of the part. This is very important for rapidly developing new products. The IDFM system can be called up to analyse the current state of their designs, point out where the design is too complicated, and indicate possible areas of improvement. This is critical at the product design stage, especially in the earlier design stage. The DFM is best performed at the conceptual product design stage, before major decisions about the product and process characteristics have been finalised. At this early stage of design, there may not be much information to work with, but the IDFM systems will ensure that all the existing information can be made available to a product design team, which may be globally distributed.

14.4.2 IDFC for Economical Mould Development

This section presents an IDFC system developed to help mould designers predict the costs of manufacturing injection moulds and enable them to make more cost-effective design decisions. The main goal of the IDFC system is to estimate the PD cost during the early design stages through achieving the following objectives:

1. to determine the segment of a part model that may cause high PD cost;

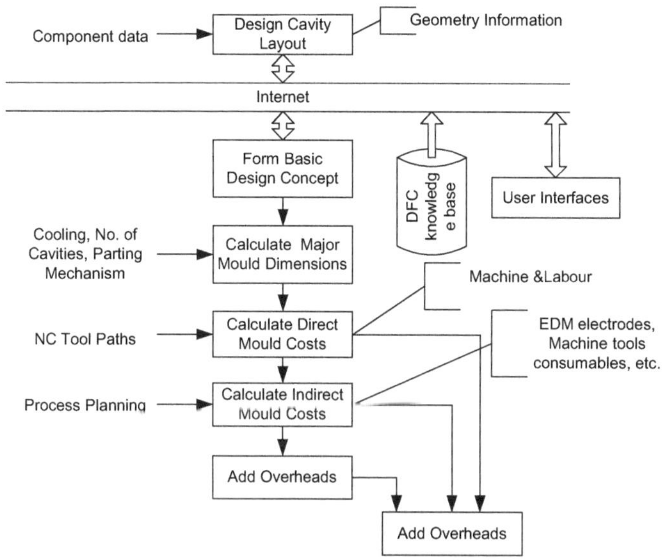

Figure 14.4 Basic structure of IDFC for developing plastic

2. to provide an environment to estimate alternative costs of possible designs;
3. to automatically calculate and optimise the PD cost and help decision-making at the early product stages.

In order to develop an IDFC software tool for early cost estimation, the steps required to estimate the manufacturing cost need to be identified. Figure 14.4 shows the cost estimation steps of developing an injection mould. Different from what is discussed in Chapter 6, the IDFC discussed in this chapter is more specific in cost estimation and optimisation, developing a mould that normally consists of basic common components. Its cost optimisation and estimation methods are also different from those used in sheet metal part development as described in Chapter 6.

In the IDFC system, the costing process normally consist of the following steps. First, to estimate the mould development cost, the mould-making processes need to be decided. The mould-making processes are decided according to the mould design specifications. Then the cost of making a mould is estimated by adding up the costs incurred in these mould-making process as well as the material costs. Usually, the cost estimate of making a plastic injection mould requires the incorporation of raw material, labour, machine setup and operating costs, and overheads. In real life, mould cost is estimated according to rough plastic part features or merely a mockup (*i.e.*, without the details of mould design) at the early design stages for the evaluation of different design alternatives or for providing a customer quotation.

To support RPD, in the cost estimation process, a number of product and process requirements and constraints in the areas of part features, mould design, mould-making processes, economics of mould-making and part moulding costs, *etc.* should be considered simultaneously. Hence, the IDFC system is designed to be used in a concurrent engineering environment to provide board density information, scrap and rework cost estimates and a breakdown of setup, labour and material costs for each step of the injection moulds development process. As shown in Figure 14.4, the IDFC system includes the following entities:

1. User and design interface: interfaces can be used by design teams/designers through the Internet.
2. Costing algorithms: developed to perform cost calculation and estimation (Gregory 1998, Tu *et al.* 2007, Xie 2001).
3. IDFC knowledge base/database: designed based on the breakdown costs and the detailed steps in the mould development process.
4. IDFC tools: designed to perform functions, *e.g.*, gather and present facts, provide suggestions for possible cost reduction, diagnose costly designs, *etc.*

As shown in Figure 14.5, the IDFC system also contains:

1. an Internet-based DFC data/knowledge base that contains specific guidelines organised in an appropriate form for easy access;

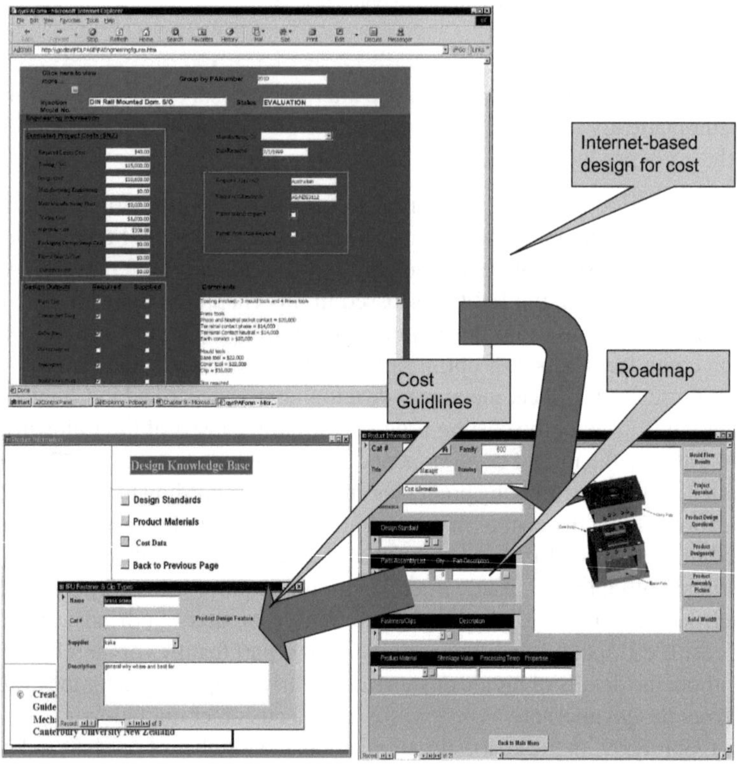

Figure 14.5 DFC via the Internet

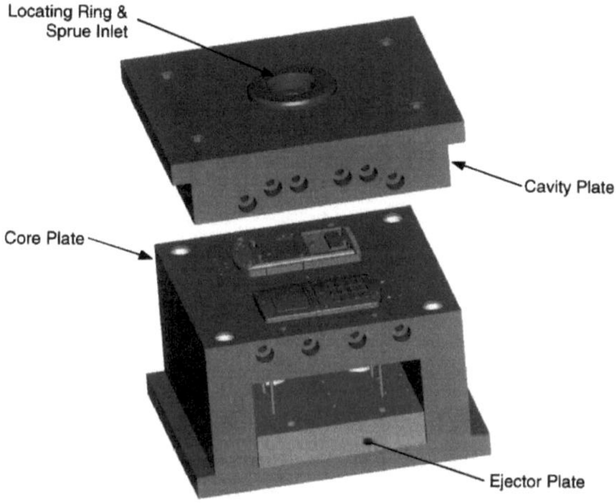

Figure 14.6 Two-cavity, two-plate mould for front and rear panels

2. Internet-based tools for cost calculation, data operations (*e.g.*, edit, search and retrieve) of the DFC data/knowledge base and other databases;
3. DFC checklists and guidelines, which provide examples of good economical design, and remind the designer to check various costly designs. Product designers can easily obtain knowledge of product design and cost by checking the detailed cost information from the mould product ID and feature ID.

The DFC guidelines can be automatically reorganised and reformed according to the subcomponents of a new mould (or product) and the generic DFX model. This will allow the manufacturing company to meet the dynamic changes in its global market by providing products with suitable prices.

To test the IDFC system with these cost functions, a case study has been carried out to determine the cost of making an injection mould, which is used to make the front and back panels of a home power manager as shown in Figure 14.6. In order to reduce the costs, the front and rear panels were moulded together in a two-cavity family mould. One of each is therefore produced with every injection cycle. The injection mould is shown in Figure 14.6. It is a two-way split injection mould with a 'Z gate' runner system. This runner system carries the molten polymer to the back of the component so that the injection point blemishes end up on the inside of the final assembly, unseen by users. The information in the pricing schedule makes it possible to form cost estimates for the various components of an average moulding tool.

As discussed in Section 14.3, a DFC knowledge base was built to model the mould PD information and cost information. The pricing schedule of the moulding tool provided by the manufacturing companies is modelled in the DFC knowledge base. These cost information data in the DFC knowledge base are used for the cost estimation

For the home power manager, the manufacturing costs can be estimated with the costing methods above, and are shown in Table 14.1. These values can be seen from

Table 14.1 Costs for manufacture of the injection mould

Item	Total cost
Raw mterials	$2308.00
Mould base cost	$1100.00
Heat treatment	$200.00
Freight	$300.00
Labour @ $53/h	$28,408.00
	$32,316.00

the Web page in Figure 14.5. The mould cost generated by the IDFC software tool is NZ$32,316. In order to calculate the labour cost for machining EDM electrodes and the cavity with the equations created above, the relevant ratios for estimating machining costs have been identified. A more detailed report for the machining costs is presented by Gregory (1998).

14.5 Conclusion

In this chapter, a new approach for developing IDFX systems is introduced for rapidly and economically developing plastic injection moulds. The structures of the systems are presented. A generic DFX model is introduced for developing various IDFX systems. Based on this data model, DFX knowledge bases for plastic injection mould development and Internet-based DFX (IDFM and IDFC) systems have been built. These tools utilise algorithms that predict the level of activity at each step in the manufacturing system, as driven by decisions about the design and production of injection moulds. The IDFM/IDFC systems allow engineers to try design alternatives, with immediate feedback given, to help lead them to cost-effective and rapid designs. These systems have several important features. First, the IDFM and IDFC systems can be used in concurrent engineering environments. Second, the developed IDFM and IDFC systems are based on a generic DFX model. Various software tools that are employed at different stages of the mould development can easily access this model without a middle data translator. Third, the management of the DFX data/knowledge base is relatively easy. Fourth, the DFX guidelines are organised in an appropriate form and can be automatically reorganised for modular products. Finally, the IDFX system can be easily integrated with other Internet-based PD software tools or systems.

References

Fang, N. S., Jiang, H., Luo, J. Z., 2000, Application of CORBA/SNMP gateway in Enterprise Intranet management, the International Conference on Advanced Manufacturing Systems and manufacturing Automation, Guangzhou, P.R.China.

Gregory, 1998, Project Report, An investigation into variant design of plastic injection moulded components. Department of Mechanical Engineering, University of Canterbury, Christchurch, New Zealand. Project No. 71

Huang, G. Q., 1996, Design for X. Concurrent Engineering Imperatives (London: Chapman & Hall).

Kim, Y., Kang, S.-H., Lee, S.-H., Yoo, S. B., 2001, A distributed, open, intelligent product data management system. International Journal of Computer Integrated Manufacturing, 14(2):224–235.

Mangesh, P., Bhandarkar, B. D., Hardwick, M., Nagi, R., 2000, Migrating from IGES to STEP: one to one translation of IGES drawing to STEP drafting data. International Journal of Computer in Industry, 41:261–277.

Olesen, J., 1992, Concurrent development in manufacturing-based on dispositional mechanisms. PhD thesis, Technical University of Denmark.

Tu, Y. L., Chu, X. L., Yang, W. Y., 2000, Computer-aided process planning in virtual one-of-a-kind production. International Journal of Computer in Industry, 41:99–110.

Tu, Y. L. and Xie, S. Q., Zhou, Z. D., 2001, An Information Modelling Framework to Concurrent Product Design and Manufacturing. IAM'2001, March 17–21, 2001, American University in Dubai, U.A.E.

Tu, Y. L., Xie, S. Q., Fung, R. Y. K., 2007, Product Development Cost Estimation in Mass Customization. IEEE Trans on Engineering Management, Vol. 54, No. 1, pp.29–40.

Wagner, R., Castanotto, G., Goldberg, K., 1997, FixtureNet: interactive computer aided design via theWWW. International Journal of Human–Computer Studies, 46(6):773–788.

Xie, S. Q., Huang, H., Tu, Y. L., 2002, A WWW-based information management system for rapid and integrated mould product development. The International Journal of Advanced Manufacturing Technology, 20(1):50–57.

Xie S. Q., Tu Y. J., Liu J. Q., Zhou Z. D., 2001, An integrated and Concurrent Approach for Compound Sheet Metal Cutting and Punching. International Journal of Product Research, 39(6):1095–1112.

Xie, S. Q., Tu, Y. J., Liu, J. Q., Zhou, Z. D., 2001, An integrated and concurrent approach for compound sheet metal cutting and punching. International Journal of Product Research, 39(6):1095–1112.

Yang, C. O., Pei, H. N., 1999, Developing a STEP-based integration environment to evaluate the impact of an engineering change on MRP. The International Journal of Advanced Manufacturing Technology, 15:769–779.

Zhang, Y. P., Zhang, C.,Wang, H. P., 2000, An internet-based STEP data exchange framework for virtual enterprises. International Journal of Computer in Industry, 41:51–63.

Chapter 15
Optimal Process Planning for Compound Laser Cutting and Punching Using Genetic Algorithms

Abstract Process planning has become an important stage for OKP companies as global competition has intensified the need to reduce the cost of products. This chapter uses sheet metal work as an example to discuss how new algorithms can be used to improve efficiency and to reduce cost. An optimal process planner can maximise the utilisation of costly raw material resources, improve machining efficiency, and hence reduce product cost. However, two problems must be overcome before such an optimal process planner can be developed; nesting and machining path planning. The nesting requirement is to maximise sheet metal material utilisation ratio by nesting parts of various shapes into the sheet. The path planning requirement is to optimise machining sequence so that the total machining path distance and machining time are minimised. This work investigates the two problems using genetic algorithms. The proposed genetic algorithm approach uses a genetic encoding scheme and a genetic reproduction strategy to reach an optimum solution. Case studies are carried out to test the genetic algorithms. The effectiveness of the genetic algorithm path planning approach is compared with that of the "ant colony" algorithm (Wang and Xie 2005). The results show that the genetic algorithm achieves better performances for path planning than the ant colony algorithm.

15.1 Introduction

In the sheet metal industry, there are many of small or medium sized enterprises (SMEs). These manufacturing companies have been facing competitive pressures in the marketplace. This pressure has forced them to make significant effort in shortening product development (PD) lead time, improve production efficiency, approach high quality standards, but at the same time cut down costs.

Compound machines (Xie *et al.* 2001), or combined punch and laser cutting machines (Clark and Carbone 1980, Wang *et al.* 2010), have been developed to increase the functionality and efficiency of sheet metal machining. By using this compound method, the cutting process and punching process can be carried out sequentially or concurrently on the same CNC sheet metal compound machine without altering the fixtures. This compound manufacturing method uses the high efficiency and low cost advantages of CNC punching as well as the high flexibility of CNC cutting for complex contour cutting (Xie *et al.* 2001).

Normally, sheet metal product design, process planning, and manufacturing are achieved using different computer aided software tools. They normally include a computer aided design (CAD) system (which takes care of product design), and computer aided process planning (which translates design information into process steps and instructions to efficiently and effectively manufacture products). Computer aided process planning includes a computer aided path planning (CAPP) system responsible for generating optimal tool paths and a computer aided nesting (CAN) system for optimal nesting of 2D parts with regular and complicated shapes (in order to effectively improve the utilisation ratio of sheet metal materials), as well as a computer aided manufacturing (CAM) system generating G and M-codes for different sheet metal processing machines (Xie and Xu 2006).

The machining path planning and the nesting problems are both combinatorial optimisation problems, which have been proven to have high computational complexity. In literature, the optimum path planning problem is traditionally addressed as a "Travelling Salesman Problem" (TSP), which has been the subject of research for many decades (Hwang and Ahuja 1992). The 2D sheet metal path planning problem is similar to the traditional TSP problem. It can be described as follows: given a set of starting and end points of the machining operations, such as cutting or punching, the objective is to find the shortest path of all of the points of a cutting process or a punching process. The distance between each pair of points is symmetric in sheet metal path planning optimisation.

The nesting problem is defined as the problem of finding an efficient layout of products to be cut in a containing region without overlapping. Its main objective is to maximise the use of material. The nesting problem is characterised by the intrinsic difficulty of dealing with geometry, satisfaction of the no-overlapping and containment constraints, and complex computation. Currently, there is a lack of practical algorithms in industry to nest complex and multiplex parts, which impedes the realisation of effective automatic nesting (Xie *et al.* 2001).

Although process planning tools have been used on general sheet metal cutting or punching machines, as well as compound machines, the optimisation of process planning dedicated to compound machines is limited. This work addresses the process planning problems for the compound punch–laser machine using genetic algorithms. The genetic algorithms are developed for optimisation of both the cutting and punching tool path as well as and the nesting of 2D parts with regular and complicated shapes. This enables future work on the efficient integration of the two algorithms for finding a global optimal solution for both nesting and path planning.

15.2 Literature Review

The branch and bound algorithm is an insertion algorithm (Nee *et al.* 1986) which does a truncated search on the entire solution space. The branching generates all the possible solutions available while bounding limits the search by not expanding a partial tour if it is already longer than the best solution. Computational experience with this method shows that there is a difficulty in setting the bound that limits the search without compromising optimality. The Clarke–Wright saving heuristic (Albano and Sapuppo 1980) is derived from a more general vehicle routing algorithm by choosing a point as a hub. The initial solution starts with the salesman returning after visiting every other point. The construction terminates when the hub is connected to only two other points. The best performance of this algorithm is better than that of the greedy algorithm.

Tabu search (Han and Na 1996) is based on the assumption that all locally optimal solutions are not good global solutions. Therefore, by minimising the randomised starting heuristic using a tabu list (a list close to the solution just found), a global optimum can be found. It is more effective than the original 2-opt and 3-opt since it considers only a tabu list instead of random starting points. However, the use of a tabu list in preference to the random starting heuristic restricts the algorithm, which in some cases "misses" the optimum path.

Meeran and Shafie (1997) implemented the convex hull boundary into a system for the given set of points as its initial subtour. Then a local search heuristic is applied successively until all the given points are included in the path. Every point is identified in a family hierarchy, hence the relationship between each point inside the convex hull boundary and the convex edges can be established without using a combinatorial search.

Considering the contours of the sheet products, the sheet metal nesting problem can be classified into two types: regular nesting and irregular nesting. The regular nesting problems typically considered is the 2D rectangular nesting problem. Lesh and Marks (2000) presented a "bottom-left-decreasing" (BLD) algorithm that includes successive random perturbations of the original four decreasing orderings. The bottom-left heuristic sorted the rectangles by decreasing width, but the heuristic is not competitive when sorted by decreasing height. Hopper and Turton (1999) solved a 2D packing problem frequently encountered in the wood, glass and paper industry, which consists of nesting rectangular shaped parts onto a rectangular object while minimising the object space used. The nesting process has to ensure that there is no overlap between the rectangular parts, which are allowed to rotate by 90°.

Qu and Sanders (1987) discussed a heuristic nesting algorithm for irregular parts and the factors affecting trim loss. They represented irregular shapes according to a set of non-overlapping rectangles. The system places each part in an orientation such that its length is larger than its height and always facing towards the bottom-left most direction. The parts are then sorted by non-increasing part height. The shapes are packed into a rectangular scene in a raster fashion, building up layers of intermeshed packed shapes.

Dori and Ben-Bassat (1984) were the first to investigate the nesting of shapes within a polygon rather than a rectangle. They made the assumption that the packing plane is infinite. The algorithm is only applicable to the nesting of congruent convex shapes. The problem involves cutting a number of similar but irregular pieces from a steel board. This is referred to as the template-layout problem.

Qu and Sanders (1987) proposed a heuristic nesting algorithm for irregular parts and the factors affecting trim loss. They represented irregular shapes according to a set of non-overlapping rectangles. The system places each part in an orientation such that its length is larger than its height and always facing towards the bottom-left most direction. The parts are then sorted by increasing part height. The shapes are packed into a rectangular scene in a raster fashion, building up layers of intermeshed packed shapes.

Wang and Xie (2005) addressed the process-planning problem for the combined punch–laser machine by integrating knowledge, quantitative analysis, and numerical optimisation approaches. The ant colony optimisation (ACO) algorithms were developed in searching the optimal tool path. Experimental results showed that their proposed method can significantly improve the operation efficiency for the combined punch–laser machine.

Based on our literature review, the optimisation of process planning dedicated to compound laser cutting and punching machines, is still limited. The research in this area is now behind the fast development of compound machines. This has significantly influenced the efficiency and productivity of compound machines. This work attempts to develop genetic algorithms for the optimisation of sheet metal cutting and punching processes.

15.3 Genetic Approach

The products to be machined in compound machines are normally small in size. They also have completely different shapes and sizes from one another. Normally, one product could need both cutting and punching operations. According to our experience, for a product that requires both cutting and punching operations, the punching operations will be carried out first. This is due to the fact that extra fixtures are required if the product is cut first. This is one of the constraints that needs to be taken into account in process planning.

One assumption needs to be made first before discussing the proposed genetic algorithm for path planning, which is that each product is represented by a starting

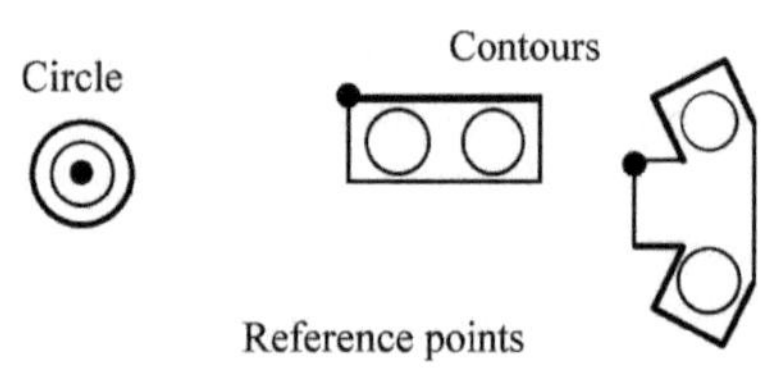

Figure 15.1 Reference points for different contours

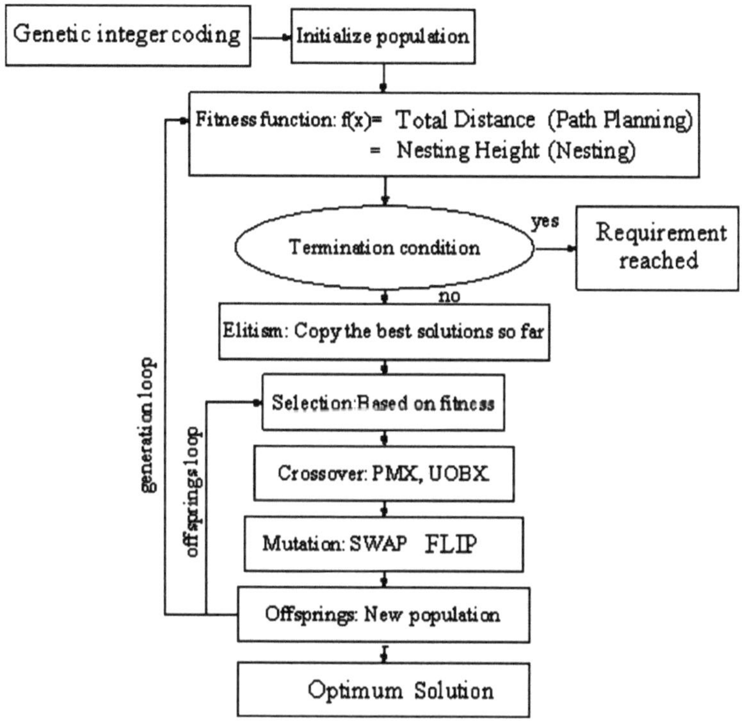

Figure 15.2 Structure of a genetic algorithm for path planning and nesting

point for the cutting operation (Xie *et al.* 2001) as shown in Figure 15.1. For punching operations, the centre point of the feature to be punched is used as the reference point.

When nesting the contour of a product in a computer, the contour is represented by a reference point, and the other vertex of this contour is calculated according to the reference point. When manufacturing a sheet metal product, the machining tool first goes to the reference point of this product. Then, different machining processes are carried out, such as auxiliary cutting path design. Figure 15.1 illustrates the reference point for three different product contours.

The main structure of the proposed genetic algorithm for path planning is showed in Figure 15.2. A genetic integer coding scheme creates an initial population for the genetic algorithm to find the optimum path. The generation loop records the best results found in each generation, while the offspring loop reproduces at each generation to generate the next generation. The process is terminated in two different ways.

1. Manipulated stop. The optimum requirement is reached. For example, in some cases, it is not necessary to have the optimal path since the computational complexity is proportional to the accuracy of the optimum.

2. Auto stop. The algorithm stops when it is designed to stop. For instance, if the optimum result is estimated to appear after around the 90th generation, then we can design the algorithm to stop at the 100th generation.

The evolving process of the proposed genetic algorithm contains two loops, the generation loop and offspring loop. The generation loop includes the offspring loop and it records the optimum path found in each generation and terminates the genetic algorithm when an ideal solution appears. On the other hand, the offspring loop evolves the solution path in each generation in an improving means using several operators, such as selection, crossover, mutation and replacement operators. In addition, an elitism scheme copies the optimum solution found in each generation to the next generation population, which ensures the genetic algorithm evolves towards the optimum.

15.3.1 Genetic Coding String

Each product is represented by an integer number. All the numbers combined form a genetic string. The genetic string contains the sequence of the product machining process. That is, which product is manufactured, which second, and so on.

Figure 15.3 shows an example representing a string of six products. The product numbers "4, 1, 6, 3, 5, 2" are considered in the same sequence for manufacturing these six products.

15.3.2 Initial Population

The size of the population significantly influences the search space and the computation time needed to reach an optimal solution. According to the literature, the population size is considered to be given by the length of the string (Hopper and Turton 1999).

15.3.3 Fitness Function

Since the objective of path planning is to find the shortest path for the machining operation, the fitness function equation is defined as follows:

$$1/F = D_{1,2} + D_{2,3} + \ldots + D_{i,i+1} + \ldots + D_{n-1,n}$$

4	1	6	3	5	2

Figure 15.3 A genetic coding string for manufacturing six products

where $D_{i,i+1}$ is the distance between the ith and $(i + 1)$th product to be manufactured, n is the total number of products to be manufactured.

On the other hand, the fitness function for genetic sheet metal nesting is

$$1/F = Height$$

where *Height* is the final height of the nested area (we assume that the nesting direction is along the height).

The fitness function for any genetic string represents the effectiveness of the string in the manufacturing process. During the genetic evolution process, strings with higher fitness values will survive and those with lower fitness values will be eliminated by the "survival of the fittest" principle in the genetic algorithm.

15.3.4 Genetic Reproduction

The main purpose of reproduction is to preserve the good strings in the population and try to generate better strings in the population. To achieve the first task in the genetic reproduction process, an elitism scheme is incorporated in the proposed genetic algorithm. The elitism scheme passes the fittest string into the next generation without any changes. However, the second task is achieved by using different genetic operators.

15.3.4.1 Selection

Two selection schemes are set to generate a mating pool from the previous generation but using different mechanisms. "Tournament" methods are more objective than "roulette wheel" methods as they choose a small group first, and then pick up the highest fitness within the small group.

In the roulette wheel method, a real-valued interval, *Sum*, is determined as either the sum of the individuals' expected selection probabilities or the sum of the raw fitness values over all the individuals in the current population. Individuals are then mapped one-to-one into contiguous intervals in the range [0, *Sum*]. The size of each individual interval corresponds to the fitness value of the associated individual. For example, in Figure 15.4 the circumference of the roulette wheel is the sum of all six individual's fitness values. Individual 5 is the fittest individual and occupies the largest interval, whereas individuals 1 and 4 are the least fit individuals and have correspondingly smaller intervals within the roulette wheel. To select an individual, a random number is generated in the interval [0, *Sum*] and the individual whose segment spans the random number is selected.

The tournament method work in three steps:

Step 1: Set a random number t as the tournament size.
Step 2: Choose t individuals from the population.
Step 3: Return the fittest individual of these t.

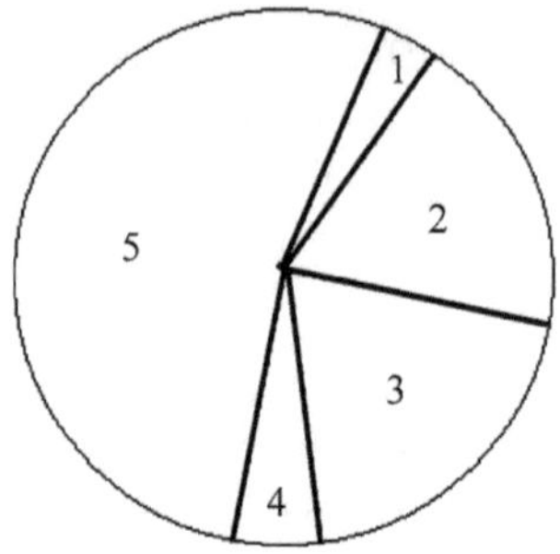

Figure 15.4 Roulette wheel selection

15.3.4.2 Crossover

Two alternative types of crossover operator are the "uniform order-based" crossover and the "partially matched" crossover. Each crossover operator is subjected to the probability of crossover.

Uniform order-based crossover (UOX): two parent paths are selected according to the probability of crossover. The UOX operator creates a mask of length equal to the parent whose position value is '1' or '0' generated randomly. Starting from the first position on the mask, if the value is '1', the two children inherit the same gene from the same position from the two parents; when the value is '0', the first child receives the corresponding gene from the second parent and the second child receives the corresponding gene from the first parent. For example, as shown in Table 15.1, the middle part of two paths is crossover by UOX, while creating new chromosome 1 and 2, the first, third and sixth genes are inherited from parents, chromosome 1 and 2, respectively, without change; while the second, fourth and fifth are switched. According to the values of the positions in the mask, the first value of the mask is 1, so the 10th gene value of the new path 1 is a copy the 10th gene value of path 1, while the 10th gene of new path 2 gets the 10th gene value of path 2. The second value of the mask is 0, then, the 11th gene value of the new path 1 is the same as the 11th gene value of path 2.

Partially matched crossover (PMX): Instead of a mask, two crossover points are generated randomly in PMX. First, PMX proceeds by position-wise exchanges between the two points. Then it maintains the crossed parts and transfers the rest that

Table 15.1 An example of applying uniform order-based crossover

Position of gene	1st	...	10th	11th	12th	13th	14th	15th	...
Chromosome 1	1	...	16	7	22	31	10	25	...
Chromosome 2	1	...	25	31	7	16	22	10	...
mask			1	0	1	0	0	1	
New_Chrom 1	1	...	16	31	22	7	10	25	...
New_Chrom 2	1	...	25	16	7	22	31	10	...

Table 15.2 An example of applying partially matched crossover

Position of gene	...	10th	11th	12th	13th	14th	15th	...
Chromosome 1	...	16	7	22	31	10	25	...
Chromosome 2	...	25	31	7	16	22	10	...
Step 1	...			7	16	22		...
	...			22	31	10		...
New_Chrom 1	...	31	10	7	16	22	25	...
New_Chrom 2	...	25	16	22	31	10	7	...

have the same value as genes in the crossed parts in each chromosome into the gene lost during the crossing operation.

After the gene positions are picked up for crossover, such as the 11th, 13th and 14th genes, the PMX operator changes the values between them as a first step, following an elimination step, which checks the same genes in one chromosome and replaces them with missing gene during the crossover process.

While shown in Table 15.2, the 12th and 14th positions are selected to be crossover points. The PMX operator processes in two steps. The integers between the 12th and 14th positions of chromosome 1 and 2 are exchanged in the first step. In the second step, since the 10th and 11th gene has the same integer as the 13th and 12th, respectively, the 10th and 11th location is changed to values which different from the integer value of crossing parts.

15.3.4.3 Mutation

While the crossover operation operates on two strings and changes the sequence in which the products are to be manufactured, the mutation operator operates on one string. Two types of mutation operators are applied in the proposed genetic algorithm.

Flip operator: This operates within two flip sites of a chromosome. The values insides the sites are reordered (inversed). For example, in a path chromosome, from 10th to 15th genes are reversed by the flip mutator as shown in Table 15.3.

Swap mutation: The values of two positions are switched under the swap operator. Table 15.3 shows the swap mutation, which exchanges 11th and 14th gene values.

Table 15.3 An example of applying two mutation operators

Position of gene	1st	...	10th	11th	12th	13th	14th	15th	...
Parent 1	1	...	16	7	22	31	10	25	...
Flip mutation	1	...	25	10	31	22	7	16	...
Swap mutation	1	...	16	10	22	31	7	25	...

15.4 Case Study

Case studies were carried out to test the feasibility of the proposed genetic algorithms for the two optimisation issues. The performance of the algorithms were compared with some existing algorithms described in the literature.

15.4.1 Sheet Metal Path Planning Problem

The purpose of this case study was to apply the proposed genetic path planning algorithm to an industrial case and to demonstrate the performance of the proposed algorithm in a sheet metal path planning process. Wang and Xie (2005) proposed an ant colony algorithm for solving the path planning issue for combined sheet metal machines. The results showed that the ant algorithm achieved a better performance than the normally used intuitive method. In this research, the ant algorithm is used as a comparison with the proposed genetic path planning algorithm.

Workpiece description. Figure 15.5 shows an example selected from Wang and Xie (2005). It includes a batch of workpieces used for the case study. There are four different types of components to be cut or punched: 104 small holes of diameter ø50, 31 holes of ø60, 31 ø100 contouring, 22 rectangular block contours and 30 clip contours.

Two machining methods were used in the case study: punching and cutting. Following the defined cutting and punching rules in Wang and Xie (2005), the 22 rectangular blocks and the 30 clips were to be cut, while the rest of the features were

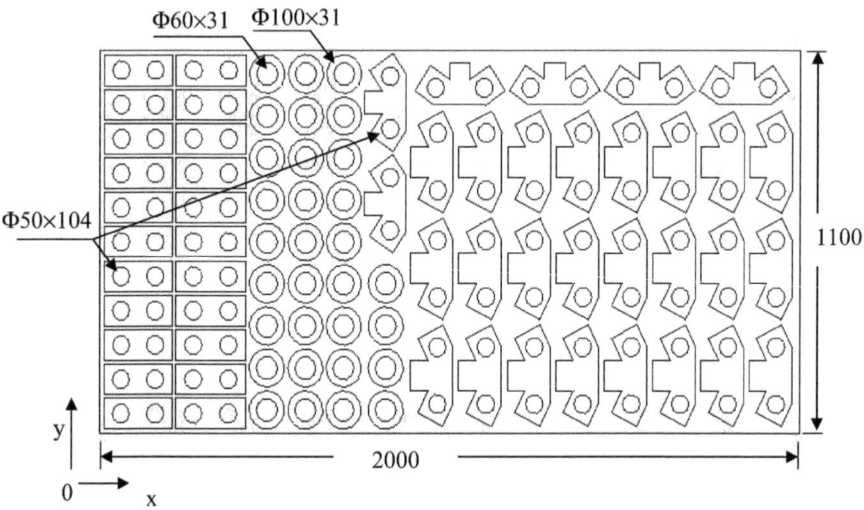

Figure 15.5 A batch of workpieces

to be punched. Since there are three sizes of hole features ø50, ø60 and ø100, three types of punch tool were used.

The case study investigates which algorithm of the two gives the shortest cutting/punching path. The optimisation result and the time that it takes to arrive at the solution are compared.

15.4.1.1 Genetic Path Planning

A genetic algorithm is developed to find the shortest machining path for the example as illustrated in Section 15.2. This algorithm includes the following key modules/steps: genetic encoding, fitness function definition, genetic reproduction and population replacement.

Genetic encoding. The coding strategy takes into consideration the notion of a point and a path. For example, a point is represented as a gene in a genetic coding string, which encodes to a positive integer. The points chosen depend on the types of features. However, a path is comprised of these reference points and is encoded into a chromosome, which is subsequently an ordered set of integers. Following the chromosome string from left to right, the order of integers in a chromosome is the order of the machining sequence.

Figure 15.6 illustrates the genetic path planning encoding according to the particular case. The name of each hole is represented as in the graph with integers from

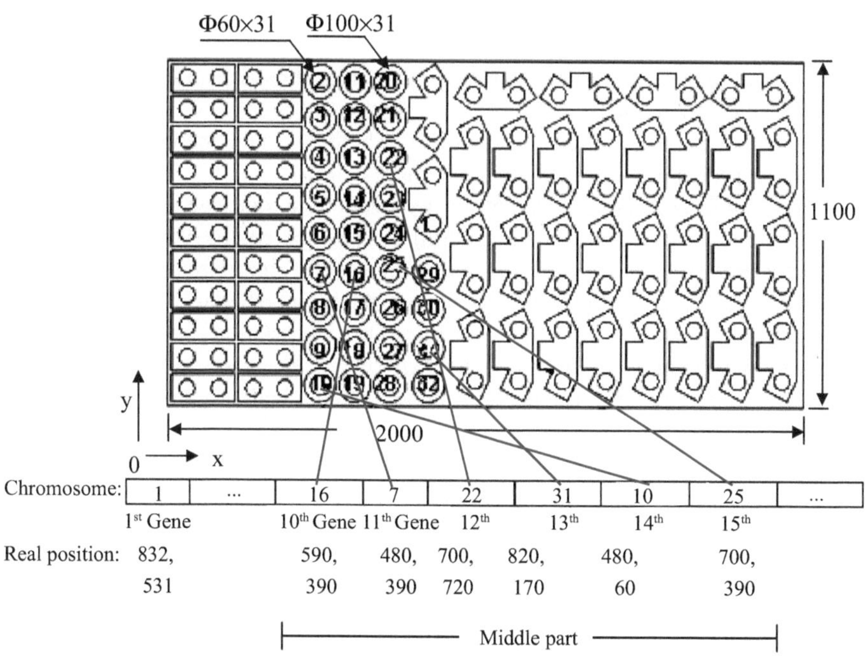

Figure 15.6 Genetic integer encoding scheme of path planning

1 to 32 as part identity integers. The chromosome in Figure 15.6 lists the possible machining sequence before optimisation is carried out. Each feature in the chromosome is represented as a gene. The genes are represented by 32 integers. Furthermore, the location of an integer in a chromosome string indicates the sequence of the machining operation. For instance, part of this chromosome is defined as a middle path for explanation, 16, 7, 22, 31, 10 and 25. Integer 16 places at the position of 10th gene, which means feature 16 is the 10th feature to be machined in terms of this path chromosome. In a similar manner, feature 7 is the 11th feature to be machined, then feature 22, 31 and 10. Feature 25 thus is the last one to be machined.

Fitness function. According to the fitness equation defined in Section 15.2, the chromosome illustrated before the fitness function of the middle parts is calculated as an example.

$$f(x) = 1/(d_{16,7} + d_{7,22} + d_{22,31} + d_{31,10} + d_{10,25})$$

where $d_{16,7}$ is the distance between feature 16 and 7.

Genetic reproduction. Several genetic operators are set up for reproduction. Roulette wheel selection and tournament selection are alternative methods. Partially matched crossover and uniform crossover with high crossover probability are two options for genetic crossover. Two points swap mutation and flip mutation operators are designed to avoid convergence in the proposed genetic algorithm.

15.4.1.2 Experimental Results and Discussion

ø60 punching path optimisation. Figure 15.7 shows an optimal path (solid line) found by the genetic algorithm, while the ant algorithm gives a different path as shown by the dotted line.

The genetic algorithm used consists of 81 individuals in the mating pool initially, a crossover probability of 0.8 and a mutation rate of 0.1.

Table 15.4 shows a comparison of the two machining routes generated by the ant algorithm and the proposed genetic algorithm. The total lengths of the two machining routes are given, and it can be seen that the genetic algorithm produces a better search result. The algorithm is able to find a machining path that is 169 mm shorter than that found by the ant colony algorithm. The computation time is also shorter than with the ant colony algorithm. In the experiment, the genetic algorithm runs 10 times with 500 loops. The results are recorded together with those of the ant colony algorithm in Table 15.5.

Table 15.4 Performance comparison between the genetic algorithm and the ant colony algorithm

	ø60 punching Time (s)	Path length (mm)
Genetic algorithm	15.51	3546
Ant colony algorithm	23	3715

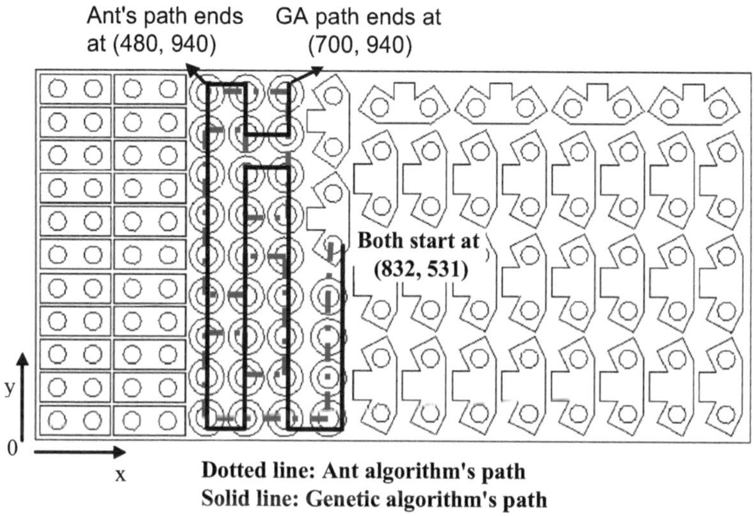

Figure 15.7 Optimum paths found by the genetic algorithm and the ant algorithm

As shown in Table 15.5, the average performance of the proposed genetic algorithm is much better than that of the ant colony algorithm. Although both algorithms are evolutionary algorithms, the genetic algorithm is superior to the ant colony algorithm with regard to time in this path planning task. This is important, especially when the optimisation issue becomes more complex.

Genetic parameter investigation. An ideal choice of the genetic parameters guarantees better exploration of the solution space and quicker convergence towards the optimal solution. A good balance between crossovers and mutations is needed. Crossover allows the exploration of a wider neighbourhood of solutions, while mutation allows diversification of the population. Similarly, a tradeoff between the size of the population and the number of generations is necessary to guarantee good quality solutions in short run times.

In this case study, there are in total 32 parts to be punched. The optimisation of the path which a punch tool travels involves a large number of combinations. The results of different combinations of crossover and mutation operators, crossover possibility and mutation possibility, generation loops are listed in Appendix A.7. The results indicate the combination of a partially matched crossover with crossover possibility

Table 15.5 Optimum punching path results using the genetic algorithm (GA) and ant colony algorithm (AC)

AC	Time (s)	22.89	22.87	22.83	22.73	22.66	22.72	22.83	23	22.89	23
	L (mm)	3611	3674	3542	3993	3827	3545	3611	3591	3452	3715
GA	Time (s)	16.14	15.95	15.89	15.75	16.15	15.95	15.78	15.67	15.92	15.51
	L (mm)	3499	3477	3540	3545	3560	3542	3568	3499	3495	3546

Table 15.6 Optimum cutting path results using both ant colony algorithm and genetic algorithm

Ant colony algorithm	Loops=60	Loops=80	Loops=100	Loops=120
Time (s) Length (mm)	156.547 9663.64	227.422 9716.586	286.094 9501.618	396.047 9492.532
Genetic algorithm	pop = 81, Loops = 3000	pop = 81, Loops = 5000	pop = 151, Loops = 3000	pop = 151, Loops = 2000
Time (s) Length (mm)	216.844 8914.5	544.688 8225.5	197.016 8637.7	288.016 8278.4

Table 15.7 Optimum paths results of punching and cutting

Cutting optimisation	Time (s)	Path length (mm)
Genetic algorithm	321.234	7911.7
Ant colony algorithm	–	9348 – 212.64 = 9135.36

Note: 212.64 mm is the distance between last second point (1856, 110) and end point (2060, 50). The end point means the machining header's final position after the whole manufacturing process.

of 0.9, a flip mutation with mutation possibility of 0.1 and initial population size of 61 yields the optimum path at around the 20,000th generation.

Cutting path optimisation. The encoding scheme and genetic operators are set up as for the punching path optimisation, but this genetic algorithm has a large size of initial population, 151, since it is a large problem size. Table 15.6 presents the different numbers of generations (loops) and population sizes that are tested and summarised.

Table 15.6 shows that the proposed genetic algorithm has better results than the ant algorithm in terms of time and quality.

Paths integration. It is time to consider the punching and cutting machining process at the same time, since in real manufacturing settings the two processes are operated sequentially. The starting point is the same in the two algorithms to enable accurate comparison. The optimum paths are plotted on the same graph in Figure 15.8.

The total distances of the cutting process using the two algorithms are shown in Table 15.7.

It is undeniable that a significant improvement is achieved using the genetic algorithm. A total shortening in the machining path of 1223.66 mm is obtained in this example. The genetic algorithm is able to deliver the result within a fixed time frame (321.234) while ant colony algorithm has no guarantees. Even running the ant algorithm 10 times did not provide results as good as the genetic algorithm result. During the investigation of both algorithms, the genetic algorithm finds the optimum quickly while the ant algorithm improves smoothly with time.

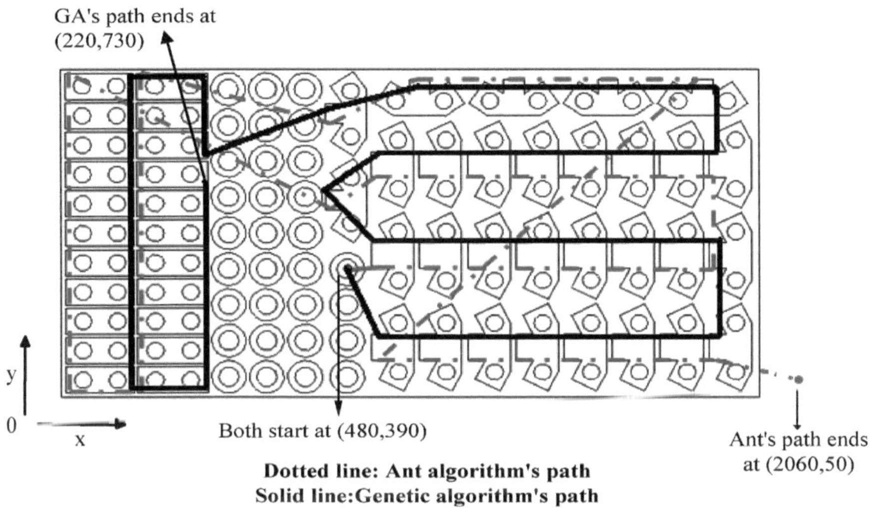

Figure 15.8 Optimum paths of punching and cutting machining process generated by ant algorithm and genetic algorithm

The results found in this case study reported in this section suggest that the proposed genetic algorithm could be an effective optimisation methodology: it has proved its performance in a sheet metal path planning optimisation problem.

15.4.2 Sheet Metal Nesting

The difference between sheet metal path planning and nesting is the fitness function. A bottom left heuristic algorithm is applied in the sheet metal nesting fitness function. In this chapter, we are concerned only with the rectangular sheet metal nesting problem.

The case study is carried out nesting six rectangular items of various sizes. The genetic coded nesting string contains only integers, which is a sequence of parts coding identifications. A typical string is formulated as shown in Table 15.8.

In this particular string, the six elements represent the six nested parts. In view of the complexity of the optimal placement of parts, the present approach considers nesting the parts in a sequential manner. Part 5 with size 15×25 is the first one placed

Table 15.8 A string coded by the genetic algorithm and corresponding dimensions of the nested parts

5	4	1	2	3	6
15 × 25	20 × 10	10 × 30	15 × 35	25 × 20	20 × 10

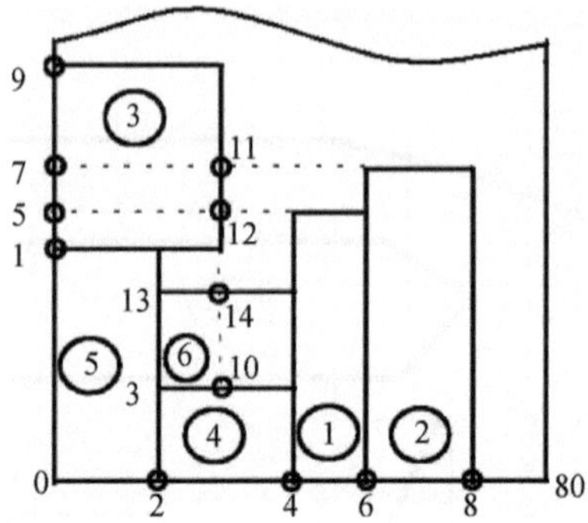

Figure 15.9 Heuristic placement considering the coding string: 5 4 1 2 3 6

in the heuristic algorithm, with Part 4 placed second. Following the sequence, Part 6 is the last to be nested into the sheet.

The heuristic algorithm considers each rectangular part in the same order that appears in the string to generate a nesting pattern. To nest any particular parts, the parts are arranged from the bottom left position of the sheet. It is ensured that the part does not overlap previous parts or cross the boundary of the sheet. After placing each part on the sheet, new positions for the next part to be placed are identified. The positioning of parts on the sheet is based on a 2D translation. For translating the part to any node, the bottom-left corner of the part, chosen as the reference point, is made to coincide with that node. The following procedure is adopted for the generation of the placement coding string shown in Table 15.8.

Step 1. The first rectangular part in the sequence, *i.e.*, Part 5 (15 × 25 mm), is translated to the bottom-left corner of the first rectangular sheet in the sequence, *i.e.*, the third sheet, in such a way that the first dimension (15 mm) is positioned along the x-axis and the second dimension (25 mm) along the y-axis. After translating this part, positions, *e.g.*, left-top corner and right-bottom corner of the part are obtained to position the next part. These two nodes are represented as nodes 1 and 2 in Figure 15.9.

Step 2. The next part in the sequence is Part 2 (20 × 10 mm), which is translated to node 2 since this particular node is located at the bottom-left position on the remaining sheet. This part is positioned with 20 mm side along the x-axis and 10 mm side parallel to the y-axis, as shown in Figure 15.9. Nodes 3 and 4 are identified as new nodes to translate the next part on the sheet. In a similar manner, Parts 1 and 2 are also nested, and nodes 5, 6, 7 and 8 are identified as the new nodes to arrange the next part. Nodes 5 and 7 are obtained by projecting the top horizontal edges of Parts 1 and 7 onto the vertical edge of the sheet.

Step 3. The next part in the string sequence is Part 3. It is translated to the first node where the part does not overlap with those that are already nested and ensures the bottom-left-most position. After translating the part, nodes 9, 10, 11 and 12 are identified as new nodes. Node 10 is obtained by projecting the right vertical edge of Part 5 onto the top horizontal edge of Part 2. Nodes 11 and 12 are projections of horizontal edges of Parts 1 and 7 on the right vertical edge of Part 3. In a similar way, the next and the last part in the sequence, is translated to the third node. Node 14 is obtained by projecting the right vertical edge of Part 5 onto the top horizontal edge of Part 3.

With the bottom left heuristic, the example given attains a material utilisation of 75 % using the proposed genetic nesting approach.

15.4.3 Discussion

The case studies show that the proposed genetic algorithms are able to provide good solutions to sheet metal machining problems. It has been shown that the genetic algorithm is able to produce better results than the ant colony algorithm that was originally developed in our research group. The experimental results using the proposed genetic algorithm show that the algorithm is able to find a better machining path. This is achieved by designing effective genetic operators, which can improve algorithm performance towards the optimum.

A genetic nesting algorithm is also proposed for solving a rectangular nesting problem. However, this area requires further work to accommodate the nesting of parts with irregular shapes (Xie and Xu 2006), which is more often encountered in the sheet metal manufacturing process. The problem becomes more complex as the shape of the parts becomes more complicated. Future studies will explore how a genetic algorithm can be developed for this nesting issue. The ant colony algorithm will be considered an alternative methodology for the nesting of irregular parts.

It is well known that the sheet metal path planning process is conducted after a computer aided nesting process, which generates a compact layout based on minimising the wastage of sheet materials. However, this does not take into consideration the efficiency of manufacturing processes. It is better, from the operation efficiency point of view, to put products that have the same types of operations together. Future study will be made of the global optimisation of both path planning and nesting integration.

15.5 Conclusion

A new search methodology based on an evolutionary process is introduced and studied in this work and its application to the solution of a classical optimisation problem in the sheet metal manufacturing industry, sheet metal machining path planning and sheet metal nesting issues.

This research investigates optimal process planning issues in sheet metal PD. Genetic algorithms are proposed and developed for sheet metal part path planning and nesting. Case studies are carried out to demonstrate the performance of the proposed algorithms used for path planning optimisation issues.

For the sheet metal path planning optimisation problem, the proposed genetic algorithms are tested in a sheet metal industrial case against the ant colony algorithm. The proposed genetic algorithms are programmed in Matlab. The experimental results are compared with results using the ant colony algorithm. The performance of the proposed genetic algorithm is better than that of the ant colony algorithm.

References

Albano, A. and Sapuppo, G., 1980, Optimal location of two-dimensional irregular shapes using heuristic search methods, IEEE Transactions on Systems, Man, and Cybernetics, 10(5), 242–248.

Clark, S. C. and Carbone, V. T., 1980, Laser Cutting Head Attachment for Punch Press. United States Patent and Trademark Office, Houaille Industries, Inc., United States, Patent No. 4201905.

Dori, D. and Ben-Bassat, M., 1984, Efficient nesting of congruent convex figures. Communications of the ACM, 27, 228–235.

Han, G. C. and Na, S. J., 1996, Two-stage approach foresting in two-dimensional cutting problems using neural network and simulated annealing, Proceedings of Institution of Mechanical Engineers, Part:B Journal of Engineering Manufacture, 210, 509–519.

Hopper, E. and Turton, B., 1999, A genetic algorithm for A 2D industrial packing problem, Computers & Industrial Engineering 37, 375–378.

Hwang, Y. and Ahuja, N., 1992, Gross motion planning: a survey, ACM Computing Surveys, 24(3), 219–291.

Lesh, N. and Marks, J., 2000, Human guided simple search. A Mitsubishi Electric Research Laboratory, http://www.merl.com.

Meeran, S. and Shafie, A., 1997, Optimum path planning using convex hull and local search heuristic algorithms, Mechatronics, 7(8), 737–756.

Nee, Y. C., Seow, K. W. and Long, V. C., 1986, Designing algorithm for nesting irregular shapes with and without boundary constraints, Annals of CIRP, 35(1), 107–110.

Qu, W. and Sanders, J. L., 1987, A nesting algorithm for irregular parts and factors affecting trim losses, International Journal of Production Research, 25, 381–397.

Wang, G. G. and Xie, S. Q. 2005, Optimal process planning for a combined punch and laser cutting machine using ant colony optimisation, International Journal of Production Research, 43(11), 2195–2216.

Wang, Q., Teng, N., Cai, D. M., Zhang, S. H., 2010, Study on Deformation Mechanism of Tube Hydropiercing. Advanced Materials Research, Vols. 97–101, pp. 166–169.

Xie, S. Q. and Xu, X. 2006, A STEP-compliant integrated system for sheet metal parts, International Journal of Integrated Manufacturing Technology, 19(6), 627–638.

Xie, S. Q., Tu, Y. L., Liu, J. Q. and Zhou, Z. D., 2001, Integrated and concurrent approach for compound sheet metal cutting and punching. International Journal of Production Research, 39(6), 1095–1112.

Chapter 16
Conclusion and Future Work

Abstract The last chapter starts with a brief summary of the goal of the book and the future research problems faced by OKP researchers. With respect to the research objectives as stated in Chapter 1, the major achievements or research findings are summarised in Section 16.2. Section 16.3 discusses possible future research directions and topics.

16.1 The Goal of this Book

The primary goal of this book is to discuss the possible technologies for improving OKP companies' rapid PD ability so that they can remain profitable in the globally and dynamically competitive market. In order to realise this goal, it is necessary to develop new OKP systems, strategies, methods and algorithms for OKP companies to handle these globally and dynamically competitive issues. The development and integration of new technologies and methodologies under the global heterogeneous environment have been the main problems faced by OKP researchers. This problem can be addressed by an Internet-based integrated PD platform, as proposed in this book. To apply this PD system in OKP sheet metal part/product or tool/mould development, modules, which are important entities of the system, and relevant methods, algorithms and concepts, have been developed through this work. These modules, methods, algorithms and concepts make good examples and will be further summarised in the following section.

16.2 Major Achievements and Research Findings

The work reported in this book is funded by the International Investment Opportunity Fund (2008) and the Public Good Science Funding (1999) of the Foundation for Research, Science and Technology (FoRST) in New Zealand. The highlights and achievements of this research work through government funded research projects include development of reference system architecture, methods, algorithms, concepts and the necessary computer software tools and interfaces for Internet-based rapid OKP development, and four software packages. The main findings of this research have been published in international journals and conference proceedings. The source codes of the main software tools and interfaces are attached in the Appendices and include:

1. A software platform for integrating different engineering software systems on the intra/Internet. This software platform was developed using computer intra/Internet communication and interfacing technology. It was designed as a window-based and user-friendly software system. It can be used to integrate different kinds of computer aided engineering software packages, *e.g.* CAD system, CAPP system, cost analysis system, manufacturing process simulation system, and shop floor operation simulation system, to effectively support the global PD practices of OKP companies. This software platform includes relevant databases and knowledge bases as presented in this book.
2. A computer aided product design, process planning and manufacturing software package for intelligent and economical PD.
3. A WWW based software package for communication and data management on the intra/Internets during a collaborative product design and manufacturing process between an OKP company and its partners/customers.

The main research findings and contributions presented in this book are summarised below.

1. *Current rapid OKP PD systems, methods and technology review and comparison.* An extensive literature review was provided in Chapter 2 on the historical background of OKP and Internet-based product design and manufacture systems. From this literature review, the current approaches to rapid OKP PD were evaluated and discussed against the needs for recently emerging global OKP PD practices. The gaps and problems between the current technological state and the practical needs have been identified. These gaps can be filled and problems can be solved through the development of an Internet-based computer aided PD system and a global collaborative PD concept and method. A reference system architecture for the Internet-based rapid OKP PD system was developed and proposed at the end of Chapter 2 of this book.
2. *Internet-based integrated PD system for rapidly developing OKP parts/product.* This is an application of the proposed reference system architecture to the rapid development of OKP sheet metal parts/products. From this successful industrial test and implementation, the feasibility of the proposed reference system architecture has been validated. To meet some special needs for sheet metal part/product development, a number of technical modules and computer interface were developed as entities of the system. These include system structure, a RTCAPP (real time computer aided process planning) module, customer interfaces, a simulation platform and a design/manufacturing knowledge bases

module for supporting concurrent product design and manufacturing. The definition and the structure of the integrated data structure for concurrent sheet metal part/product design and manufacturing was also given in Part II of this book. This Internet-based rapid OKP sheet metal part/product development system has been tested with several industrial cases. These industrial implementations clearly show the advantages this Internet-based OKP development system has over traditional systems.

3. *Information modelling framework for intelligent and concurrent OKP parts design and manufacturing.* As information integration is an important issue in supporting integrated and concurrent PD, *i.e.* rapid PD, the structure of an information framework for concurrent design and manufacturing of OKP sheet metal parts/products was developed and presented in Part II of this book. This information framework was developed to build an information bridge to fill the electronic data communication gaps among the sheet metal part/product design system, process planning system and manufacturing/production system. The principles of zero thickness and zero bend radius were proposed, which can be used to abstract the geometry entities of sheet metal parts in order to facilitate part modelling and information modelling. A tree-based information modelling methodology was also presented in Chapter 4 as an important concept and method to support the implementation of this information framework. This information framework enhances the overall system integration of the Internet-based rapid OKP sheet metal part/product development system. Its feasibility was tested by an industrial case study as reported in Part II of the book.
4. *Compound OKP product manufacturing process.* To speed up an OKP sheet metal part/product development, a compound cutting and punching method was developed and tested in an industrial case study. This compound sheet metal manufacturing process can simplify the sheet metal cutting and punching process planning and thereby simplify the part/product design. This simplification can significantly reduce a sheet metal part development cycle time. To implement this compound sheet metal manufacturing process, an integrated CAD/CAPP/CAM system was developed and presented in Chapter 5 owing to the fact that existing commercial CAD/CAM systems are not suitable for this manufacturing method, especially under a concurrent and global PD strategy. Some problems have to be solved before these existing commercial CAD/CAM systems can be employed and integrated for this compound manufacturing method. Some solutions to solve these problems were suggested. These include an integrated data exchange platform developed on the Pro/INTRALINK of the Pro/ENGINEER (a commercial CAD/CAM package) and in the format of STEP, and a knowledge-based RTCAPP system for compound sheet metal cutting and punching. In addition to these, some modules were also developed as entities of the integrated CAD/CAPP/CAM system. These include an automatic tool selection and manufacturing process planning module, a shortest tool path optimisation module, and an automatic insertion of auxiliary path module based on knowledge bases.

5. *Cost estimation and optimisation for the rapid development of sheet metal parts/ products.* To provide in-time cost estimates and optimisation results to support decision-making of product developers in RPD, a feasible computer aided automatic cost estimation and optimisation system was developed. This system was built on a Cost Index Structure, which is generic and can be applied to all kinds of PD. The cost estimate methods and optimisation algorithms are effective. They can be easily understood by and implemented in industry. This concurrent approach enables product developers to design the product, plan the processes, schedule operations on the shop floor, manage logistics, and estimate and optimally control the cost feature by feature.
6. *Integrated product data model for PD life cycle.* An integrated product data model for the PD life cycle was developed. The emphasis of this research work was placed on integrated data management and knowledge reuse to support a company's practice to shorten its PD cycle. The integrated data structure was modelled utilising EXPRESS of STEP. In terms of the data structure, a design/manufacturing knowledge base was developed as a major part of the WWW-based PD system. This data structure was successfully applied to support the rapid development of an injection mould.
7. *Internet-based product information management system for rapid OKP product development.* To facilitate the complicated data exchange and communications between a master OKP company and its branches or partners via intra/Internet, an Internet-based product information management system was developed. This information management system includes a WWW-based, distributed and object oriented databases and knowledge bases for product design and manufacturing; a WWW-based database management system; Web browser interface to link a local database to a web page browser; and an integrated software platform for the integration of CAD, CAPP, CAM and WWW based information management systems. This Internet-based information management system was implemented in a New Zealand manufacturing company, which has its headquarter in Christchurch and a number of branches or partners in other places in New Zealand and overseas.
8. *Internet-based DFX for rapid OKP product development.* The application of DFX, which is an effective method in concurrent engineering for RPD, to the global and collaborative OKP PD was tested in this work. An Internet-based DFX or IDFX system was developed for rapid tool/mould making. The structures of the system were presented and discussed in Chapter 9. A generic IDFX system structure was further introduced in Chapter 9 for the development of various IDFX systems. Based on this system structure, IDFX knowledge bases for a plastic injection mould development were created and the Internet-based DFX (*i.e.* Internet-based design for manufacturing (IDFM) and Internet-based design for cost (IDFC)) systems were built. The IDFM/IDFC systems allow engineers to evaluate alternative designs with immediate feedback on manufacturability and cost. This allows them to quickly come to an effective and economical design.

16.3 Future Work

The increasing level of customisation and the demand for individualised products has been identified as a clear trend in the international manufacturing market, due to this, more and more manufacturing companies are being forced to move towards OKP and to develop OKP products at a faster pace. Hence, the Internet-based rapid OKP product development as presented in this book will become an important research area in the development of a new generation of manufacturing enterprises and systems.

It can be seen from this book that the development of suitable systems, methods and technology for rapid OKP PD in a global and collaborative environment is a large and extensive research task. Although much research effort has been reported in this book, there are still unsolved problems. Research work as reported in this book can be further extended and some possible research directions have been identified and summarised below.

16.3.1 Extending the Internet-based Rapid OKP Product Development System

In Chapter 2, the structure and major components of the Internet-based rapid OKP product development system have been defined. However, for the development of different OKP products, some of the components may need to be different. As illustrated in Chapter 3, the PD system for sheet metal part/product development contains some specific modules such as nesting, process planning and costing. Hence, to be able to use the integrated PD system for rapidly developing other OKP products, further research should be carried out on the compatibility and interoperability of the proposed reference system architecture.

16.3.2 Continually Upgrading the Internet-based Rapid OKP Product Development System

As the PD system depends on new technologies and theories, the proposed reference system architecture has to be continually upgraded with the development of modern technology. New technology may affect data standards and modelling methodologies, knowledge modelling methods and AI support, global optimisation algorithms, and Internet communication. Only by continuous upgrades, can the integrated OKP PD system keep its advantage over other systems and help companies face the increasing market competition.

16.3.3 Finding New Implementing Methodologies

Internet technology can be used by most global manufacturing companies. Research efforts as reviewed in Chapter 2, as well as the research work reported in this book, have been aimed at the development of various kinds of RPD systems. However, it is still not clear how effectively these systems and methods can be implemented to support rapid OKP PD in a global and collaborative environment. In particular, some software interfacing and development tools and methods are needed to efficiently develop and effectively implement this type of Internet-based PD system. These tools and methods will be able to provide a good connection between modern computer technology and a rapid OKP PD process. They can be used to decrease the time spent on programming work and shorten the time used to update an existing Internet-based rapid OKP product development system. This research requires cooperation from both computer science and manufacturing areas.

16.3.4 Achieving High Level System Integration

As IRPD requires the integration of people, business processes and information technology through a whole PD cycle, a high level system integration across all the organisational functions in an OKP company as well as its suppliers and customers is required. The information of a product on its design, manufacture, use, maintenance and disposal is not only shared between the departments within a company, but also shared between the "partner" companies around the world (Chiu *et al.* 2006, Wagner *et al.* 1997). This wider information sharing practice in today's manufacturing approach requires system integration that covers not only the information sharing through a central data base in tradition, but also the technologies for product design, process planning, computer simulation, manufacturing, knowledge reuse and optimisation algorithms applied in a PD. To meet this demand of high level system integration, new data structures, models and software developing tools need to be developed. This includes: 1. life-time product data structure and models for integrating various computer aided engineering systems in a PD cycle (Tu *et al.* 2001); 2. definition of generic attributes and consistent data models for PD processes; 3. further development of STEP and improving its standardisation and system interoperability.

16.3.5 Developing New CSCW and Interfacing Techniques

Since an IRPD system requires the cooperation of various engineering systems over a network environment, CSCW and interfacing techniques are important for developing the IRPD systems. They directly affect the accuracy and efficiency of the data management through a whole PD cycle. However, early as well as recent experience has shown that CSCW technologies have not had much success in its application for the development of Internet-based PD systems (Lubich 1995). Critical problems result mainly from the limitations of the technology itself and the compatibility with other technical systems and social norms.

On the other hand, the interfacing techniques in IRPD systems should have the following capabilities: 1. workbench (common users interface to multiple PD applications); 2. process modelling and workflow management; 3. application encapsulation and invocation; and 4. application data translation and transfer and multi-vender networking and platform support. The common data exchange standards used, *e.g.* CAD-NT, CGI, CGM, SET, PDES, STEP, and IGES, do not have the functionalities to directly support interfacing via variant application software packages. Although great efforts have been made by us and many other researchers, *e.g.* Wang and Bourne (1995), Lau and Jiang (1998), Dorador and Young (1999), Tu and Xie (2000), Tu *et al.* (2001), Huaglory (2001), to resolve this problem, the research into CSCW and interfacing techniques for the purpose of RPD is still in an early development stage and further research is still needed. Interesting research topics for IRPD include the use of STEP to represent non-geometric information (Wang 1995) and the combination of Internet modelling methodology (*e.g.* HTML, CORBA, etc.) with STEP (Hardwick *et al.* 1996), and Internet-based data translators between various data formats (Heck *et al.* 2009, Zhang *et al.* 2000). These would be interesting research topics for IRPD.

16.3.6 Expanding the Information Integration Framework

The information integration framework as proposed in Chapter 4 is an important technology to improve system integration and was successfully used to fill the information gap between the design and other downstream sheet metal part/product development processes. Although this information integration framework was developed for rapid OKP sheet metal part/product development, we believe it can be further developed as a generic information integration framework for other OKP product development.

16.3.7 Improving the Cost Estimation and Optimisation Algorithm

From the foregoing discussion, it is clear that cost estimation is an area of great research interest. New technologies such as neural networks (Ju and Xi 2008, Ehrlenspiel and Schaal 1992; Bode 1998, 2000), activity based costing (Tamas *et al.* 2000), log–linear and non-linear learning curve model (Timothy 1999) AI and mathematical models (Xie *et al.* 2001) have been widely researched. However, a widely accepted costing system or a system with wide applicability has not been found. Chapter 6 proposed a cost estimate and control methodology for rapid sheet metal part/product development. The cost optimisation algorithm as presented in Chapter 6 was based on a multi-objective algorithm with the support of cost resources. This algorithm works well when all the needed information is available. However, at the very early design stage, there is usually not enough design or downstream information available for the estimation algorithm. New and intelligent cost estimation and optimisation algorithms need to be further researched.

16.3.8 Further Development of Internet-based DFX

In Chapter 14, a prototype IDFX system was introduced for rapid tool-/mould-making. To further apply this prototype IDFX system for rapidly developing other types of OKP products, research is needed in the following areas:

1. Detailed DFX guidelines should be drawn up for various functions of "X" and compiled in the IDFX systems.
2. Standard interfaces of the IDFX systems with product designers/design team should be created for the purpose of providing the right information first time.
3. IDFX systems need to provide tools for building new DFX functions and criteria for setting up the relevant guidelines.
4. Further research into generic DFX model for developing various IDFX systems is needed.

References

Bode, J., 1998, Neutral networks for cost estimation. Journal of Cost Engineering, 40, 25–30.

Bode, J., 2000, Neutral networks for cost estimation: simulations and pilot applications. International Journal of Production Research, 38, 1231–1254.

Chiu, C. M., Hsu, M. H. and Wang, E. T. G., 2006, Understanding knowledge sharing in virtual communities: An integration of social capital and social cognitive theories. Decision Support Systems, 42(3), 1872–1888.

Dorador, J. M. and Young, R. L .M., 1999, Information Models to support the interaction between Design for Assembly and Assembly Process Planning. Paper presented at the 1999 IEEE International Symposium on Assembly and Task Planning, Porto, Portugal.

Ehrlenspiel, K. and Schaal, S., 1992, In CAD integrierte Kostenkalkulation. Konstruktion, 44(12), 407–414.

Hardwick, M., Spooner, D., Rando, T. and Morris, K., 1996, Sharing manufacturing information in virtual enterprises. Communications of the ACM, 39, 46–54.

Heck, B. S., Wills, L. M. and Vachtsevanos, G. J., 2009, Software Technology for Implementing Reusable, Distributed Control Systems. Intelligent Systems, Control and Automation: Science and Engineering, Volume 39, Part VI, 267–293, DOI: 10.1007/978-90-481-3018-4_11.

Huaglory, T., 2001, Advanced life-cycle model for complex product development via stagealigned information-substitutive concurrency and detour. International Journal of Computer Integrated Manufacturing, 14, 281–303.

Ju, B. and Xi, L.F., 2008, A product cost estimation for the early design of sedans using neural networks. International Journal of Automotive Technology and Management, 8(3), 331–349.

Lau, H. and Jiang, B., 1998, A Generic Integrated System from CAD to CAPP: A Neural File-Cum-GT Approach, International Journal of Computer Integrated Manufacturing Systems, Vol.11, No.1–2, pp. 67–75.

Lubich, H. P., 1995, Towards a CSCW Framework for Scientific Cooperation in Europe. Lecture Notes in Computer Science 889, Springer-Verlag, ISBN 3-540-58844-2.

Smunt, T. L., 1999, Log-linear and non-log-linear learning curve models for production research and cost estimation. International Journal of Production Research, 37(17), 3901–3911.

Tamas, K., Lozano, S., Guerrero, F. and Onieva, L., 2000, A flexible costing system for flexible manufacturing systems using activity based costing. International Journal of Production Research, 38, 1615–1630.

Tu, Y. L. and Xie, S. Q., 2001, An information modelling framework for sheet metal parts intelligent and concurrent design and manufacturing. International Journal of Advanced Manufacturing Technology, 18(12), 873–883.

Tu, Y. L. and Xie, S. Q., 2000, A WWW-based Integrated Product Development Information Management System. IFAC Symposium on Manufacturing, Modeling, Management and Control (MIM2000), 12–14 July 2000, Patras, Greece.

Tu, Y. L. and Xie, S. Q. and Zhou, Z. D., 2001, An Information Modelling Framework to Concurrent Product Design and Manufacturing. IAM'2001, March 17–21, 2001, American University in Dubai, U.A.E.

Wang, C. H. and Bourne, D. A., 1995, Using features and their constraints to aid process planning of sheet metal parts. Proceedings of IEEE International Conference on Robotics and Automation, pp. 1020–1026.

Wagner, R., Castanotto, G. and Goldberg, K., 1997, FixtureNet: interactive computer aided design via the WWW. International Journal of Human–Computer Studies, 46(6), 773–788.

Xie, S. Q., Tu, Y. L., Liu, J. Q. and Zhou, Z. D., 2001, An integrated and concurrent approach for compound sheet metal cutting and punching. International Journal of Production Research, 39(6), 1095–1112.

Zhang, Y. P., Zhang, C. C. and Wang, H. P. B., 2000, An Internet-based STEP data exchange framework for virtual enterprises. International Journal of Computers in Industry, 41, 51–63.

Appendix

A.1 Examples of the Internet-based Product Development Systems

Some Internet-based PD products and research projects	Purpose	Data formats and compatible softwares	Key implementation technology	Stage of the PD process	Developer	Key functionalities
WinCHILL	Life cycle management	Compatible with CAD softwares, *e.g.* AutoCAD, Pro/ENGINEER. Support standard data formats, *e.g.* STEP, IGES, DXF and VRML and Internet technologies, *e.g.* CORBA, OLE.	RAD Tool, JAVA platform	Design, Process planning and manufacturing	WinCHILL Corporation	Product definition Product planning Customer management Maintenance and repair Production Manufacturing Planning Sourcing Product modelling
SMARTTEAM	Product data management	Connect with CAD, MRP through API function	Smart Basic/ Visual Basic	Product design and planning.	Smart solution	Document management and workflow management
QUEST	Process simulation and optimisation	Provides standard interface with CAD and supports IGES	Similar with Pascal programming language	Process planning and optimisation	Deneb robotics	Virtual simulation

CVW (Collaborative Virtual Workspace)	Communication with partners	Internet, Video conference software	Java	Earlier design stage	The MITRE Corporation	Audio/video conferencing Conference servers Text chat and instant messaging Data conferencing Place-based collaboration environments
CIMOSA project	CIM Open Structure	Internet, enterprise object, process modelling	Enterprise modelling and integration	The whole PD processes	http://cimosa.cnt.pl/	Enterprise integration, product management Enterprise structure Enterprise engineering
FixtureNet	Fixture design	Interactive CAD via the WWW	HTML, Java	Design		Design stage
OPDX	Collaboration design and information sharing to develop competitive products	PD with sharing information	Web Collaboration Tools,	Partner tools Product Information management Workflow and document management	ORACLE corporation	Design and product data management
Testbeds	Collaborative research	STEP	Visualisation tools, STEP, information modelling, etc.	Test platform	NIST	Collaborative design, manufacturing, process planning, etc.
METEOR2	Workflow management	Internet based workflow management	CORBA, HTML, JAVASCRIPT and CGI	Design and planning	http://www.scra.org/hiit.html	Internet-based planning and management

MatrixOne	PD integration	Various CAD, CAM, ERP Systems integration	Innovative Adaplet™ technology, Web Interface (CWI)	Design, planning and integration	http://www.matrix-one.com/products/integration_products.html	Internet-based systems integration platform
CyberSystem	Supply chain planning and enterprise resource planning	ERP systems and supply chain systems	Not mention		http://www.intramalls.com/	Seamless integration of supply chain planning and enterprise resource planning
Knowledge base.net	Knowledge sharing and management		Enterprise wide integration	The whole PD process	http://www.knowledgebase.net/	Self-help customer support Help Desk knowledge base FAQ management Document management policies and procedure sharing Product/Project documentation
ProcessPoint	Collaborative design			Design and planning stage	http://www.processpoint.com/	Collaborative design, deployment and manufacturing
WFMC	Workflow management	PDM, knowledge management		Planning stage	http://www.wfmc.org	Internet-based WFM Process management and optimisation

ISCM	Supply chain management	Supplier's resource management systems	Agent technology	Supply management	Integration laboratory University of Toronto	Information agents Temporal reasoning Scheduling Resource management
ABMA	Enterprise Integration				Budenske *et al.* 1998 Architecture Tech. Co.	Middleware architecture
AARIA	Manufacturing, scheduling and control				Parunrak *et al.* 1998	Using autonomous agents to represent physical entities, processes and operations
IA framework	Enterprise Integration				Pan and Tenenbaum, 1991	Using large number of intelligent agents for integration
IAO	Intelligent Manufacturing				Kwok and Norrie, 1994 U. of Calgary	A rule-based object system for developing intelligent manufacturing software
iDCSS	Concurrent Engineering				Klein 1995	Propose an integrated model to combine existing coordination technologies
MetaMorph II	Intelligent manufacturing and supply chain management				Shen *et al.* 1998, University of Calgary	

Madefast	Collaborative Engineering		Cutkosky *et al.* 1996, standford	Using Internet as a platform for collaboration
AMC	Manufacturing Scheduling		Goldsmith and Interrante 1998	Using physical agents: part agents and machine agents
CIIMPLEX	Enterprise Integration		Peng *et al.* 1998 UMBC EECOMS	Service agents (Name Server, Facilitator gent, Gateway Agent)
ADE	Supply Chain Management		Mehra and Nissen 1998	Using delegation based event handling similar to JavaBeans
VPD	Virtual product development	The whole PD process	NIST	Redefining PD Virtual Enterprise Service Centre Virtual Enterprise Architecture

Detailed information concerning the references: See reference list at the end of Chapter 2.

A.2 Current Product Modelling Methods and Tools

References	Modelling methodology	Data format	Main features	Tools	Purpose	Advantages
Pilz *et al.* 1989	Solid product modelling	Geometric	CSG model		Visualisation of a given CSG-object for verification purposes	Successfully creates a variety of renderings of various CSG-models
Kodiyalam *et al.* 1990	Solid product modelling	Geometric	Geometric model, automatic mesh generation module, shape sensitivity analysis and approximation module		For generic 3D shape optimisation	Successfully integrate CSG and B-rep for geometry definition and shape control
Shapiro *et al.* 1993	Solid product modelling	Geometric	Boundary-based separation half-spaces method	Parasolid^{TM}, PADL-2	B-Rep->CSG conversion for a curved solid object	Successfully convert natural quadric B-Reps to efficient CSG representations.
Xue *et al.* 1994	Feature-based modelling	Feature	A fuzzy c-means clustering algorithm, a fast feature searching scheme, a design feature and manufacturing feature coding method		Concurrent engineering design	[1]Support quick feasible design identification [2]Support the generation of multiple design solutions and joint evaluation of design performance and manufacturing costs
Gu *et al.* 1995	Integrated modelling (integrates CAX)	STEP	A generic product model, a product model database, a GPM/AutoSolid interface, a GPM/AutoCAD interface, a user interface.	SMALLTALK AutoCAD, AutoSolid	Integrate a variety of manufacturing activities in a concurrent engineering environment.	[1]Support a complete product information structure and standard data format [2]Can integrate a variety of manufacturing activities in CIM environment. [3]High level of interaction with user.

Chen *et al.* 1997	Feature-based modelling		Features, feature relationship		Design for net shape manufacturing	Facilitate the integration of geometry, design and manufacturing knowledge.
Meng *et al.* 1997	Feature-based modelling	STEP	User interface of feature modelling, feature recognition module, information matching module	Pro/E system	To realise the CAD/CAM/CAPP integration	Realise primary integration of CAD/ CAPP / CAM
Shaharoun *et al.* 1998	Integrated modelling	STEP	Geometric data, plastic material property, machining and grinding parameters, plastic injection process variables	ST-Devoloper V1.5	To support the plastic injection moulding industry	Improve current design and manufacturing practices in the mould and die industry
Roy *et al.* 1999	Integrated modelling (integrates VRML standard for 3D geometry with conventional CAD packages)	VRML	Graphical layer of product, technological information layer of product	VRML Version 1.0	To develop an open collaborative design environment though WWW technology	[1]Facilitates human-assisted or automated interpretation of product model for process planning, manufacturability analysis and so on [2]Having open architecture and interoperability.
Song *et al.* 1999	Integrated modelling (integrates part feature and manufacturing knowledge)		Integrated product model, expert system of product manufacturability evaluation, controller of concurrent design		Real-time concurrent product and process design	[1]Optimise design process [2]Realise information and function integration [3]Reduce time to market and cost
Gao *et al.* 2000	Feature-based modelling		Application features, system features,	ANSI C	Intelligent product modelling	New complicated features of a product could be developed utilising the system features.

Van Holland *et al.* 2000	Integrated modelling (integrates single parts and assemblies modelling and also integrates modelling and planning)		Single part feature model, assembly feature model		For assembly modelling and planning	[1]Integrating single-part and assembly modelling in one system [2]Bridging the gap between functional and geometric modellers
Tang *et al.* 2001	Integrated modelling	STEP	Application model, uniform model (DPIIM), resource model, product data, physical format		Concurrent stamped part and die development	[1]Ensuring data specification consistency [2]Can be integrated with other STEP-based or STEP-compliant data models and databases
Xie *et al.* 2001	Integrated modelling		Customer interface, a simulation platform, a cost estimation module, an information integration framework	Visual C++, VBSCRIPT, JAVASCRIPT	Sheet metal parts intelligent concurrent design and manufacturing	[1]Share product design and manufacturing knowledge in real-time [2] Realise intelligent decision support at the earlier design stage [3]Support sheet metal parts' concurrent design and manufacturing in a distributed computing environment; [4]Multi-vendor software compatible.

Chin *et al.* 2002	Integrated modelling	STEP	A geometric model, a feature model, a product definition model (including the domain product model), an integrated core model.	Concurrent engineering	All the information required for the product development can be integrated. Enhances the effectiveness and efficiency of the product development process.
Zha *et al.* 2002	Knowledge-based modelling	STEP	Feature-based CAD system, product modelling system, assembly planning system, assembly evaluation system	Concurrent integrated design and assembly planning	[1]Enhancing the communication ability between systems [2]Simplifying the assembly planning of the complex product [3]Improving design process [4]Eliminating human interpretation, errors and reworks in data exchanging [5]Supporting integration of new product
Lee *et al.* 2004	Feature-based modelling	UML	Graphic User Interface, semantic product model data structure, geometric modelling kernel, graphic library	To support rapid product modelling of the initial hull structure in shipbuilding	[1] 3D solid model and the production material information can be generated automatically for CAPP [2] Can be efficiently used for overall initial ship design environment.

Chen, L. *et al.* 2005	Integrated modelling (integrates physical structure, design semantic, collaboration management data)		Physical structure data, design semantic data, relation data, constraint data, collaboration data, process data.	Java, XML	Multidisciplinary collaborative design	Good flexibility, expansibility, and evolution ability.
Lin *et al.* 2005	Integrated modelling (integrates static and dynamic evolvement information)	STEP, XML	A master model, application models		Remote collaborative product development	[1]High reusability of product information [2]Good information consistency
Zhang *et al.* 2005	Knowledge-based modelling		Assembly model, function model, configuration model, knowledge presentation model	PowerBuilder 7.0, SQL Server 2000	Configuration modelling for MTO-ME	[1]Product knowledge can be easily managed [2]Product configuration can be effectively supported
Li *et al.* 2006	Integrated modelling	STEP, XML, UML	Product requirement model, product Concept design model, product Engineering design model, product Manufacturing model and product Service model	J2EE, Java, Jbuilder8.0, SQL Server2000	Collaborative design in product life cycle management	Shorten the PD time

Detailed information concerning the references: See reference list at the end of Chapter 3.

A.3 EDM Schemas

```
SCHEMA assembly_information_schema;

REFERENCE FROM supporting_schema
(identifier, label, text);

REFERENCE FROM product_definition_schema
(product);

TYPE super_item_type = SELECT
(assembly_product, subassembly);
END_TYPE;

ENTITY assembly_product SUBTYPE OF (product);
INVERSE
sub_parts: SET [0:?] OF part FOR of_assembly;
sub_assemblies: SET [0:?] OF subassembly FOR of_assembly;
connection_information: SET [1:?] OF connector FOR of_super_item;
WHERE
WR1: SELF.sub_parts.level_in_assembling_hierarchy = 2;
WR2: SELF.sub_assemblies.level_in_assembling_hierarchy = 2;
WR3: SELF.connection_information.level_in_assembling_hierarchy = 2;
END_ENTITY;

ENTITY part;
id: identifier;
name: label;
description: OPTIONAL text;
of_assembly: assembly_product;
super_item: super_item_type;
level_in_assembling_hierarchy: INTEGER;
INVERSE
connection_information: SET [1:?] OF connector FOR relating_parts;
WHERE
WR1: SELF.level_in_assembling_hierarchy >=2;
END_ENTITY;

ENTITY subassembly;
id: identifier;
name: label;
description: OPTIONAL text;
of_assembly: assembly_product;
super_item: super_item_type;
```

```
level_in_assembling_hierarchy: INTEGER;
INVERSE
sub_parts : SET [0:?] OF part FOR super_item;
sub_subassemblies: SET [0:?] OF subassembly FOR super_item;
connection_information: SET [1:?] OF connector FOR of_super_item;
WHERE
WR1: EXISTS (SELF.sub_parts) OR EXISTS (SELF.sub_assemblies);
WR2: SELF.level_in_assembling_hierarchy >=2;
WR3: SELF.sub_parts.level_in_assembling_level:=:
   SELF.level_in_assembling_hierarchy + 1;
WR4: SELF.sub_sub_assemblies.level_in_assembling_level:=:
   SELF.level_in_assembling_hierarchy + 1;
WR5: SELF.connection_information.level_in_assembling_level:=:
SELF.level_in_assembling_hierarchy + 1;
END_ENTITY;

ENTITY connector;
id: identifier;
name: label;
description: OPTIONAL text;
of_super_item: super_item_type;
connecting_method_used: connecting_method;
level_in_assembling_hierarchy: INTEGER;
relating_parts: SET [0:?] OF part;
relating_subassemblies: SET [0:?] OF subassembly;
WHERE
WR1: SELF.relating_parts.level_in_assembling_hierarchy:=:
SELF.level_in_assembling_hierarchy;
WR2: SELF.relating_subassemblies.level_in_assembling_hierachy
   :=:SELF.level_in_assembling_hierarchy;
WR3: SELF.level_in_assembling_hierarchy >= 2;
END_ENTITY;

ENTITY connecting_method;
id: identifier;
name: label;
description: OPTIONAL text;
constrain: OPTIONAL text;
END_ENTITY;

END_SCHEMA;

SCHEMA bill_of_material_information_schema;

REFERENCE FROM supporting_schema
```

```
(identifier, label, text, person, amount_of_day, source, un_known);
REFERENCE FROM assembly_information_schema
(part, subassembly);

REFERENCE FROM product_definition_schema
(product);

REFERENCE FROM material_information_schema
(material);

REFERENCE FROM supplier_information_schema
(supplier);

TYPE bom_item_type = SELECT
(subassembly, part, material);
END_TYPE;

TYPE order_type = SELECT
(external_ordering_information, internal_ordering_information, un_known);
END_TYPE;

ENTITY bom_item;
id: identifier;
name: label;
description: OPTIONAL text;
of_bom_list: bill_of_material_list;
quantity: INTEGER;
item_type: bom_item_type;
item_source: source;
DERIVE
item_ordering_information: order_type:=
   order_type_selection(SELF.item_source);
END_ENTITY;

ENTITY bill_of_material_list;
id: identifier;
of_product: product;
INVERSE
bom_list: SET [1:?] OF bom_item FOR of_bom_list;
END_ENTITY;

ENTITY external_ordering_information;
id: identifier;
supplier_identifier: supplier;
ordered_product: bom_item;
```

```
ordered_quantity: INTEGER;
person_in_charge: SET [1:?] OF person;
delivery_time: amount_of_day;
END_ENTITY;

ENTITY internal_ordering_information;
id: identifier;
ordered_product: bom_item;
ordered_quantity: INTEGER;
person_in_charge: SET [1:?] OF person;
delivery_time: amount_of_day;
END_ENTITY;

FUNCTION order_type_selection
(item_source : source) : order_type;
LOCAL x : item_source;
END_LOCAL;
IF x :=:source.bought THEN RETURN (external_ordering_information);END_IF;
IF x:=:source.made THEN RETURN (internal_ordering_information); END_IF;
IF x:=: source.un_known THEN RETURN (un_known);
END_IF;
END_FUNCTION;

END_SCHEMA;

SCHEMA cost_information_schema;

REFERENCE FROM supporting_schema
(identifier, text, label, money);

REFERENCE FROM product_definition_schema
(product);

TYPE fixed_cost_type = ENUMERATION OF
(depreciation_on_capital_investment,
interest_on_capital_investment_and_inventory,
property_taxes,
insurance,
technical_service,
product_design_and_development,
nontechnical_service,
general_supplies,
rental_of_equipment,
share_of_corporate_executive_staff,
legal_staff,
```

```
share_of_corporate_research_and_devlopment_staff,
maket_staff,
sales_force,
technical_service_staff);
END_TYPE;

TYPE variable_cost_type = ENUMERATION OF
(material,
direct_label,
maitenance_costs,
power_and_utilities,
quality_control_staff,
royalty_payment,
packaging_and_storage_costs,
scrap_loss_and_spoilag);
END_TYPE;

ENTITY cost ABSTRACT SUPERTYPE OF (ONEOF (fixed_cost, variable_cost));
id: identifier;
name: label;
description: OPTIONAL text;
END_ENTITY;

ENTITY fixed_cost SUBTYPE OF (cost);
belonging: fixed_cost_association;
cost_type: fixed_cost_type;
cost_amount: money;
WHERE
WR1: SELF.cost_amount.amount >= 0;
END_ENTITY;

ENTITY fixed_cost_association;
of_product: product;
sum: REAL;
INVERSE
cost_association: SET [1:?] OF fixed_cost FOR belonging;
END_ENTITY;

ENTITY variable_cost
SUBTYPE OF (cost);
belonging: variable_cost_association;
cost_type: variable_cost_type;
rate: variable_cost_rate;
quantity: REAL;
DERIVE
```

```
cost_amount: REAL := SELF.rate.rate * SELF.quantity;
WHERE
WR1: SELF.rate.rate >= 0;
WR2: SELF.quantity >= 0;
END_ENTITY;

ENTITY variable_cost_association;
of_product: product;
sum: REAL;
INVERSE
ssociation: SET [1:?] OF variable_cost FOR belonging;
END_ENTITY;

ENTITY variable_cost_rate;
id: identifier;
name: text;
rate: OPTIONAL REAL;
unit: label;
END_ENTITY;

ENTITY sum_of_cost;
of_product: product;
DERIVE
sum_cost: REAL := SELF.fixed_cost_association.sum +
   SELF.varial_cost_association.sum;
WHERE
WR1: SELF.sum_of_cost.of_product.id :=:
   SELF.fixed_cost_association.of_product.id;
WR2: SELF.sum_of_cost.of_product.id :=:
   SELF.variable_cost_association.of_product.id;
END_ENTITY;

END_SCHEMA;

SCHEMA product_document_schema;

REFERENCE FROM supporting_schema
(identifier, label, text);

REFERENCE FROM product_definition_schema
(product);

ENTITY document SUPERTYPE OF (ONE OF (facility_document,
   inspection_standard,
material_document,bom_document, manufacturing_process_document,
```

```
    manufacturing_process_action_document));
id: identifier;
name: label;
description: OPTIONAL text;
kind: document_type;
INVERSE
representation_types: SET [0:?] OF document_representation_type FOR
    represented_document;
END_ENTITY;

ENTITY document_product_association;
name: label;
description: OPTIONAL text;
relating_document: document;
related_product: product;
END_ENTITY;

ENTITY document_relationship;
name: label;
description: OPTIONAL text;
relating_document: document;
related_document: document;
END_ENTITY;

ENTITY document_representation_type;
name: label;
represented_document: document;
END_ENTITY;

ENTITY document_type;
product_data_type: label;
END_ENTITY;

ENTITY document_usage_constraint;
source: document;
subject_element: label;
subject_element_value: text;
END_ENTITY;

ENTITY facility_document
SUBTYPE OF (document);
END_ENTITY;

ENTITY material_document
SUBTYPE OF (document);
```

```
END_ENTITY;

ENTITY inspection_standard
SUBTYPE OF (document);
END_ENTITY;

ENTITY bom_document
SUBTYPE OF (document);
END_ENTITY;

ENTITY manufacturing_process_document
SUBTYPE OF (document);
END_ENTITY;

ENTITY manufacturing_process_action_document
SUBTYPE OF (document);
END_ENTITY;

END_SCHEMA;

SCHEMA inspection_information_schema;

REFERENCE FROM supporting_schema
(identifier, label, text);

REFERENCE FROM product_definition_schema
(product);

REFERENCE FROM product_document_schema
(inspection_standard);

TYPE result = SELECT
(pass, fail);
END_TYPE;

TYPE pass = STRING;
END_TYPE;

TYPE fail = STRING;
END_TYPE;

ENTITY product_inspections;
of_product: product;
inspecetions_list: SET [0:?] OF inspection_test;
WHERE
```

```
WR1: SELF.product_inspection.of_product
    :=:SELF.product_inspections.inspections_list.test_product;
END_ENTITY;

ENTITY inspection_test;
id: identifier;
name: label;
description: OPTIONAL text;
testing_product: product;
INVERSE
inspection_pass_requirements: SET [1:?] OF
    inspection_pass_requirement FOR of_inspection;
inspection_test_result : result_value FOR of_inspection;
END_ENTITY;

ENTITY inspection_pass_requirement;
name: label;
description: OPTIONAL text;
of_inspection: inspection_test;
frame_of_reference: SET [1:?] OF inspection_standard;
END_ENTITY;

ENTITY result_value;
of_inspection: inspection_test;
value_amount: OPTIONAL REAL;
text_description: OPTIONAL STRING;
test_result : result;
WHERE
WR1: EXISTS (value_amount) OR EXISTS (text_description);
END_ENTITY;

END_SCHEMA;

SCHEMA manufacturing_facility_schema;

REFERENCE FROM supporting_schema
(identifier, label, text, property_definition, source);

REFERENCE FROM supplier_information_schema
(facility_supplier);

REFERENCE FROM product_definition_schema
(product);

REFERENCE FROM product_document_schema
```

```
(facility_document);

ENTITY facility SUPERTYPE OF (substitute_facility);
id: identifier;
name: label;
description: OPTIONAL text;
facility_type: facility_category;
facility_source: source;
supplier: facility_supplier;
related_documents: SET [0:?] OF facility_document;
END_ENTITY;

ENTITY facility_property SUBTYPE OF (property_definition);
of_facility : facility;
END_ENTITY;

ENTITY facility_category;
id: identifier;
name: label;
category_definition: text;
END_ENTITY;

ENTITY facility_relationship;
id: identifier;
name: label;
description: text;
relating_facility: facility;
reltaed_facility: facility;
END_ENTITY;

ENTITY substitute_facility SUBTYPE OF (facility);
of_facility : facility;
requirements: text;
WHERE
WR1: SELF.id :<>: SELF.of_facility.id;
END_ENTITY;

ENTITY facility_usage_information;
of_facility : facility;
for_product: product;
related_facilities: SET [1:?] OF facility;
usage_description: text;
substitution_information: SET [0:?] OF substitute_facility;
END_ENTITY;
```

END_SCHEMA;

SCHEMA material_information_schema;

REFERENCE FROM supporting_schema
(identifier, label, text, property_definition);

REFERENCE FROM supplier_information_schema
(material_supplier);

REFERENCE FROM manufacturing_facility_schema
(facility);

ENTITY material SUPERTYPE OF (substitute_material);
id: identifier;
name: label;
description: OPTIONAL text;
supplier: SET [1:?] OF material_supplier;
INVERSE
characteristics: SET [1:?] OF material_property FOR of_material;
subsitution_select: SET [0:?] OF substitute_material FOR of_material;
fabrication_information: material_fabrication_information FOR of_material;
END_ENTITY;

ENTITY material_property SUBTYPE OF (property_definition);
of_material: material;
UNIQUE
UR1: SELF

Property 16.1. _definition.name, SELF

Property 16.2. _definition.id;
END_ENTITY;
ENTITY substitute_material SUBTYPE OF (material);
of_material: material;
END_ENTITY;

ENTITY material_fabrication_information;
of_material: material;
relating_manufacturing_facility: SET [1:?] OF facility;
fabraication_requirements: SET [1:?] OF text;
END_ENTITY;

END_SCHEMA;

```
SCHEMA process_planning_schema;

REFERENCE FROM supporting_schema
(identifier, label, text);

REFERENCE FROM product_document_schema
(manufacturing_process_document, manufacturing_process_action_document);

REFERENCE FROM product_definition_schema
(product);

TYPE tolerance_type = ENUMERATION OF
(straightness, flatness, circularity, cylindericity,
   profile_of_a_line, profile_of_a_surface, all_around_profile, angularity,
   perpendicularity, parallelism, position, concentricity_or_coaxiality, symmetry,
   circular_runout, total_runout);
END_TYPE;

TYPE process_method = SELECT
(sequential_action_method, concurrent_action_method);
END_TYPE;

TYPE characterized_action_definition = SELECT
(action, action_method, action_method_relationship, action_relationship);
END_TYPE;

ENTITY action ABSTRACT SUPERTYPE OF (manufacturing_process_action);
id: identifier;
name: label;
description: OPTIONAL text;
END_ENTITY;

ENTITY action_method SUPERTYPE OF
   (ONEOF (sequential_action_method, concurrent_action_method));
name: label;
description: OPTIONAL text;
consequence: text;
purpose: text;
END_ENTITY;

ENTITY action_method_relationship;
name: label;
description: OPTIONAL text;
related_method: action_method;
relating_method: action_method;
```

```
END_ENTITY;

ENTITY action_relationship;
name: label;
description: OPTIONAL text;
relating_action: action;
related_action: action;
END_ENTITY;

ENTITY versioned_action_request;
id: identifier;
version: label;
purpose: text;
description: OPTIONAL text;
END_ENTITY;

ENTITY manufacturing_feature;
id: identifier;
name: label;
description: OPTIONAL text;
manufacturing_method: action_method;
END_ENTITY;

ENTITY sequential_action_method SUBTYPE OF (action_method_relationship);
sequence_position: INTEGER;
END_ENTITY;

ENTITY concurrent_action_method SUBTYPE OF (action_method_relationship);
END_ENTITY;

ENTITY action_property;
id: identifier;
name: label;
description: text;
definition: characterized_action_definition;
END_ENTITY;

ENTITY action_resource;
id: identifier;
name: label;
description: OPTIONAL text;
END_ENTITY;

ENTITY manufacturing_process_action SUPERTYPE OF (surface_finish)
    SUBTYPE OF (action);
```

```
type_of_process: process_method;
desired_result: versioned_action_request;
manufacturing_process_action_property: SET [0:?] OF action_property;
manufacturing_process_action_time: REAL;
manufacturing_process_action_resources: SET [1:?] OF action_resource;
relating_manufacturing_feature: OPTIONAL manufacturing_feature;
documents: SET [1:?] OF manufacturing_process_action_document;
WHERE
WR1: manufacturing_process_action_time >= 0;
END_ENTITY;

ENTITY manufacturing_process;
id: identifier;
of_product: product;
process_action_list: LIST [1:?] OF manufacturing_process_action;
process_cycle_time: cycle_time;
END_ENTITY;

ENTITY cycle_time;
id: identifier;
name: label;
description: OPTIONAL text;
cycle_time_amount: REAL;
unit: label;
WHERE
WR1: SELF.cycle_time_amount >=0;
WR2: SELF.cycle_time_amount >= ('PROCESS_PLANNING_SCHEMA.' +
   'MANUFACTURING_PROCESS_ACTION.' +
   'MANUFACTURING_PROCESS_ACTION_TIME');
END_ENTITY;

ENTITY surface_finish SUBTYPE OF (manufacturing_process_action);
tolerance_requirement: SET [1:?] OF tolerance;
END_ENTITY;

ENTITY tolerance;
id: identifier;
name: label;
description: OPTIONAL text;
type_select: tolerance_type;
tolerance_value: REAL;
tolereance_value_unit: label;
reference_description : LIST [1:?] OF text;
END_ENTITY;
```

```
END_SCHEMA;

SCHEMA product_definition_schema;

REFERENCE FROM supporting_schema
(label, text, identifier, bag_to_set, property_definition);

ENTITY product;
id: identifier;
name: label;
description: OPTIONAL text;
of_category: product_category;
INVERSE
property_information : SET [1:?] OF product_property FOR of_product;
END_ENTITY;

ENTITY product_category;
name: label;
description: OPTIONAL text;
id: identifier;
END_ENTITY;

ENTITY product_category_relationship;
name: label;
description: OPTIONAL text;
category: product_category;
ub_category: product_category;
WHERE
WR1: acyclic_product_category_relationship (SELF, [SELF.sub_category]);
END_ENTITY;

ENTITY product_relationship;
id: identifier;
name: label;
description: OPTIONAL text;
relating_product: product;
related_product: product;
END_ENTITY;

ENTITY product_property SUBTYPE OF (property_definition);
of_product : product;
END_ENTITY;

FUNCTION acyclic_product_category_relationship
(relation : product_category_relationship;
```

```
children : SET OF product_category) : BOOLEAN;
LOCAL
x: SET OF product_category_relationship;
local_children: SET OF product_category;
END_LOCAL;
REPEAT i:= 1 to HIINDEX (children);
IF relation.category :=: children [i] THEN
RETURN (FALSE);
END_IF;
END_REPEAT;
x := bag_to_set (USEDIN (relation.category,
    'PRODUCT_DEFINITION_SCHEMA.' +
    'PRODUCT_CATEGORY_RELATIONSHIP.SUB_CATEGORY'));
local_children := children +relation.category;
IF SIZEOF(x) > 0 THEN
REPEAT i:= 1 to HIINDEX(x);
IF NOT acyclic_product_category_relationship(x[i], local_children) THEN
RETURN (FALSE);
END_IF;
END_REPEAT;
END_IF;
RETURN(TRUE);
END_FUNCTION;

END_SCHEMA;

SCHEMA supplier_information_schema;

REFERENCE FROM supporting_schema
(label, identifier, text, organisation, person, address, amount_of_day, money);

REFERENCE FROM product_definition_schema
(product);

REFERENCE FROM manufacturing_facility_schema
(facility);

REFERENCE FROM assembly_information_schema
(part, subassembly);

REFERENCE FROM material_information_schema
(material);

TYPE supplier_type = SELECT
(facility, part, subassembly, material);
```

```
END_TYPE;

ENTITY supplier SUPERTYPE OF (ONEOF(material_supplier, facility_supplier,
part_supplier, subassembly_supplier)) SUBTYPE OF (organisation);
supplier_type_selection: supplier_type;
contact_information: supplier_contact;
INVERSE
products_range: SET [1:?] OF product_supplying FOR of_supplier;
END_ENTITY;

ENTITY supplier_contact;
contact_person: person;
contact_person_address: address;
END_ENTITY;

ENTITY product_supplying SUBTYPE OF (product,material);
of_supplier: supplier;
leading_time: amount_of_day;
product_unit_price: money;
INVERSE
other_fees : SET [0:?] OF fee FOR of_product;
END_ENTITY;

ENTITY fee;
id: identifier;
name: label;
description: OPTIONAL text;
of_product: product_supplying;
fee_information: money;
END_ENTITY;

ENTITY material_supplier SUBTYPE OF (supplier);
WHERE
WR1: SELF.supplier_type_selection :=: material;
END_ENTITY;

ENTITY facility_supplier SUBTYPE OF (supplier);
WHERE
WR1: SELF.supplier_type_selection :=: facility;
END_ENTITY;

ENTITY part_supplier SUBTYPE OF (supplier);
WHERE
WR1: SELF.supplier_type_selection :=: part;
```

```
END_ENTITY;

ENTITY subassembly_supplier SUBTYPE OF (supplier);
WHERE
WR1: SELF.supplier_type_selection :=: subassembly;
END_ENTITY;

END_SCHEMA;

SCHEMA supporting_schema;

TYPE amount_of_day = REAL;
WHERE
WR1: SELF>=0;
END_TYPE;

TYPE identifier = STRING;
END_TYPE;

TYPE label = STRING;
END_TYPE;

TYPE text = STRING;
END_TYPE;

TYPE source = ENUMERATION OF
(made, bought, not_known);
END_TYPE;

ENTITY address;
name: label;
internal_location: OPTIONAL label;
street_number: OPTIONAL label;
street: OPTIONAL label;
postal_box: OPTIONAL label;
town: OPTIONAL label;
region: OPTIONAL label;
postal_code: OPTIONAL label;
country: OPTIONAL label;
facsimile_number: OPTIONAL label;
telephone_number: OPTIONAL label;
electronic_mail_address: OPTIONAL label;
telex_number: OPTIONAL label;
url: identifier;
WHERE
```

```
WR1: EXISTS (internal_location) OR EXISTS (street_number) OR
EXISTS (street) OR EXISTS (postal_box) OR EXISTS (town) OR
EXISTS (region) OR EXISTS (postal_code) OR EXISTS (country) OR
EXISTS (facsimile_number)OR EXISTS (telephone_number) OR
EXISTS (electronic_mail_address) OR EXISTS (telex_number);
END_ENTITY;

ENTITY money;
amount: REAL;
unit: label;
WHERE
WR1: SELF.amount >= 0;
END_ENTITY;

ENTITY organisation;
id: OPTIONAL identifier;
name: label;
description: OPTIONAL text;
END_ENTITY;

ENTITY person;
id: identifier;
last_name: OPTIONAL label;
first_name : OPTIONAL label;
middle_names: OPTIONAL LIST[1:?] OF label;
prefix_titles: OPTIONAL LIST[1:?] OF label;
suffix_titles: OPTIONAL LIST[1:?] OF label;
company: OPTIONAL organisation;
position: OPTIONAL label;
WHERE
WR1: EXISTS (last_name) OR EXISTS (first_name);
END_ENTITY;

ENTITY property_definition;
id: identifier;
name: label;
description: OPTIONAL text;
END_ENTITY;

ENTITY un_known;
END_ENTITY;

FUNCTION bag_to_set
(the_bag :BAG OF GENERIC : intype) : SET OF GENERIC : intype;
LOCAL the_set : SET OF GENERIC : intype := [];
```

```
END_LOCAL;
IF SIZEOF (the_bag) > 0 THEN
REPEAT i:= 1 to HIINDEX (the_bag);
the_set := the_set + the_bag [i];
END_REPEAT;
END_IF;
RETURN (the_set);
END_FUNCTION;

END_SCHEMA;
```

A.4 QFD Program

A.4.1 Programs for QED Interface and Interactions with Microsoft Access Database

```
Private Sub BoxA_Click()

End Sub

Private Sub cmdClearAll1_Click()

BoxA.Clear
End Sub

Private Sub cmdClearAll2_Click()
BoxB.Clear
End Sub

Private Sub cmdDeleteA_Click()
BoxA.RemoveItem (BoxA.ListIndex)
End Sub

Private Sub cmdDeleteB_Click()
BoxB.RemoveItem (BoxB.ListIndex)
End Sub

Private Sub cmdEnter_Click()

Dim IRow As Integer
Dim INumRow As Integer

Application.ScreenUpdating = False

ListSizeA = BoxA.ListCount
  For iRow1 = 1 To ListSizeA
    With Worksheets("QFD Table")
      .Range("B3").Cells(iRow1, 1).Value = BoxA.List(iRow1 - 1)
      .Activate
    End With
  Next iRow1

ListSizeB = BoxB.ListCount
  For iRow2 = 1 To ListSizeB
    With Worksheets("QFD Table")
```

```
    .Range("D2").Cells(1, iRow2).Value = BoxB.List(iRow2 - 1)
  End With
Next iRow2

For iRow2 = 0 To ListSizeB
  With Worksheets("QFD Table")
    .Range("C2").Cells(1, iRow2 + 1).Borders(xlEdgeTop).Weight = xlMedium
    .Range("C3").Cells(1, iRow2 + 1).Borders(xlEdgeTop).Weight = xlMedium
    .Range("A3:B3").Borders(xlEdgeTop).Weight = xlMedium
  End With
Next iRow2

For iRow1 = 1 To 34
  For iRow2 = 0 To ListSizeB + 1
    For IBor = -1 To 0
      With Worksheets("QFD Table").Range("D3")
        .Cells(iRow1, iRow2 - 1).Borders(xlEdgeRight).Weight = xlThin
        .Cells(iRow1, iRow2 - 1).Borders(xlEdgeBottom).Weight = xlThin
        .Cells(iRow1, 0).Borders(xlEdgeBottom).Weight = xlThin
        .Cells(1, IBor).Borders(xlEdgeTop).Weight = xlMedium
        .Cells(0, iRow2).Borders(xlEdgeLeft).Weight = xlThin
      End With
    Next IBor
  Next iRow2
Next iRow1

For iRow2 = 1 To ListSizeB
  With Worksheets("QFD Table").Range("D1")
    .Cells(1, iRow2).Borders(xlEdgeTop).Weight = xlMedium
  End With
Next iRow2

For iRow2 = -2 To ListSizeB
  For INumRow = -1 To ListSizeB
  With Worksheets("QFD Table")
    .Cells(33, iRow2 + 3).Borders(xlEdgeBottom).Weight = xlMedium
    .Cells(36, INumRow + 3).Borders(xlEdgeBottom).Weight = xlMedium
  End With
  Next INumRow
Next iRow2

sBorderLeft = Cells(3, 1).Address & ":" & Cells(33, 1).Address
sBorderCenter0 = Cells(3, 1).Address & ":" & Cells(36, 1).Address
sBorderCenter1 = Cells(1, 3).Address & ":" & Cells(36, 3).Address
sBorderRight = Cells(1, BoxSizeB + 3).Address & ":"
```

```
        & Cells(36, 3 + ListSizeB).Address
With Worksheets("QFD Table")
  .Range(sBorderLeft).Borders(xlEdgeLeft).Weight = xlMedium
  .Range(sBorderCenter0).Borders(xlEdgeRight).Weight = xlMedium
  .Range(sBorderCenter1).Borders(xlEdgeRight).Weight = xlMedium
  .Range(sBorderRight).Borders(xlEdgeRight).Weight = xlMedium
End With
```

'Merge cells

```
sMerge1 = Cells(3, 1).Address & ":" & Cells(33, 1).Address
sMerge2 = Cells(1, 4).Address & ":" & Cells(1, 3 + ListSizeB).Address
 With Worksheets("QFD Table")
   .Range(sMerge1).Merge
   .Range(sMerge1).Value = "Voice of Customer"
   .Range(sMerge2).Merge
   .Range(sMerge2).Value = "Voice of Engineer"
   .Cells(2, 3).Value = "Customer Importance"
   .Cells(34, 2).Value = "Absolute Importance"
   .Cells(35, 2).Value = "Customer Satisfaction Performance"
   .Cells(36, 2).Value = "Competitive Benchmark"
   .Cells(38, 2).Value = "*Note: All information are based"
   .Cells(39, 2).Value = "on product code PA337D"
   .Cells(41, 2).Value = "*Note: All values for Competitive"
   .Cells(42, 2).Value = "Benchmark based on 1-5 Rating"
   .Cells(43, 2).Value = "Customer Performance are based"
   .Cells(44, 2).Value = "on 1-9 rating"
   .Cells(46, 2).Value = "*Note: All values in Relation Matrix"
   .Cells(47, 2).Value = "are of <>, 1, 3 & 9 rating"

 End With
 ActiveSheet.Columns("A:A").Select
 With Selection
   .HorizontalAlignment = xlCenter
   .VerticalAlignment = xlCenter
   .WrapText = False
   .Orientation = 90
   .ShrinkToFit = False
 End With
 ActiveSheet.Columns("B:B").Select
 With Selection
   .HorizontalAlignment = xlLeft
```

```
    .VerticalAlignment = xlCenter
    .WrapText = False
    .Orientation = 0
    .ShrinkToFit = False
  End With
  ActiveSheet.Rows("2:2").Select
  Selection.Orientation = 90
  Selection.Rows.AutoFit
  Selection.HorizontalAlignment = xlCenter
  Selection.VerticalAlignment = xlBottom
  ActiveSheet.Columns("B:B").Select
  Selection.Columns.AutoFit
  ActiveSheet.Range("A1").Select
```

'Technical Correlation

```
ListSizeB = BoxB.ListCount
  For iRow2 = 1 To ListSizeB
    With Worksheets("Tech Correlation")
      .Range("B3").Cells(iRow2, 1).Value = BoxB.List(iRow2 - 1)
      .Range("C2").Cells(1, iRow2).Value = BoxB.List(iRow2 - 1)
      .Activate
    End With
  Next iRow2

sHighlight = Cells(2, 4).Address & ":" & Cells(2, 3 + ListSizeB).Address

sColumn = Cells(1, 2).Address & ":" & Cells(60, 3 + ListSizeB).Address
sRelation = Cells(2, 2).Address & ":" & Cells(2 + ListSizeB, ListSizeB + 2).Address
sBorder1 = Cells(2, 2).Address & ":" & Cells(2 + ListSizeB, 2).Address
sBorder2 = Cells(2, 2).Address & ":" & Cells(2, 2 + ListSizeB).Address

    ActiveSheet.Range(sHighlight).Select
    Selection.Copy
    ActiveSheet.Range(sColumn).EntireColumn.AutoFit
    Application.CutCopyMode = False
    With Selection

      .HorizontalAlignment = xlLeft
      .VerticalAlignment = xlBottom
      .WrapText = False
      .Orientation = 0
      .IndentLevel = 0
```

```
    .ShrinkToFit = False
    .MergeCells = False
  End With

  ActiveSheet.Rows("2:2").Select
  Selection.Orientation = 90
  Selection.Rows.AutoFit
  Selection.HorizontalAlignment = xlCenter
  Selection.VerticalAlignment = xlBottom
  ActiveSheet.Columns("B:B").Select
  Selection.Columns.AutoFit
  ActiveSheet.Range("A1").Select
```

'Technical Correlation's Borders

```
 ActiveSheet.Range(sRelation).Select
 Selection.Borders(xlDiagonalDown).LineStyle = xlNone
 Selection.Borders(xlDiagonalUp).LineStyle = xlNone
 With Selection.Borders(xlEdgeLeft)

   .LineStyle = xlContinuous
   .Weight = xlThin
   .ColorIndex = xlAutomatic
 End With
 With Selection.Borders(xlEdgeTop)

   .LineStyle = xlContinuous
   .Weight = xlThin
   .ColorIndex = xlAutomatic
 End With
 With Selection.Borders(xlEdgeBottom)

   .LineStyle = xlContinuous
   .Weight = xlThin
   .ColorIndex = xlAutomatic
 End With
 With Selection.Borders(xlEdgeRight)

   .LineStyle = xlContinuous
   .Weight = xlThin
   .ColorIndex = xlAutomatic
 End With
```

```
With Selection.Borders(xlInsideVertical)

  .LineStyle = xlContinuous
  .Weight = xlThin
  .ColorIndex = xlAutomatic
End With
With Selection.Borders(xlInsideHorizontal)

  .LineStyle = xlContinuous
  .Weight = xlThin
  .ColorIndex = xlAutomatic
End With

Selection.Borders(xlDiagonalDown).LineStyle = xlNone
Selection.Borders(xlDiagonalUp).LineStyle = xlNone
With Selection.Borders(xlEdgeLeft)

  .LineStyle = xlContinuous
  .Weight = xlMedium
  .ColorIndex = xlAutomatic
End With
With Selection.Borders(xlEdgeTop)

  .LineStyle = xlContinuous
  .Weight = xlMedium
  .ColorIndex = xlAutomatic
End With
With Selection.Borders(xlEdgeBottom)

  .LineStyle = xlContinuous
  .Weight = xlMedium
  .ColorIndex = xlAutomatic
End With
With Selection.Borders(xlEdgeRight)

  .LineStyle = xlContinuous
  .Weight = xlMedium
  .ColorIndex = xlAutomatic
End With
With Selection.Borders(xlInsideVertical)

  .LineStyle = xlContinuous
  .Weight = xlThin
  .ColorIndex = xlAutomatic
End With
```

```
With Selection.Borders(xlInsideHorizontal)

  .LineStyle = xlContinuous
  .Weight = xlThin
  .ColorIndex = xlAutomatic
End With

ActiveSheet.Range(sBorder1).Select
Selection.Borders(xlDiagonalDown).LineStyle = xlNone
Selection.Borders(xlDiagonalUp).LineStyle = xlNone
With Selection.Borders(xlEdgeLeft)

  .LineStyle = xlContinuous
  .Weight = xlMedium
  .ColorIndex = xlAutomatic
End With
With Selection.Borders(xlEdgeTop)

  .LineStyle = xlContinuous
  .Weight = xlMedium
  .ColorIndex = xlAutomatic
End With
With Selection.Borders(xlEdgeBottom)

  .LineStyle = xlContinuous
  .Weight = xlMedium
  .ColorIndex = xlAutomatic
End With
With Selection.Borders(xlEdgeRight)

  .LineStyle = xlContinuous
  .Weight = xlMedium
  .ColorIndex = xlAutomatic
End With
With Selection.Borders(xlInsideHorizontal)

  .LineStyle = xlContinuous
  .Weight = xlThin
  .ColorIndex = xlAutomatic
End With

ActiveSheet.Range(sBorder2).Select
Selection.Borders(xlDiagonalDown).LineStyle = xlNone
Selection.Borders(xlDiagonalUp).LineStyle = xlNone
```

```
  With Selection.Borders(xlEdgeLeft)

    .LineStyle = xlContinuous
    .Weight = xlMedium
    .ColorIndex = xlAutomatic
  End With
  With Selection.Borders(xlEdgeTop)

    .LineStyle = xlContinuous
    .Weight = xlMedium
    .ColorIndex = xlAutomatic
  End With
  With Selection.Borders(xlEdgeBottom)

    .LineStyle = xlContinuous
    .Weight = xlMedium
    .ColorIndex = xlAutomatic
  End With
  With Selection.Borders(xlEdgeRight)

    .LineStyle = xlContinuous
    .Weight = xlMedium
    .ColorIndex = xlAutomatic
  End With
```

'Merge Technical Correlation

```
sMerge3 = Cells(1, 3).Address & ":" & Cells(1, 2 + ListSizeB).Address
  ActiveSheet.Range(sMerge3).Select
  With Selection
    .HorizontalAlignment = xlCenter
    .VerticalAlignment = xlCenter
    .WrapText = False
    .Orientation = 0
    .ShrinkToFit = False
    .MergeCells = False
  End With
  Selection.Merge
  ActiveSheet.Range(sMerge3).Value = "Technical Correlation"
  ActiveSheet.Range("B2").Select

End Sub
```

```
Private Sub image1_Click()

End Sub

Private Sub InsertA_Click()

  BoxA.AddItem txtBoxA.Text
  txtBoxA.Text = ""
End Sub

Private Sub InsertB_Click()

  BoxB.AddItem txtBoxB.Text
  txtBoxB.Text = ""
End Sub

Private Sub lblCustomer_Click()

End Sub

Private Sub lblEngineer_Click()

End Sub

Private Sub BoxB_Click()

End Sub

Private Sub txtBoxA_KeyDown(ByVal KeyCode As MSForms.ReturnInteger, By-
Val Shift As Integer)
  If KeyCode = vbKeyReturn Then
    If txtBoxA.Text <> "" Then
      BoxA.AddItem txtBoxA.Text
      txtBoxA.Text = ""
    End If
  End If
End Sub

Private Sub txtBoxB_KeyDown(ByVal KeyCode As MSForms.ReturnInteger, By-
Val Shift As Integer)
  If KeyCode = vbKeyReturn Then
    If txtBoxB.Text <> "" Then
      BoxB.AddItem txtBoxB.Text
      txtBoxB.Text = ""
    End If
```

```
  End If
End Sub

Private Sub txtTitle_Change()

End Sub
```

A.4.2 Programs for QFD Calculations and Table Operations

```
Private Sub cmdCalculate_Click()
ListSizeB = Worksheets("Input").BoxB.ListCount
sCalc = Cells(34, 4).Address & ":" & Cells(34, ListSizeB + 3).Address
sEndCalc = Cells(34, ListSizeB + 3).Address
  ActiveSheet.Range("D34").Select
  ActiveCell.FormulaR1C1 = _
    "=R3C3*R[-31]C+R4C3*R[-30]C+R5C3*R[-29]C+R6C3*R[-28]C+R7C3*
R[-27]C+R8C3*R[-26]C+R9C3*R[-25]C+R10C3*R[-24]C+R11C3*R[-23]C+
R12C3*R[-22]C+R13C3*R[-21]C+R14C3*R[-20]C+R15C3*R[-19]C+R16C3*
R[-18]C+R17C3*R[-17]C+R18C3*R[-16]C+R19C3*R[-15]C+R20C3*R[-14]C+
R21C3*R[-13]C+R22C3*R[-12]C+R23C3*R[-11]C+R24C3*R[-10]C+R25C3*
R[-9]C+R26C3*R[-8]C+R27C3*R[-7]C+R28C3*R[-6]C+R29C3*R[-5]C+
R30C3*R[-4]C+R31C3*R[-3]C+R32C3*R[-2]C+R33C3*R[-1]C"
    Selection.AutoFill Destination:=Range(sCalc), Type:=xlFillDefault
  ActiveSheet.Range(sCalc).Select
    With Selection.Borders(xlInsideVertical)
      .LineStyle = xlContinuous
      .Weight = xlThin
      .ColorIndex = xlAutomatic
    End With
  ActiveSheet.Range(sEndCalc).Borders(xlEdgeRight).Weight = xlMedium
  ActiveSheet.Range("B34").Select
End Sub

Private Sub cmdClear_Click()
Cells.Select
  Selection.Borders(xlDiagonalDown).LineStyle = xlNone
  Selection.Borders(xlDiagonalUp).LineStyle = xlNone
  Selection.Borders(xlEdgeLeft).LineStyle = xlNone
  Selection.Borders(xlEdgeTop).LineStyle = xlNone
  Selection.Borders(xlEdgeBottom).LineStyle = xlNone
  Selection.Borders(xlEdgeRight).LineStyle = xlNone
  Selection.Borders(xlInsideVertical).LineStyle = xlNone
  Selection.Borders(xlInsideHorizontal).LineStyle = xlNone
  Selection.ColumnWidth = 4
  Selection.RowHeight = 13.2
  With Selection
    .HorizontalAlignment = xlCenter
    .VerticalAlignment = xlCenter
    .Orientation = 0
    .WrapText = False
    .ShrinkToFit = False
```

```
    .MergeCells = False
  End With
  Selection.ClearContents
  Range("F15").Select
  With Worksheets("Input")
    .Activate
  End With
End Sub

Private Sub cmdGraph_Click()
ListSizeB = Worksheets("Input").BoxB.ListCount
sPlotvalue = Cells(34, 4).Address & ":" & Cells(34, 3 + ListSizeB).Address
sXvalue = Cells(2, 4).Address & ":" & Cells(2, 3 + ListSizeB).Address

ActiveWindow.ScrollRow = 1
  Range(sPlotvalue).Select
  Charts.Add
  ActiveChart.ChartType = xl3DColumnClustered
  ActiveChart.SetSourceData Source:=Sheets("QFD Table").Range(sPlotvalue),
  PlotBy:=xlColumns_
  ActiveChart.SeriesCollection(1).XValues = Worksheets("QFD Table").
  Range(sXvalue)
  ActiveChart.Location Where:=xlLocationAsNewSheet, Name:="QFD Chart"
  ActiveChart.Axes(xlCategory).Select
  Selection.TickLabels.Orientation = xlUpward
  With ActiveChart
    .RightAngleAxes = True
    .HasTitle = True
    .ChartTitle.Characters.Text = "Absolute Importance for Engineer"
    .Axes(xlCategory, xlPrimary).HasTitle = True
    .Axes(xlCategory, xlPrimary).AxisTitle.Characters.Text = "Voice of Engineer"
    .Axes(xlValue, xlPrimary).HasTitle = True
    .Axes(xlValue, xlPrimary).AxisTitle.Characters.Text = "Rating"
    .Axes(xlValue, xlPrimary).AxisTitle.Orientation = 90
  End With
  With ActiveChart.Axes(xlCategory)
    .HasMajorGridlines = False
    .HasMinorGridlines = False
  End With
  With ActiveChart.Axes(xlValue)
    .HasMajorGridlines = False
    .HasMinorGridlines = False
  End With
  ActiveChart.HasLegend = False
  ActiveChart.PlotArea.Select
```

```
Selection.Top = 31
Selection.Height = 387
With Selection.Border
  .ColorIndex = 30
  .Weight = xlThin
  .LineStyle = xlContinuous
End With
Selection.Interior.ColorIndex = 19

ActiveChart.SeriesCollection(1).Select
With Selection.Border
  .Weight = xlThin
  .LineStyle = xlAutomatic
End With

Selection.Shadow = True
Selection.InvertIfNegative = False
With Selection.Interior
  .ColorIndex = 30
  .Pattern = 13
End With
Sheets("QFD Chart").Select
Sheets("QFD Chart").Move After:=Sheets("QFD Table")
End Sub

Private Sub cmdGraph2_Click()
ListSizeB = Worksheets("Input").BoxB.ListCount
sPlotvalue = Cells(35, 4).Address & ":" & Cells(35, 3 + ListSizeB).Address
sXvalue = Cells(2, 4).Address & ":" & Cells(2, 3 + ListSizeB).Address

ActiveWindow.ScrollRow = 1
  Range(sPlotvalue).Select
  Charts.Add
  ActiveChart.ChartType = xl3DColumnClustered
  ActiveChart.SetSourceData Source:=Sheets("QFD Table").Range(sPlotvalue),
     PlotBy:=xlColumns_
  ActiveChart.SeriesCollection(1).XValues = Worksheets("QFD Table").
     Range(sXvalue)
  ActiveChart.Location Where:=xlLocationAsNewSheet, Name:="QFD Chart2"
  ActiveChart.Axes(xlCategory).Select
  Selection.TickLabels.Orientation = xlUpward
  With ActiveChart
    .RightAngleAxes = True
    .HasTitle = True
    .ChartTitle.Characters.Text = "Customer Satisfaction Performance"
```

```
    .Axes(xlCategory, xlPrimary).HasTitle = True
    .Axes(xlCategory, xlPrimary).AxisTitle.Characters.Text = "Voice of Engineer"
    .Axes(xlValue, xlPrimary).HasTitle = True
    .Axes(xlValue, xlPrimary).AxisTitle.Characters.Text = "Customer Satisfaction
       Rating"
    .Axes(xlValue, xlPrimary).AxisTitle.Orientation = 90
    End With
  With ActiveChart.Axes(xlCategory)
    .HasMajorGridlines = False
    .HasMinorGridlines = False
  End With
  With ActiveChart.Axes(xlValue)
    .HasMajorGridlines = False
    .HasMinorGridlines = False
  End With
  ActiveChart.HasLegend = False
  ActiveChart.PlotArea.Select
  Selection.Top = 31
  Selection.Height = 387
  With Selection.Border
    .ColorIndex = 30
    .Weight = xlThin
    .LineStyle = xlContinuous
  End With
  Selection.Interior.ColorIndex = 19
  ActiveChart.SeriesCollection(1).Select
  With Selection.Border
    .Weight = xlThin
    .LineStyle = xlAutomatic
  End With
  Selection.Shadow = False
  Selection.InvertIfNegative = False
  With Selection.Interior
    .ColorIndex = 30
    .Pattern = xlSolid
  End With
  Sheets("QFD Chart2").Select
  Sheets("QFD Chart2").Move After:=Sheets("QFD Table")
End Sub

Private Sub CmdCalculate1_Click()
ListSizeB = Worksheets("Input").BoxB.ListCount
sCalc = Cells(35, 4).Address & ":" & Cells(35, ListSizeB + 3).Address
sEndCalc = Cells(35, ListSizeB + 3).Address
  ActiveSheet.Range("D35").Select
```

```
    ActiveCell.FormulaR1C1 = _
      "=(R3C3*R[-32]C+R4C3*R[-31]C+R5C3*R[-30]C+R6C3*R[-29]C+
        R7C3*R[-28]C+R8C3*R[-27]C+R9C3*R[-26]C+R10C3*R[-25]C+
        R11C3*R[-24]C+R12C3*R[-23]C+R13C3*R[-22]C+R14C3*R[-21]C+
        R15C3*R[-20]C+R16C3*R[-19]C+R17C3*R[-18]C+R18C3*R[-17]C+
        R19C3*R[-16]C+R20C3*R[-15]C+R21C3*R[-14]C+R22C3*R[-13]C+
        R23C3*R[-12]C+R24C3*R[-11]C+R25C3*R[-10]C+R26C3*R[-9]C+
        R27C3*R[-8]C+R28C3*R[-7]C+R29C3*R[-6]C+R30C3*R[-5]C+R31C3*
        R[-4]C+R32C3*R[-3]C+R33C3*R[-2]C)/(R[-32]C+R[-31]C+R[-30]C+
        R[-29]C+R[-28]C+R[-27]C+R[-26]C+R[-25]C+R[-24]C+R[-23]C+
        R[-22]C+R[-21]C+R[-20]C+R[-19]C+R[-18]C+R[-17]C+R[-16]C+
        R[-15]C+R[-14]C+R[-13]C+R[-12]C+R[-11]C+R[-10]C+R[-9]C+
        R[-8]C+R[-7]C+R[-6]C+R[-5]C+R[-4]C+R[-3]C+R[-2]C)"
      Selection.AutoFill Destination:=Range(sCalc), Type:=xlFillDefault
    ActiveSheet.Range(sCalc).Select
      With Selection.Borders(xlInsideVertical)
        .LineStyle = xlContinuous
        .Weight = xlThin
        .ColorIndex = xlAutomatic
      End With
    ActiveSheet.Range(sEndCalc).Borders(xlEdgeRight).Weight = xlMedium
    ActiveSheet.Range("B35").Select
End Sub

Private Sub Worksheet_SelectionChange(ByVal Target As Excel.Range)

End Sub
```

A.4.3 Programs for Tech Correlation

```
Private Sub cmdClear_Click()
Cells.Select
  Selection.Borders(xlDiagonalDown).LineStyle = xlNone
  Selection.Borders(xlDiagonalUp).LineStyle = xlNone
  Selection.Borders(xlEdgeLeft).LineStyle = xlNone
  Selection.Borders(xlEdgeTop).LineStyle = xlNone
  Selection.Borders(xlEdgeBottom).LineStyle = xlNone
  Selection.Borders(xlEdgeRight).LineStyle = xlNone
  Selection.Borders(xlInsideVertical).LineStyle = xlNone
  Selection.Borders(xlInsideHorizontal).LineStyle = xlNone
  Selection.ColumnWidth = 4
  Selection.RowHeight = 13.2
With Selection

    .MergeCells = False
  End With
  Selection.ClearContents
  Range("A1").Select
  With Worksheets("Input")
    .Activate
  End With
End Sub
```

A.5 The Integrated CAD/CAPP/CAM System Program

A.5.1 Source Code for Process Planning for Sheet Metal Cutting

```
          /* add assistant path */
#include "lc.h"
#include "user1.h"
#include "user2.h"
void inner_path_2(struct RINGS *info)
{
  struct ELEMENTS *p1,*p2,*p0,*top;
  float xs,ys,xe,ye,ea,sa,r,xc,yc,ra,da,dx,dy;
  float k,k0,xm,ym,x1,y1,x2,y2,sa1,ea2;
  float len_i,len_o,ALFA1,ALFA2;
  int kk,clock,shape,mangle;

clock=info->CLOCK;
kk=info->members;
dx=info->x_max-info->x_min;
dy=info->y_max-info->y_min;
ALFA1=120;ALFA2=60;
len_i=3;len_o=3;
if((dx<2*len_i)&&(dy<2*len_i)) len_i=len_i/2;
p1=info->p_elem;
if((p1->add==1)&&(p1->type!=CIRCLE))
{
  out_path_1(info);
  return;
}
p0=(struct ELEMENTS *)malloc(LEN1);
if(!p0)
{
  MessageBox(hWnd,"Memory Error!!","",MB_ICONSTOP|MB_OK);
  exit(0);
}
info->p_elem=p0;
p0->prior_e=NULL;
if(p1->type==CIRCLE)
{
  xc=p1->graph.circle.x_center;
  yc=p1->graph.circle.y_center;
  ra=p1->graph.circle.radius;
  info->p_elem=p0;
  if(ra>4)
```

```
 {
 // if(ra>10) r=len_i;
 // else r=ra/2;
// if(r>10)
   r=16;
   if(p1->add==2)
   {
   xm=xc+ra-r;
   sa=360-ALFA1;
   ea=360;
                xe=xc+ra;
 }
 else{
 xm=xc-ra+r;
 sa=180-ALFA1;
 ea=180;
 xe=xc-ra;
 }
 ym=yc;
 x1=xm+r*cos(sa*PI/180);
 y1=ym+r*sin(sa*PI/180);
 p0->num=0;
 p0->add=0; /*half circular*/
 p0->type=ARC;
 p0->graph.arc.aradius=r;
 p0->graph.arc.x_acenter=xm;//xc-ra+r;
 p0->graph.arc.y_acenter=ym;
 p0->graph.arc.x_end=xe;//xc-ra;
 p0->graph.arc.y_end=yc;
 p0->graph.arc.eangle=ea;//180;
 p0->graph.arc.sangle=sa;//180-ALFA1;
 p0->graph.arc.dangle=ALFA1;
 p0->graph.arc.shape=3;
 p0->graph.arc.x_start=x1;
 p0->graph.arc.y_start=y1;
 p0->next_e=p1;
 p1->prior_e=p0;
 kk++;
 p0=(struct ELEMENTS *)malloc(LEN1);
 if(!p0)
 {
   MessageBox(hWnd,"memory error!","",MB_ICONSTOP|MB_OK);
   exit(0);
 }
 p0->num=kk;
```

```
  p0->add=0;
  p0->type=ARC;
  p0->graph.arc.aradius=r;
  if(p1->add==2)
  {
  sa=0;
  ea=ALFA2;
  }
  else{
  sa=180;
  ea=180+ALFA2;
  }
  x2=xm+r*cos(ea*PI/180);          /* other half circular*/
  y2=ym+r*sin(ea*PI/180);
  p0->graph.arc.x_acenter=xm;
  p0->graph.arc.y_acenter=ym;
  p0->graph.arc.x_start=xe;
  p0->graph.arc.y_start=yc;
  p0->graph.arc.eangle=ea;
  p0->graph.arc.sangle=sa;
  p0->graph.arc.dangle=ALFA2;
  p0->graph.arc.shape=3;
  p0->graph.arc.x_end=x2;
  p0->graph.arc.y_end=y2;
  p1->next_e=p0;
  p0->prior_e=p1;
  p0->next_e=NULL;
  kk++;
}
else {
  p0->num=0;
  p0->add=0;
  p0->type=LINE;
  p0->graph.line.x_start=xc;
  p0->graph.line.y_start=yc;
  if(p1->add==2) p0->graph.line.x_end=xc+ra;
  else p0->graph.line.x_end=xc-ra;
  p0->graph.line.y_end=yc;
  p0->next_e=p1;
  p1->prior_e=p0;
  kk++;
  p0=(struct ELEMENTS *)malloc(LEN1);
  if(!p0) {
    MessageBox(hWnd, “memory error!!","",MB_ICONSTOP|MB_OK);
    exit(0);
```

```
    }
    p0->num=kk;
    p0->add=0;
    p0->type=LINE;
    if(p1->add==2) p0->graph.line.x_start=xc+ra;
    else p0->graph.line.x_start=xc-ra;
    p0->graph.line.y_start=yc;
    p0->graph.line.x_end=xc;
    p0->graph.line.y_end=yc;
    p1->next_e=p0;
    p0->prior_e=p1;
    p0->next_e=NULL;
    kk++;
  }
  info->members=kk;
  return;
}
if(p1->type==LINE) /*start with line*/
{
  xs=p1->graph.line.x_start;
  ys=p1->graph.line.y_start;
  xe=p1->graph.line.x_end;
  ye=p1->graph.line.y_end;
  xm=0.5*(xs+xe);
  ym=0.5*(ys+ye);
  // r=len_i;
  r=16.;
  k=slope_angle(xs,ys,xe,ye);
  if(clock==1)
  {
    xc=xm-r*sin(k);
    yc=ym+r*cos(k);
    k0=slope_angle(xc,yc,xm,ym);
    sa1=k0-ALFA1*PI/180;
    if(sa1<0) sa1=sa1+2*PI;
    ea2=k0+ALFA2*PI/180;
    if(ea2>2*PI) ea2=ea2-2*PI;
  }
  if(clock==0)
  {
    xc=xm+r*sin(k);
    yc=ym-r*cos(k);
    k0=slope_angle(xc,yc,xm,ym);
    sa1=k0+ALFA1*PI/180;
    if(sa1>2*PI) sa1=sa1-2*PI;
```

```
    ea2=k0-ALFA2*PI/180;
    if(ea2<0) ea2=ea2+2*PI;
  }
  x1=xc+r*cos(sa1);
  y1=yc+r*sin(sa1);
  x2=xc+r*cos(ea2);
  y2=yc+r*sin(ea2);
  p0->num=0;
  p0->add=0;
  p0->type=ARC;
  p0->graph.arc.aradius=r;
  p0->graph.arc.x_acenter=xc;
  p0->graph.arc.y_acenter=yc;
  p0->graph.arc.x_end=xm;
  p0->graph.arc.y_end=ym;
  p0->graph.arc.eangle=k0*180/PI;
  p0->graph.arc.sangle=sa1*180/PI;
  p0->graph.arc.dangle=ALFA1;
  if(clock==0) p0->graph.arc.shape=2;
  else p0->graph.arc.shape=3;
  p0->graph.arc.x_start=x1;
  p0->graph.arc.y_start=y1;
  end=p0;
// new_store(p0);
  kk++;
  p2=(struct ELEMENTS *)malloc(LEN1);
  if(!p2)
  {
    MessageBox(hWnd ,“memory error!","",MB_ICONSTOP|MB_OK);
    exit(0);
  }
  p2->add=1;
  p2->num=1;
  p2->type=LINE;
  p2->graph.line.x_start=xm;
  p2->graph.line.y_start=ym;
  p1->graph.line.x_end=xm;
  p1->graph.line.y_end=ym;
  p2->graph.line.x_end=xe;
  p2->graph.line.y_end=ye;
  break_LA(p1,p2,p0);
  kk++;
  p0=(struct ELEMENTS *)malloc(LEN1);
  if(!p0)
  {
```

```
    MessageBox(hWnd," memory error!","",MB_ICONSTOP|MB_OK);
    exit(0);
  }
  p0->num=kk+2;
  p0->add=0;
  p0->type=ARC;
  p0->graph.arc.aradius=r;
  p0->graph.arc.x_acenter=xc;
  p0->graph.arc.y_acenter=yc;
  p0->graph.arc.x_start=xm;
  p0->graph.arc.y_start=ym;
  p0->graph.arc.x_end=x2;
  p0->graph.arc.y_end=y2;
  p0->graph.arc.eangle=ea2*180/PI;
  p0->graph.arc.sangle=k0*180/PI;
  p0->graph.arc.dangle=ALFA2;
  if(clock==0) p0->graph.arc.shape=2;
  else p0->graph.arc.shape=3;
  p1->next_e=p0;
  p0->prior_e=p1;
  p0->next_e=NULL;
// new_store(p0);
  kk++;
  info->members=kk;
}
if(p1->type==ARC)
{
  xs=p1->graph.arc.x_start;
  ys=p1->graph.arc.y_start;
  xe=p1->graph.arc.x_end;
  ye=p1->graph.arc.y_end;
  xc=p1->graph.arc.x_acenter;
  yc=p1->graph.arc.y_acenter;
  ra=p1->graph.arc.aradius;
  sa=p1->graph.arc.sangle;
  ea=p1->graph.arc.eangle;
  da=p1->graph.arc.dangle;
  shape=p1->graph.arc.shape;
  if(shape==3){
    xm=xc+ra*cos((sa+da/2)*PI/180);
    ym=yc+ra*sin((sa+da/2)*PI/180);
    mangle=sa+da/2;
  }
  if(shape==2){
    xm=xc+ra*cos((sa-da/2)*PI/180);
```

```
  ym=yc+ra*sin((sa-da/2)*PI/180);
  mangle=sa-da/2;
}
k0=slope_angle(xc,yc,xm,ym);
if(clock==1){
  k=k0+PI/2;
  if(k>2*PI) k=k-2*PI;
}
if(clock==0){
  k=k0-PI/2;
  if(k0<0) k=k+2*PI;
}
if(((clock==0)&&(shape==2))||((clock==1)&&(shape==3)))
{
// r=len_i;
r=16.;
if(clock==1)
{
  xc=xm-r*sin(k);
  yc=ym+r*cos(k);
  sa1=k0-ALFA1*PI/180;
  if(sa1<0) sa1=sa1+2*PI;
  ea2=k0+ALFA2*PI/180;
  if(ea2>2*PI) ea2=ea2-2*PI;
}
if(clock==0)
{
  xc=xm+r*sin(k);
  yc=ym-r*cos(k);
  sa1=k0+ALFA1*PI/180;
  if(sa1>2*PI) sa1=sa1-2*PI;
  ea2=k0-ALFA2*PI/180;
  if(ea2<0) ea2=ea2+2*PI;
}
x1=xc+r*cos(sa1);
y1=yc+r*sin(sa1);
x2=xc+r*cos(ea2);
y2=yc+r*sin(ea2);
p0->num=0;
p0->add=0;
p0->type=ARC;
p0->graph.arc.aradius=r;
p0->graph.arc.x_acenter=xc;
p0->graph.arc.y_acenter=yc;
p0->graph.arc.x_end=xm;
```

```
  p0->graph.arc.y_end=ym;
  p0->graph.arc.eangle=k0*180/PI;
  p0->graph.arc.sangle=sa1*180/PI;
  p0->graph.arc.dangle=ALFA1;
  if(shape==3) p0->graph.arc.shape=3;
  else p0->graph.arc.shape=2;
  p0->graph.arc.x_start=x1;
  p0->graph.arc.y_start=y1;
// new_store(p0);
  kk++;
  p2=(struct ELEMENTS *)malloc(LEN1);
  if(!p2)
  {
    MessageBox(hWnd, "memory error!","",MB_ICONSTOP|MB_OK);
    exit(0);
  }
  p2->add=1;
  p2->num=1;
  p2->type=ARC;
  p2->graph.arc.x_start=xm;
  p2->graph.arc.y_start=ym;
  p2->graph.arc.x_end=xe;
  p2->graph.arc.y_end=ye;
  p2->graph.arc.sangle=mangle;
  p2->graph.arc.eangle=ea;
  p2->graph.arc.dangle=da/2;
  p2->graph.arc.aradius=ra;
  p2->graph.arc.x_acenter=p1->graph.arc.x_acenter;
  p2->graph.arc.y_acenter=p1->graph.arc.y_acenter;
  p2->graph.arc.shape=p1->graph.arc.shape;
  p1->graph.arc.x_end=xm;
  p1->graph.arc.y_end=ym;
  p1->graph.arc.eangle=mangle;
  p1->graph.arc.dangle=da/2;
  break_LA(p1,p2,p0);
// A_ins_L(p1,p2,start);
  kk++;
  p0=(struct ELEMENTS *)malloc(LEN1);
  if(!p0)
  {
    MessageBox(hWnd, " memory error!","",MB_ICONSTOP|MB_OK);
    exit(0);
  }
  p0->num=kk;
  p0->add=0;
```

```
  p0->type=ARC;
  p0->graph.arc.aradius=r;
  p0->graph.arc.x_acenter=xc;
  p0->graph.arc.y_acenter=yc;
  p0->graph.arc.x_start=xm;
  p0->graph.arc.y_start=ym;
  p0->graph.arc.x_end=x2;
  p0->graph.arc.y_end=y2;
  p0->graph.arc.eangle=ea2*180/PI;
  p0->graph.arc.sangle=k0*180/PI;
  p0->graph.arc.dangle=ALFA2;
  if(shape==3) p0->graph.arc.shape=3;
  else p0->graph.arc.shape=2;
  p1->next_e=p0;
  p0->prior_e=p1;
  p0->next_e=NULL;
// new_store(p0);
kk++;
}
if(((clock==0)&&(shape==3))||((clock==1)&&(shape==2)))
{
  len_o=ra/2;
  sa1=k;
  if(clock==1) ea2=sa1-PI;
  else ea2=sa1+PI;
  if(ea2>2*PI) ea2=ea2-2*PI;
  if(ea2<0) ea2=ea2+2*PI;
  x2=xm+len_o*cos(ea2);
  y2=ym+len_o*sin(ea2);
  x1=xm+len_o*cos(sa1);
  y1=ym+len_o*sin(sa1);
  p0->num=0;
  p0->add=0;
  p0->type=LINE;
  p0->graph.line.x_start=x1;
  p0->graph.line.y_start=y1;
  p0->graph.line.x_end=xm;
  p0->graph.line.y_end=ym;
// new_store(p0);
  kk++;
  p2=(struct ELEMENTS *)malloc(LEN1);
  if(!p2)
  {
    MessageBox(hWnd, “memory error!”,"",MB_ICONSTOP|MB_OK);
    exit(0);
```

```
  }
  p2->add=1;
  p2->num=1;
  p2->type=ARC;
  p2->graph.arc.aradius=ra;
  p2->graph.arc.x_acenter=p1->graph.arc.x_acenter;
  p2->graph.arc.y_acenter=p1->graph.arc.y_acenter;
  p2->graph.arc.x_start=xm;
  p2->graph.arc.y_start=ym;
  p2->graph.arc.sangle=mangle;
  p2->graph.arc.eangle=ea;
  p2->graph.arc.dangle=da/2;
  p2->graph.arc.shape=p1->graph.arc.shape;
  p2->graph.arc.x_end=xe;
  p2->graph.arc.y_end=ye;
  p1->graph.arc.x_end=xm;
  p1->graph.arc.y_end=ym;
  p1->graph.arc.eangle=mangle;
  p1->graph.arc.dangle=da/2;
  break_LA(p1,p2,p0);
// A_ins_L(p1,p2,top);
  kk++;
  p0=(struct ELEMENTS *)malloc(LEN1);
  if(!p0)
  {
    MessageBox(hWnd, "memory error!","",MB_ICONSTOP|MB_OK);
    exit(0);
  }
  p0->num=kk+2;
  p0->add=0;
  p0->type=LINE;
  p0->graph.line.x_start=xm;
  p0->graph.line.y_start=ym;
  p0->graph.line.x_end=x2;
  p0->graph.line.y_end=y2;
  p1->next_e=p0;
  p0->prior_e=p1;
  p0->next_e=NULL;
// new_store(p0);
  kk++;
  }
  info->members=kk;
}
  }
  void out_path_2(struct RINGS *info)
```

```
{
  struct ELEMENTS *p1,*p2,*p0,*p3,*top;
  float xs,ys,xe,ye,ra,xc,yc,r,sa1,ea2,da,sa,ea;
  float k,k0,x1,y1,x2,y2,xm,ym,mangle;
  float ALFA1,ALFA2,len_o=12,len_i=12;
  int kk,clock,shape;
 p1=info->p_elem;
 kk=info->members;
 clock=info->CLOCK;
 ALFA1=120;ALFA2=60;
 if((info->include==-1)||(info->include==0))
 {
  if(solid==1){ len_o=15; len_i=15;}
  else{ len_o=10; len_i=10;}
 }
 p0=(struct ELEMENTS *)malloc(LEN1);
 if(!p0)
 {
  MessageBox(hWnd, "memory error!","",MB_ICONSTOP|MB_OK);
  exit(0);
 }
 info->p_elem=p0;
 p0->prior_e=NULL;
 if(p1->type==CIRCLE)
 {
  xc=p1->graph.circle.x_center;
  yc=p1->graph.circle.y_center;
  ra=p1->graph.circle.radius;
  if(ra<10) r=len_o;
  else r=0.5*ra;
// if(r>10) r=10;
 p0->num=0;
 p0->add=0;
 p0->type=LINE;
 p0->graph.line.x_end=xc+ra;
 p0->graph.line.y_end=yc;
 p0->graph.line.x_start=xc+ra;
 p0->graph.line.y_start=yc-r;
 kk++;
 p0->next_e=p1;
 p1->prior_e=p0;
 p0=(struct ELEMENTS *)malloc(LEN1);
 if(!p0)
 {
  MessageBox(hWnd , "memory error!","",MB_ICONSTOP|MB_OK);
```

```
    exit(0);
  }
  p0->num=kk;
  p0->add=0;
  p0->type=LINE;
  p0->graph.line.x_end=xc+ra;
  p0->graph.line.y_end=yc+r;
  p0->graph.line.x_start=xc+ra;
  p0->graph.line.y_start=yc;
  p1->next_e=p0;
  p0->prior_e=p1;
  p0->next_e=NULL;
  kk++;
  info->members=kk;
}
if(p1->type==LINE)
{
  xe=p1->graph.line.x_end;
  ye=p1->graph.line.y_end;
  xs=p1->graph.line.x_start;
  ys=p1->graph.line.y_start;
    xm=0.5*(xs+xe);
    ym=0.5*(ys+ye);
  // r=len_i;
    r=16.;
    k=slope_angle(xs,ys,xe,ye);
    if(clock==0)
    {
      xc=xm-r*sin(k);
      yc=ym+r*cos(k);
      k0=slope_angle(xc,yc,xm,ym);
      sa1=k0-ALFA1*PI/180;
      if(sa1<0) sa1=sa1+2*PI;
      ea2=k0+ALFA2*PI/180;
      if(ea2>2*PI) ea2=ea2-2*PI;
    }
    if(clock==1)
    {
      xc=xm+r*sin(k);
      yc=ym-r*cos(k);
      k0=slope_angle(xc,yc,xm,ym);
      sa1=k0+ALFA1*PI/180;
      if(sa1>2*PI) sa1=sa1-2*PI;
      ea2=k0-ALFA2*PI/180;
      if(ea2<0) ea2=ea2+2*PI;
```

```
}
x1=xc+r*cos(sa1);
y1=yc+r*sin(sa1);
x2=xc+r*cos(ea2);
y2=yc+r*sin(ea2);
p0->num=0;
p0->add=0;
p0->type=ARC;
p0->graph.arc.aradius=r;
p0->graph.arc.x_acenter=xc;
p0->graph.arc.y_acenter=yc;
p0->graph.arc.x_end=xm;
p0->graph.arc.y_end=ym;
p0->graph.arc.eangle=k0*180/PI;
p0->graph.arc.sangle=sa1*180/PI;
p0->graph.arc.dangle=ALFA1;
if(clock==1) p0->graph.arc.shape=2;
else p0->graph.arc.shape=3;
p0->graph.arc.x_start=x1;
p0->graph.arc.y_start=y1;
kk++;
p2=(struct ELEMENTS *)malloc(LEN1);
if(!p2)
{
MessageBox(hWnd , “memory error!”,"",MB_ICONSTOP|MB_OK);
exit(0);
}
p2->add=1;
p2->num=1;
p2->type=LINE;
p2->graph.line.x_start=xm;
p2->graph.line.y_start=ym;
p1->graph.line.x_end=xm;
p1->graph.line.y_end=ym;
p2->graph.line.x_end=xe;
p2->graph.line.y_end=ye;
break_LA(p1,p2,p0);
kk++;
p0=(struct ELEMENTS *)malloc(LEN1);
if(!p0)
{
  MessageBox(hWnd , “memory error!”,"",MB_ICONSTOP|MB_OK);
  exit(0);
}
p0->num=kk+2;
```

```
    p0->add=0;
    p0->type=ARC;
    p0->graph.arc.aradius=r;
    p0->graph.arc.x_acenter=xc;
    p0->graph.arc.y_acenter=yc;
    p0->graph.arc.x_start=xm;
    p0->graph.arc.y_start=ym;
    p0->graph.arc.x_end=x2;
    p0->graph.arc.y_end=y2;
    p0->graph.arc.eangle=ea2*180/PI;
    p0->graph.arc.sangle=k0*180/PI;
    p0->graph.arc.dangle=ALFA2;
    if(clock==1) p0->graph.arc.shape=2;
    else p0->graph.arc.shape=3;
    p1->next_e=p0;
    p0->prior_e=p1;
    p0->next_e=NULL;
    kk++;
  info->members=kk;
}
if(p1->type==ARC)
{
  xs=p1->graph.arc.x_start;
  ys=p1->graph.arc.y_start;
  xe=p1->graph.arc.x_end;
  ye=p1->graph.arc.y_end;
  xc=p1->graph.arc.x_acenter;
  yc=p1->graph.arc.y_acenter;
  ra=p1->graph.arc.aradius;
  sa=p1->graph.arc.sangle;
  ea=p1->graph.arc.eangle;
  da=p1->graph.arc.dangle;
  shape=p1->graph.arc.shape;
  if(shape==3){
    xm=xc+ra*cos((sa+da/2)*PI/180);
    ym=yc+ra*sin((sa+da/2)*PI/180);
    mangle=sa+da/2;
  }
  if(shape==2){
    xm=xc+ra*cos((sa-da/2)*PI/180);
    ym=yc+ra*sin((sa-da/2)*PI/180);
    mangle=sa-da/2;
  }
  k0=slope_angle(xc,yc,xm,ym);
  if(clock==1){
```

```
   k=k0+PI/2;
   if(k>2*PI) k=k-2*PI;
  }
  if(clock==0){
   k=k0-PI/2;
   if(k0<0) k=k+2*PI;
  }
  if(((clock==0)&&(shape==2))||((clock==1)&&(shape==3)))
  {
  len_o=ra;
// if(len_o>10) len_o=10;
  ea2=k;
  if(clock==1) sa1=ea2-PI;
  else sa1=ea2+PI;
  if(sa1>2*PI) sa1=sa1-2*PI;
  if(sa1<0) sa1=sa1+2*PI;
  x2=xm+len_o*cos(ea2);
  y2=ym+len_o*sin(ea2);
  x1=xm+len_o*cos(sa1);
  y1=ym+len_o*sin(sa1);
  p0->num=0;
  p0->add=0;
  p0->type=LINE;
  p0->graph.line.x_start=x1;
  p0->graph.line.y_start=y1;
  p0->graph.line.x_end=xm;
  p0->graph.line.y_end=ym;
// new_store(p0);
  kk++;
  p2=(struct ELEMENTS *)malloc(LEN1);
  if(!p2)
  {
   MessageBox(hWnd , "memory error!","",MB_ICONSTOP|MB_OK);
   exit(0);
  }
  p2->add=1;
  p2->num=1;
  p2->type=ARC;
  p2->graph.arc.aradius=ra;
  p2->graph.arc.x_acenter=p1->graph.arc.x_acenter;
  p2->graph.arc.y_acenter=p1->graph.arc.y_acenter;
  p2->graph.arc.x_start=xm;
  p2->graph.arc.y_start=ym;
  p2->graph.arc.sangle=mangle;
  p2->graph.arc.eangle=ea;
```

```
  p2->graph.arc.dangle=da/2;
  p2->graph.arc.shape=p1->graph.arc.shape;
  p2->graph.arc.x_end=xe;
  p2->graph.arc.y_end=ye;
  p1->graph.arc.x_end=xm;
  p1->graph.arc.y_end=ym;
  p1->graph.arc.eangle=mangle;
  p1->graph.arc.dangle=da/2;
// A_ins_L(p1,p2,top);
  break_LA(p1,p2,p0);
  kk++;
  p0=(struct ELEMENTS *)malloc(LEN1);
  if(!p0)
  {
    MessageBox(hWnd , "memory error!","",MB_ICONSTOP|MB_OK);
    exit(0);
  }
  p0->num=kk+2;
  p0->add=0;
  p0->type=LINE;
  p0->graph.line.x_start=xm;
  p0->graph.line.y_start=ym;
  p0->graph.line.x_end=x2;
  p0->graph.line.y_end=y2;
  p1->next_e=p0;
  p0->prior_e=p1;
  p0->next_e=NULL;
// new_store(p0);
  kk++;
  }
  if(((clock==0)&&(shape==3))||((clock==1)&&(shape==2)))
 {
 // r=0.5*ra;
 // if(r<len_i) r=len_i;
  r=16.;
  if(clock==0)
  {
    xc=xm+r*sin(k);
    yc=ym-r*cos(k);
    sa1=k0-ALFA1*PI/180;
    if(sa1<0) sa1=sa1+2*PI;
    ea2=k0+ALFA2*PI/180;
    if(ea2>2*PI) ea2=ea2-2*PI;
  }
  if(clock==1)
```

```
{
  xc=xm-r*sin(k);
  yc=ym+r*cos(k);
  sa1=k0+ALFA1*PI/180;
  if(sa1>2*PI) sa1=sa1-2*PI;
  ea2=k0-ALFA2*PI/180;
  if(ea2<0) ea2=ea2+2*PI;
}
x1=xc+r*cos(sa1);
y1=yc+r*sin(sa1);
x2=xc+r*cos(ea2);
y2=yc+r*sin(ea2);
p0->num=0;
p0->add=0;
p0->type=ARC;
p0->graph.arc.aradius=r;
p0->graph.arc.x_acenter=xc;
p0->graph.arc.y_acenter=yc;
p0->graph.arc.x_end=xm;
p0->graph.arc.y_end=ym;
p0->graph.arc.eangle=k0*180/PI;
p0->graph.arc.sangle=sa1*180/PI;
p0->graph.arc.dangle=ALFA1;
if(shape==3) p0->graph.arc.shape=3;
else p0->graph.arc.shape=2;
p0->graph.arc.x_start=x1;
p0->graph.arc.y_start=y1;
// new_store(p0);
kk++;
p2=(struct ELEMENTS *)malloc(LEN1);
if(!p2)
{
  MessageBox(hWnd , "memory error!","",MB_ICONSTOP|MB_OK);
  exit(0);
}
p2->add=1;
p2->num=1;
p2->type=ARC;
p2->graph.arc.x_start=xm;
p2->graph.arc.y_start=ym;
p2->graph.arc.x_end=xe;
p2->graph.arc.y_end=ye;
p2->graph.arc.sangle=mangle;
p2->graph.arc.eangle=ea;
p2->graph.arc.dangle=da/2;
```

```
  p2->graph.arc.aradius=ra;
  p2->graph.arc.x_acenter=p1->graph.arc.x_acenter;
  p2->graph.arc.y_acenter=p1->graph.arc.y_acenter;
  p2->graph.arc.shape=p1->graph.arc.shape;
  p1->graph.arc.x_end=xm;
  p1->graph.arc.y_end=ym;
  p1->graph.arc.eangle=mangle;
  p1->graph.arc.dangle=da/2;
  break_LA(p1,p2,p0);
// A_ins_L(p1,p2,top);
  kk++;
  p0=(struct ELEMENTS *)malloc(LEN1);
  if(!p0)
  {
    MessageBox(hWnd , "memory error!","",MB_ICONSTOP|MB_OK);
    exit(0);
  }
  p0->num=kk;
  p0->add=0;
  p0->type=ARC;
  p0->graph.arc.aradius=r;
  p0->graph.arc.x_acenter=xc;
  p0->graph.arc.y_acenter=yc;
  p0->graph.arc.x_start=xm;
  p0->graph.arc.y_start=ym;
  p0->graph.arc.x_end=x2;
  p0->graph.arc.y_end=y2;
  p0->graph.arc.eangle=ea2*180/PI;
  p0->graph.arc.sangle=k0*180/PI;
  p0->graph.arc.dangle=ALFA2;
  if(shape==3) p0->graph.arc.shape=3;
  else p0->graph.arc.shape=2;
  p1->next_e=p0;
  p0->prior_e=p1;
  p0->next_e=NULL;
// new_store(p0);
  kk++;
  }
  info->members=kk;
 }
}
void out_path_1(struct RINGS *info)
{
  struct ELEMENTS *p1,*p2,*p0;
  float xs,ys,xe,ye,ra,xc,yc,r,sa1,ea2,da,sa,ea;
```

```
  float k1,k2,x1,y1,x2,y2,xm,ym,mangle;
  float ALFA1,ALFA2,len_o=16,len_i=16;
  int kk,clock,shape;
// top=info->p_elem;
  kk=info->members;
  clock=info->CLOCK;
  ALFA1=120;ALFA2=60;
  if((info->include==-1)||(info->include==0))
  {
    if(solid==1){ len_o=16; len_i=16;}
  }
  p1=info->p_elem;
  while(p1){
    p2=p1;
    p1=p1->next_e;
  }
  p1=info->p_elem;
// last_path=NULL;
  if(p1->type==CIRCLE)
  {
  xc=p1->graph.circle.x_center;
  yc=p1->graph.circle.y_center;
  ra=p1->graph.circle.radius;
  if(ra<10) r=len_o;
  else r=0.5*ra;
// if(r>10) r=10;
  p0=(struct ELEMENTS *)malloc(LEN1);
  if(!p0)
  {
    MessageBox(hWnd , "memory error!","",MB_ICONSTOP|MB_OK);
    exit(0);
  }
  info->p_elem=p0;
// last_path=NULL;
  p0->num=0;
  p0->add=0;
  p0->type=LINE;
  p0->graph.line.x_end=xc-ra;
  p0->graph.line.y_end=yc;
  p0->graph.line.x_start=xc-ra;
  p0->graph.line.y_start=yc+r;
// new_store(p0);
  p0->prior_e=NULL;
  p0->next_e=p1;
  p1->prior_e=p0;
```

```
  kk++;
// C_ins_ring(p1);
  p0=(struct ELEMENTS *)malloc(LEN1);
  if(!p0)
  {
    MessageBox(hWnd , “memory error!”,"",MB_ICONSTOP|MB_OK);
    exit(0);
  }
  p0->num=kk;
  p0->add=0;
  p0->type=LINE;
  p0->graph.line.x_end=xc-ra;
  p0->graph.line.y_end−yc r;
  p0->graph.line.x_start=xc-ra;
  p0->graph.line.y_start=yc;
// new_store(p0);
  p2->next_e=p0;
  p0->prior_e=p2;
  p0->next_e=NULL;
  kk++;
  info->members=kk;
  return;
}
if(p1->type==LINE)
{
  xe=p1->graph.line.x_end;
  ye=p1->graph.line.y_end;
  xs=p1->graph.line.x_start;
  ys=p1->graph.line.y_start;
  k1=slope_angle(xe,ye,xs,ys);
}
if(p1->type==ARC)
{
  xs=p1->graph.arc.x_start;
  ys=p1->graph.arc.y_start;
  xc=p1->graph.arc.x_acenter;
  yc=p1->graph.arc.y_acenter;
  shape=p1->graph.arc.shape;
  k1=slope_angle(xs,ys,xc,yc);
  if(shape==2) k1=k1-PI/2;
  else k1=k1+PI/2;
  }
  x1=xs+len_o*cos(k1);
  y1=ys+len_o*sin(k1);
  p0=(struct ELEMENTS *)malloc(LEN1);
```

```
  if(!p0)
  {
    MessageBox(hWnd, “memory error!”,"",MB_ICONSTOP|MB_OK);
    exit(0);
  }
  info->p_elem=p0;
//last_path=NULL;
  p0->num=1;
  p0->add=0;
  p0->type=LINE;
  p0->graph.line.x_end=xs;
  p0->graph.line.y_end=ys;
  p0->graph.line.x_start=x1;
  p0->graph.line.y_start=y1;
  p0->prior_e=NULL;
  p0->next_e=p1;
  p1->prior_e=p0;
// new_store(p0);
  kk++;
// L_ins_ring(p1,top);
  if(p2->type==LINE)
  {
    xe=p2->graph.line.x_end;
    ye=p2->graph.line.y_end;
    xs=p2->graph.line.x_start;
    ys=p2->graph.line.y_start;
    k2=slope_angle(xs,ys,xe,ye);
  }
  if(p2->type==ARC)
  {
    xe=p2->graph.arc.x_end;
    ye=p2->graph.arc.y_end;
    xc=p2->graph.arc.x_acenter;
    yc=p2->graph.arc.y_acenter;
    shape=p2->graph.arc.shape;
    k2=slope_angle(xe,ye,xc,yc);
    if(shape==2) k2=k2+PI/2;
    else k2=k2-PI/2;
  }
  x2=xe+len_o*cos(k2);
  y2=ye+len_o*sin(k2);

  p0=(struct ELEMENTS *)malloc(LEN1);
  if(!p0)
  {
```

```
    MessageBox(hWnd , "memory error!","",MB_ICONSTOP|MB_OK);
    exit(0);
  }
  p0->num=kk+1;
  p0->add=0;
  p0->type=LINE;
  p0->graph.line.x_end=x2;
  p0->graph.line.y_end=y2;
  p0->graph.line.x_start=xe;
  p0->graph.line.y_start=ye;
  p2->next_e=p0;
  p0->prior_e=p2;
  p0->next_e=NULL;
// new_store(p3);
  kk++;
  info->members=kk;
  return;
}
void break_LA(struct ELEMENTS *p1,struct ELEMENTS *p2,struct ELEMENTS
*p0)
/**/
{
  struct ELEMENTS *p,*pend;
  int j;
  p2->next_e=p1->next_e;
  p1->next_e->prior_e=p2;
  p2->prior_e=p0;
  p1->next_e=NULL;
  p0->next_e=p2;
  j=0;
  p=p2;
  while(p){
    p->num=++j;
    pend=p;
    p=p->next_e;
  }
  p1->prior_e=pend;
  pend->next_e=p1;
  p1->num=++j;
}
/*struct ELEMENTS *add_path(struct ELEMENTS *p1,struct RINGS *info)
{
  struct ELEMENTS *p2,*p0,*p3;
  float x11,y11,x12,y12,x22,y22;
  float xc1,yc1,xc2,yc2,k,x,y;
```

```
float LL=10.0,CL=2.0;
int num1,shape1,shape2;
p2=p1->next_e;
if(p2==NULL) p2=info->p_elem;
if((p1->type==LINE)&&(p2->type==LINE))
{
  x11=p1->graph.line.x_start;
  y11=p1->graph.line.y_start;
  x12=p1->graph.line.x_end;
  y12=p1->graph.line.y_end;
  x22=p2->graph.line.x_end;
  y22=p2->graph.line.y_end;
}
if((p1->type==LINE)&&(p2->type==ARC))
{
  x11=p1->graph.line.x_start;
  y11=p1->graph.line.y_start;
  x12=p1->graph.line.x_end;
  y12=p1->graph.line.y_end;
  xc2=p2->graph.arc.x_acenter;
  yc2=p2->graph.arc.y_acenter;
  shape2=p2->graph.arc.shape;
  if(fabs(xc2-x12)<=0.001){
    if(y12>yc2){
      if(shape2==2) x22=x12+LL;
      else x22=x12-LL;
      y22=y12;
    }
    if(y12<yc2){
      if(shape2==2) x22=x12-LL;
      else x22=x12+LL;
      y22=y12;
    }
  }
  if(fabs(x12-xc2)>0.001){
    if(y12==yc2){
      if(x12>xc2){
        if(shape2==2) y22=y12-LL;
        else y22=y12+LL;
      }
      if(x12<xc2){
        if(shape2==2) y22=y12+LL;
        else y22=y12-LL;
      }
      x22=x12;
```

```
}
else{
  k=(y12-yc2)/(x12-xc2);
  k=atan(fabs(k));
  if(((x12-xc2)>0)&&((y12-yc2)>0)){
    if(shape2==3){
      x22=x12-LL*sin(k);
      y22=y12+LL*cos(k);
    }
    if(shape2==2){
      x22=x12+LL*sin(k);
      y22=y12-LL*cos(k);
    }
  }
  if(((x12-xc2)<0)&&((y12-yc2)>0)){
    if(shape2==3){
      x22=x12-LL*sin(k);
      y22=y12-LL*cos(k);
    }
    if(shape2==2){
      x22=x12+LL*sin(k);
      y22=y12+LL*cos(k);
    }
  }
  if(((x12-xc2)<0)&&((y12-yc2)<0)){
    if(shape2==3){
      x22=x12+LL*sin(k);
      y22=y12-LL*cos(k);
    }
    if(shape2==2){
      x22=x12-LL*sin(k);
      y22=y12+LL*cos(k);
    }
  }
  if(((x12-xc2)>0)&&((y12-yc2)<0)){
    if(shape2==3){
      x22=x12+LL*sin(k);
      y22=y12+LL*cos(k);
    }
    if(shape2==2){
      x22=x12-LL*sin(k);
      y22=y12-LL*cos(k);
    }
  }
}
```

```
    }
}
if((p1->type==ARC)&&(p2->type==ARC))
{
  x12=p1->graph.arc.x_end;
  y12=p1->graph.arc.y_end;
  xc1=p1->graph.arc.x_acenter;
  yc1=p1->graph.arc.y_acenter;
  xc2=p2->graph.arc.x_acenter;
  yc2=p2->graph.arc.y_acenter;
  shape1=p1->graph.arc.shape;
  shape2=p2->graph.arc.shape;
  if(fabs(xc1-x12)<=0.001){
  if(y12>yc1){
    if(shape1==2) x11=x12-LL;
    if(shape1==3) x11=x12+LL;
    y11=y12;
  }
  if(y12<yc1){
    if(shape1==2) x11=x12+LL;
    if(shape1==3) x11=x12-LL;
    y11=y12;
  }
}
if(fabs(x12-xc1)>0.001){
  if(y12==yc1){
    if(x12>xc1){
      if(shape1==2) y11=y12+LL;
      if(shape1==3) y11=y12-LL;
    }
    if(x12<xc1){
      if(shape1==2) y11=y12-LL;
      if(shape1==3) y11=y12+LL;
    }
    x11=x12;
  }
  else {
    k=(y12-yc1)/(x12-xc1);
    k=atan(fabs(k));
    if(((x12-xc1)>0)&&((y12-yc1)>0)){
    if(shape1==3){
      x11=x12+LL*sin(k);
      y11=y12-LL*cos(k);
    }
    if(shape1==2){
```

```
      x11=x12-LL*sin(k);
      y11=y12+LL*cos(k);
     }
   }
   if(((x12-xc1)<0)&&((y12-yc1)>0)){
     if(shape1==2){
      x11=x12-LL*sin(k);
      y11=y12-LL*cos(k);
     }
     if(shape1==3){
      x11=x12+LL*sin(k);
      y11=y12+LL*cos(k);
     }
   }
   if(((x12-xc1)<0)&&((y12-yc1)<0)){
     if(shape1==2){
      x11=x12+LL*sin(k);
      y11=y12-LL*cos(k);
     }
     if(shape1==3){
      x11=x12-LL*sin(k);
      y11=y12+LL*cos(k);
     }
   }
   if(((x12-xc1)>0)&&((y12-yc1)<0)){
     if(shape1==2){
      x11=x12+LL*sin(k);
      y11=y12+LL*cos(k);
   }
     if(shape1==3){
      x11=x12-LL*sin(k);
      y11=y12-LL*cos(k);
     }
    }
   }
 }
 if(fabs(xc2-x12)<=0.001){
   if(y12>yc2){
     if(shape2==2) x22=x12+LL;
     if(shape2==3) x22=x12-LL;
     y22=y12;
   }
   if(y12<yc2){
     if(shape2==2) x22=x12-LL;
     if(shape2==3) x22=x12+LL;
```

```
    y22=y12;
   }
  }
 if(fabs(x12-xc2)>0.001){
  if(y12==yc2){
   if(x12>xc2){
    if(shape2==2) y22=y12-LL;
    if(shape2==3) y22=y12+LL;
   }
   if(x12<xc2){
    if(shape2==2) y22=y12+LL;
    if(shape2==3) y22=y12-LL;
   }
   x22=x12;
  }
  else{
   k=(y12-yc2)/(x12-xc2);
   k=atan(fabs(k));
   if(((x12-xc2)>0)&&((y12-yc2)>0)){
    if(shape2==2){
     x22=x12+LL*sin(k);
     y22=y12-LL*cos(k);
    }
    if(shape2==3){
     x22=x12-LL*sin(k);
     y22=y12+LL*cos(k);
    }
   }
   if(((x12-xc2)<0)&&((y12-yc2)>0)){
    if(shape2==2){
     x22=x12+LL*sin(k);
     y22=y12+LL*cos(k);
    }
    if(shape2==3){
     x22=x12-LL*sin(k);
     y22=y12-LL*cos(k);
    }
   }
   if(((x12-xc2)<0)&&((y12-yc2)<0)){
    if(shape2==2){
     x22=x12-LL*sin(k);
     y22=y12+LL*cos(k);
    }
    if(shape2==3){
     x22=x12+LL*sin(k);
```

```
        y22=y12-LL*cos(k);
      }
    }
    if(((x12-xc2)>0)&&((y12-yc2)<0)){
      if(shape2==2){
        x22=x12-LL*sin(k);
        y22=y12-LL*cos(k);
      }
      if(shape2==3){
        x22=x12+LL*sin(k);
        y22=y12+LL*cos(k);
      }
    }
   }
  }
}
if((p1->type==ARC)&&(p2->type==LINE))
{
  x12=p2->graph.line.x_start;
  y12=p2->graph.line.y_start;
  x22=p2->graph.line.x_end;
  y22=p2->graph.line.y_end;
  xc1=p1->graph.arc.x_acenter;
  yc1=p1->graph.arc.y_acenter;
  shape1=p1->graph.arc.shape;
  if(fabs(xc1-x12)<=0.001){
    if(y12>yc1){
      if(shape1==2) x11=x12-LL;
      if(shape1==3) x11=x12+LL;
      y11=y12;
    }
    if(y12<yc1){
      if(shape1==2) x11=x12+LL;
      if(shape1==3) x11=x12-LL;
      y11=y12;
    }
  }
  if(fabs(x12-xc1)>0.001){
    if(y12==yc1){
      if(x12>xc1){
        if(shape1==2) y11=y12+LL;
        if(shape1==3) y11=y12-LL;
      }
      if(x12<xc1){
        if(shape1==2) y11=y12-LL;
```

```
    if(shape1==3) y11=y12+LL;
  }
  x11=x12;
}
else {
  k=(y12-yc1)/(x12-xc1);
  k=atan(fabs(k));
  if(((x12-xc1)>0)&&((y12-yc1)>0)){
    if(shape1==3){
      x11=x12+LL*sin(k);
      y11=y12-LL*cos(k);
    }
    if(shape1==2){
      x11=x12-LL*sin(k);
      y11=y12+LL*cos(k);
    }
  }
  if(((x12-xc1)<0)&&((y12-yc1)>0)){
    if(shape1==3){
      x11=x12+LL*sin(k);
      y11=y12+LL*cos(k);
    }
    if(shape1==2){
      x11=x12-LL*sin(k);
      y11=y12-LL*cos(k);
    }
  }
  if(((x12-xc1)<0)&&((y12-yc1)<0)){
    if(shape1==3){
      x11=x12-LL*sin(k);
      y11=y12+LL*cos(k);
    }
    if(shape1==2){
      x11=x12+LL*sin(k);
      y11=y12-LL*cos(k);
    }
  }
  if(((x12-xc1)>0)&&((y12-yc1)<0)){
    if(shape1==3){
      x11=x12-LL*sin(k);
      y11=y12-LL*cos(k);
    }
    if(shape1==2){
      x11=x12+LL*sin(k);
      y11=y12+LL*cos(k);
```

```
        }
      }
    }
  }
}
num1=p1->num;
  p0=(struct ELEMENTS *)malloc(sizeof(struct ELEMENTS));
  if(!p0){
    MessageBox(hWnd , “memory error!”,"",MB_ICONSTOP|MB_OK);
    exit(0);
  }
  p1->next_e=p0;
  p0->add=0;
  p0->prior_e=p1;
  p0->type=LINE;
// if(p1->type==LINE) p0->num=-1;
// else { p0->num=num1+1; num0++;}
  p0->num=num1+1;
  p0->graph.line.x_start=x12;
  p0->graph.line.y_start=y12;
  if(fabs(x12-x11)<=0.001){
    if(y11<y12) y=y12+CL;
    if(y12<y11) y=y12-CL;
    x=x12;
  }
  if(fabs(x12-x11)>0.001){
    if(y12==y11){
      if(x12>x11) x=x12+CL;
      if(x12<x11) x=x12-CL;
      y=y12;
    }
    else {
      k=(y12-y11)/(x12-x11);
      k=atan(fabs(k));
      if(((x12-x11)>0)&&((y12-y11)>0))
    {
      x=x12+CL*cos(k);
      y=y12+CL*sin(k);
    }
    if(((x12-x11)<0)&&((y12-y11)>0))
    {
      x=x12-CL*cos(k);
      y=y12+CL*sin(k);
    }
    if(((x12-x11)<0)&&((y12-y11)<0))
```

```
    {
      x=x12-CL*cos(k);
      y=y12-CL*sin(k);
    }
    if(((x12-x11)>0)&&((y12-y11)<0))
    {
      x=x12+CL*cos(k);
      y=y12-CL*sin(k);
    }
  }
}
p0->graph.line.x_end=x;
p0->graph.line.y_end=y;
if(fabs(x12-x22)<=0.001){
  if(y22<y12) y=y12+CL;
  if(y12<y22) y=y12-CL;
  x=x12;
}
if(fabs(x12-x22)>0.001){
  if(y12==y22){
    if(x12>x22) x=x12+CL;
    if(x12<x22) x=x12-CL;
    y=y12;
  }
  else {
      k=(y12-y22)/(x12-x22);
      k=atan(fabs(k));
      if(((x12-x22)>0)&&((y12-y22)>0))
    {
      x=x12+CL*cos(k);
      y=y12+CL*sin(k);
    }
    if(((x12-x22)<0)&&((y12-y22)>0))
    {
      x=x12-CL*cos(k);
      y=y12+CL*sin(k);
    }
    if(((x12-x22)<0)&&((y12-y22)<0))
    {
      x=x12-CL*cos(k);
      y=y12-CL*sin(k);
    }
    if(((x12-x22)>0)&&((y12-y22)<0))
    {
      x=x12+CL*cos(k);
```

```
      y=y12-CL*sin(k);
    }
  }
}
  p3=(struct ELEMENTS *)malloc(sizeof(struct ELEMENTS));
  if(!p3){
    MessageBox(hWnd, "memory error!","",MB_ICONSTOP|MB_OK);
    exit(0);
  }
  p0->next_e=p3;
  p3->prior_e=p0;
  p3->type=LINE;
  p3->add=0;
// num0++;
// if(p1->type==LINE) p3->num=num1+1;
// else p3->num=num1+2;
  p3->num=num1+2;
  p3->graph.line.x_start=p0->graph.line.x_end;
  p3->graph.line.y_start=p0->graph.line.y_end;
  p3->graph.line.x_end=x;
  p3->graph.line.y_end=y;
  p0=(struct ELEMENTS *)malloc(sizeof(struct ELEMENTS));
  if(!p0){
    MessageBox(hWnd , "memory error!","",MB_ICONSTOP|MB_OK);
    exit(0);
  }
  p3->next_e=p0;
  p0->prior_e=p3;
  p0->next_e=p2;
  p2->prior_e=p0;
  p0->type=LINE;
  p0->add=0;
  p0->graph.line.x_start=x;
  p0->graph.line.y_start=y;
  p0->graph.line.x_end=x12;
  p0->graph.line.y_end=y12;
//if(p2->type==LINE) p0->num=-1;
// else { p0->num=num1+3; num0++;}
  p0->num=num1+3;
  p2->num=num1+4;
  p0=p1->next_e;
// while(p0){
// draw_elem(p0,8,4);
// p0=p0->next_e;
// if(p0==p2) break;
```

```
// }
  info->members=info->members+3;
  return p0;
}*/
struct ELEMENTS *add_path(struct ELEMENTS *p1,struct RINGS *info)
{
  struct ELEMENTS *p2,*p0,*p3;
  float xs,ys,xe,ye,xc,yc;
  float x1,y1,x2,y2,k1,k2,x0,y0;
  float LL=16.0;
  int num1,shape;

  p2=p1->next_e;
  if(p2==NULL) return p2;//p2=info->p_elem;
  if(p1->type==LINE)
  {
    xe=p1->graph.line.x_end;
    ye=p1->graph.line.y_end;
    xs=p1->graph.line.x_start;
    ys=p1->graph.line.y_start;
    k1=slope_angle(xs,ys,xe,ye);
  }
  if(p1->type==ARC)
  {
    xe=p1->graph.arc.x_end;
    ye=p1->graph.arc.y_end;
    xc=p1->graph.arc.x_acenter;
    yc=p1->graph.arc.y_acenter;
    shape=p1->graph.arc.shape;
    k1=slope_angle(xe,ye,xc,yc);
    if(shape==2) k1=k1+PI/2;
    else k1=k1-PI/2;
  }
  x0=xe;
  y0=ye;
  x1=xe+LL*cos(k1);
  y1=ye+LL*sin(k1);
  if(p2->type==LINE)
  {
    xe=p2->graph.line.x_end;
    ye=p2->graph.line.y_end;
    xs=p2->graph.line.x_start;
    ys=p2->graph.line.y_start;
    k2=slope_angle(xe,ye,xs,ys);
  }
```

```
if(p2->type==ARC)
{
  xs=p2->graph.arc.x_start;
  ys=p2->graph.arc.y_start;
  xc=p2->graph.arc.x_acenter;
  yc=p2->graph.arc.y_acenter;
  shape=p2->graph.arc.shape;
  k2=slope_angle(xs,ys,xc,yc);
  if(shape==2) k2=k2-PI/2;
  else k2=k2+PI/2;
}
x2=xs+LL*cos(k2);
y2=ys+LL*sin(k2);

num1=p1->num;
p0=(struct ELEMENTS *)malloc(sizeof(struct ELEMENTS));
if(!p0){
  MessageBox(hWnd , "memory error!","",MB_ICONSTOP|MB_OK);
  exit(0);
}
p1->next_e=p0;
p0->add=0;
p0->prior_e=p1;
p0->type=LINE;
p0->graph.line.x_start=x0;
p0->graph.line.y_start=y0;
p0->graph.line.x_end=x1;
p0->graph.line.y_end=y1;
p3=(struct ELEMENTS *)malloc(sizeof(struct ELEMENTS));
if(!p3){
  MessageBox(hWnd, "memory error!","",MB_ICONSTOP|MB_OK);
  exit(0);
}
p0->next_e=p3;
p3->prior_e=p0;
p3->type=LINE;
p3->add=0;
p3->graph.line.x_start=p0->graph.line.x_end;
p3->graph.line.y_start=p0->graph.line.y_end;
p3->graph.line.x_end=x2;
p3->graph.line.y_end=y2;
p0=(struct ELEMENTS *)malloc(sizeof(struct ELEMENTS));
if(!p0){
  MessageBox(hWnd,"memory error!","",MB_ICONSTOP|MB_OK);
  exit(0);
```

```
  }
  p3->next_e=p0;
  p0->prior_e=p3;
  p0->next_e=p2;
  p2->prior_e=p0;
  p0->type=LINE;
  p0->add=0;
  p0->graph.line.x_start=x2;
  p0->graph.line.y_start=y2;
  p0->graph.line.x_end=x0;
  p0->graph.line.y_end=y0;
  p0=p1->next_e;
  while(p0){
    num1++;
    p0->num=num1;
    p0=p0->next_e;
  }
  p0=p1->next_e;
  info->members=info->members+3;
  return p0;
}
```

A.6 Database Program

A.6.1 Product Database Operation

```
Option Compare Database
Option Explicit

Private Sub Mould_Flow_Results_Click()
On Error GoTo Err_Mould_Flow_Results_Click

  Dim stDocName As String
  Dim stLinkCriteria As String

  stDocName = "fPU Product Mould Flow Results"

  stLinkCriteria = "[Cat #]=" & "'" & Me![Cat #] & "'"
  DoCmd.OpenForm stDocName, , , stLinkCriteria

Exit_Mould_Flow_Results_Click:
  Exit Sub

Err_Mould_Flow_Results_Click:
  MsgBox Err.Description
  Resume Exit_Mould_Flow_Results_Click

End Sub
Private Sub Project_Appraisal_Click()
On Error GoTo Err_Project_Appraisal_Click

  Dim stDocName As String
  Dim stLinkCriteria As String

  stDocName = "fSF Project Appraisals"

  stLinkCriteria = "[PA number]=" & "'" & Me![Cat #] & "'"
  DoCmd.OpenForm stDocName, , , stLinkCriteria

Exit_Project_Appraisal_Click:
  Exit Sub

Err_Project_Appraisal_Click:
  MsgBox Err.Description
  Resume Exit_Project_Appraisal_Click
```

```
End Sub
Private Sub Product_Design_Questions_Click()
On Error GoTo Err_Product_Design_Questions_Click

  Dim stDocName As String
  Dim stLinkCriteria As String

  stDocName = "fPU Product Design Questions"

  stLinkCriteria = "[cat #]=" & "'" & Me![Cat #] & "'"
  DoCmd.OpenForm stDocName, , , stLinkCriteria

Exit_Product_Design_Questions_Click:
  Exit Sub

Err_Product_Design_Questions_Click:
  MsgBox Err.Description
  Resume Exit_Product_Design_Questions_Click

End Sub
Private Sub Product_Designers_Click()
On Error GoTo Err_Product_Designers_Click

  Dim stDocName As String
  Dim stLinkCriteria As String

  stDocName = "fPU Product Designers"

  stLinkCriteria = "[cat #]=" & "'" & Me![Cat #] & "'"
  DoCmd.OpenForm stDocName, , , stLinkCriteria

Exit_Product_Designers_Click:
  Exit Sub

Err_Product_Designers_Click:
  MsgBox Err.Description
  Resume Exit_Product_Designers_Click

End Sub
Private Sub Product_Assembly_Picture_Click()
On Error GoTo Err_Product_Assembly_Picture_Click

  Dim stDocName As String
  Dim stLinkCriteria As String
```

```
  stDocName = "f Product assembly picture"

  stLinkCriteria = "[cat #]=" & "'" & Me![Cat #] & "'"
  DoCmd.OpenForm stDocName, , , stLinkCriteria

Exit_Product_Assembly_Picture_Click:
  Exit Sub

Err_Product_Assembly_Picture_Click:
  MsgBox Err.Description
  Resume Exit_Product_Assembly_Picture_Click

End Sub
Private Sub Tool_Make_Product_Click()
On Error GoTo Err_Tool_Make_Product_Click
  Dim stDocName As String
  Dim stLinkCriteria As String

  stDocName = "fSF (Tools) Product Tool Relationship"

  stLinkCriteria = "[Cat #]=" & "'" & Me![Cat #] & "'"
  DoCmd.OpenForm stDocName, , , stLinkCriteria

Exit_Tool_Make_Product_Click:
  Exit Sub

Err_Tool_Make_Product_Click:
  MsgBox Err.Description
  Resume Exit_Tool_Make_Product_Click

End Sub
Private Sub Tool_Make_Part_Click()
On Error GoTo Err_Tool_Make_Part_Click

  Dim stDocName As String
  Dim stLinkCriteria As String

  stDocName = "fProduct Tool Relationship"

  stLinkCriteria = "[Cat #]=" & "'" & Me![Cat #] & "'"
  DoCmd.OpenForm stDocName, , , stLinkCriteria

Exit_Tool_Make_Part_Click:
  Exit Sub
```

```
Err_Tool_Make_Part_Click:
  MsgBox Err.Description
  Resume Exit_Tool_Make_Part_Click

End Sub
Private Sub Solid_Work99_Click()
On Error GoTo Err_Solid_Work99_Click

  Dim stAppName As String

  stAppName = "C:\sw99\sldworks.exe"
  Call Shell(stAppName, 1)

Exit_Solid_Work99_Click:
  Exit Sub

Err_Solid_Work99_Click:
  MsgBox Err.Description
  Resume Exit_Solid_Work99_Click

End Sub
Private Sub Back_to_Main_Menu_Click()
On Error GoTo Err_Back_to_Main_Menu_Click
  Dim stDocName As String
  Dim stLinkCriteria As String

  stDocName = "Switchboard"
  DoCmd.OpenForm stDocName, , , stLinkCriteria

Exit_Back_to_Main_Menu_Click:
  Exit Sub

Err_Back_to_Main_Menu_Click:
  MsgBox Err.Description
  Resume Exit_Back_to_Main_Menu_Click

End Sub
```

A.6.2 Interface Programs

```
Option Compare Database
Option Explicit

Private Sub Command28_Click()
On Error GoTo Err_Command28_Click

  Dim stDocName As String

  stDocName = "rProduct Assembly Detail"
  DoCmd.OpenReport stDocName, acPreview

Exit_Command28_Click:
  Exit Sub

Err_Command28_Click:
  MsgBox Err.Description
  Resume Exit_Command28_Click

End Sub
Private Sub Find_Record_Click()
On Error GoTo Err_Find_Record_Click

  Screen.PreviousControl.SetFocus
  DoCmd.DoMenuItem acFormBar, acEditMenu, 10, , acMenuVer70

Exit_Find_Record_Click:
  Exit Sub

Err_Find_Record_Click:
  MsgBox Err.Description
  Resume Exit_Find_Record_Click

End Sub
Private Sub Tool_info_for_products_Click()
On Error GoTo Err_Tool_info_for_products_Click

  Dim stDocName As String
  Dim stLinkCriteria As String

  stDocName = "fProduct Tool Relationship"

  stLinkCriteria = "[Cat #]=" & "'" & Me![Cat #] & "'"
```

```
  DoCmd.OpenForm stDocName, , , stLinkCriteria

Exit_Tool_info_for_products_Click:
  Exit Sub

Err_Tool_info_for_products_Click:
  MsgBox Err.Description
  Resume Exit_Tool_info_for_products_Click

End Sub
Private Sub Search_for_record_Click()
On Error GoTo Err_Search_for_record_Click

  Screen.PreviousControl.SetFocus
  DoCmd.DoMenuItem acFormBar, acEditMenu, 10, , acMenuVer70

Exit_Search_for_record_Click:
  Exit Sub

Err_Search_for_record_Click:
  MsgBox Err.Description
  Resume Exit_Search_for_record_Click

End Sub
Private Sub Product_mould_flow_results_Cmd_Click()
On Error GoTo Err_Product_mould_flow_results_Cmd_Click

  Dim stDocName As String
  Dim stLinkCriteria As String

  stDocName = "fPU Product Mould Flow Results"

  stLinkCriteria = "[Cat #]=" & "'" & Me![Cat #] & "'"
  DoCmd.OpenForm stDocName, , , stLinkCriteria

Exit_Product_mould_flow_results_Cmd_Clic:
  Exit Sub

Err_Product_mould_flow_results_Cmd_Click:
  MsgBox Err.Description
  Resume Exit_Product_mould_flow_results_Cmd_Clic

End Sub
Private Sub Project_Appraisal_Info_Cmd_Click()
```

```
On Error GoTo Err_Project_Appraisal_Info_Cmd_Click

  Dim stDocName As String
  Dim stLinkCriteria As String

  stDocName = "fSF Project Appraisals"

  stLinkCriteria = "[cat #]=" & "'" & Me![Cat #] & "'"
  DoCmd.OpenForm stDocName, , , stLinkCriteria

Exit_Project_Appraisal_Info_Cmd_Click:
  Exit Sub

Err_Project_Appraisal_Info_Cmd_Click:
  MsgBox Err.Description
  Resume Exit_Project_Appraisal_Info_Cmd_Click

End Sub
Private Sub Product_Design_Questions_Cmd_Click()
On Error GoTo Err_Product_Design_Questions_Cmd_Click

  Dim stDocName As String
  Dim stLinkCriteria As String

  stDocName = "fPU Product Design Questions"

  stLinkCriteria = "[cat #]=" & "'" & Me![Cat #] & "'"
  DoCmd.OpenForm stDocName, , , stLinkCriteria

Exit_Product_Design_Questions_Cmd_Click:
  Exit Sub

Err_Product_Design_Questions_Cmd_Click:
  MsgBox Err.Description
  Resume Exit_Product_Design_Questions_Cmd_Click

End Sub
Private Sub Product_Designer_s__Cmd_Click()
On Error GoTo Err_Product_Designer_s__Cmd_Click

  Dim stDocName As String
  Dim stLinkCriteria As String

  stDocName = "fPU Product Designers"
```

```
  stLinkCriteria = "[cat #]=" & "'" & Me![Cat #] & "'"
  DoCmd.OpenForm stDocName, , , stLinkCriteria

Exit_Product_Designer_s__Cmd_Click:
  Exit Sub

Err_Product_Designer_s__Cmd_Click:
  MsgBox Err.Description
  Resume Exit_Product_Designer_s__Cmd_Click

End Sub
Private Sub Assembly_picture_detail_Click()
On Error GoTo Err_Assembly_picture_detail_Click

  Dim stDocName As String
  Dim stLinkCriteria As String

  stDocName = "fProduct Assembly Detail Subform"

  stLinkCriteria = "[Cat #]=" & "'" & Me![Cat #] & "'"
  DoCmd.OpenForm stDocName, , , stLinkCriteria

Exit_Assembly_picture_detail_Click:
  Exit Sub

Err_Assembly_picture_detail_Click:
  MsgBox Err.Description
  Resume Exit_Assembly_picture_detail_Click

End Sub
Private Sub Product_Assembly_Picture_Cmd_Click()
On Error GoTo Err_Product_Assembly_Picture_Cmd_Click

  Dim stDocName As String
  Dim stLinkCriteria As String

  stDocName = "f Product assembly picture"

  stLinkCriteria = "[cat #]=" & "'" & Me![Cat #] & "'"
  DoCmd.OpenForm stDocName, , , stLinkCriteria

Exit_Product_Assembly_Picture_Cmd_Click:
  Exit Sub

Err_Product_Assembly_Picture_Cmd_Click:
```

```
  MsgBox Err.Description
  Resume Exit_Product_Assembly_Picture_Cmd_Click

End Sub
Private Sub Command67_Click()
On Error GoTo Err_Command67_Click

  Screen.PreviousControl.SetFocus
  DoCmd.FindNext

Exit_Command67_Click:
  Exit Sub

Err_Command67_Click:
  MsgBox Err.Description
  Resume Exit_Command67_Click

End Sub
Private Sub Go_Back_to_Main_Menu_Click()
On Error GoTo Err_Go_Back_to_Main_Menu_Click

  Dim stDocName As String
  Dim stLinkCriteria As String

  stDocName = "Switchboard"
  DoCmd.OpenForm stDocName, , , stLinkCriteria

Exit_Go_Back_to_Main_Menu_Click:
  Exit Sub

Err_Go_Back_to_Main_Menu_Click:
  MsgBox Err.Description
  Resume Exit_Go_Back_to_Main_Menu_Click

End Sub
Private Sub Run_Word_Click()
On Error GoTo Err_Run_Word_Click

  Dim oApp As Object

  Set oApp = CreateObject("Word.Application")
  oApp.Visible = True

Exit_Run_Word_Click:
```

```
  Exit Sub

Err_Run_Word_Click:
  MsgBox Err.Description
  Resume Exit_Run_Word_Click

End Sub
Private Sub Solid_Work99_Click()
On Error GoTo Err_Solid_Work99_Click

  Dim stAppName As String

  stAppName = "C:\sw99\sldworks.exe"
  Call Shell(stAppName, 1)

Exit_Solid_Work99_Click:
  Exit Sub

Err_Solid_Work99_Click:
  MsgBox Err.Description
  Resume Exit_Solid_Work99_Click

End Sub
```

A.6.3 Programs for Tool Database Management

```
Option Compare Database
Option Explicit

Private Sub Command22_Click()
On Error GoTo Err_Command22_Click

  Dim stDocName As String
  Dim stLinkCriteria As String

  stDocName = "fSFPU Tool Components"
  DoCmd.OpenForm stDocName, , , stLinkCriteria

Exit_Command22_Click:
  Exit Sub

Err_Command22_Click:
  MsgBox Err.Description
  Resume Exit_Command22_Click

End Sub
Private Sub Tool_manufacturing_information_Click()
On Error GoTo Err_Tool_manufacturing_information_Click

  Dim stDocName As String
  Dim stLinkCriteria As String

  stDocName = "fSFPU Tool Manufacturing Information"

  stLinkCriteria = "[Tool Code]=" & "'" & Me![Tool code] & "'"
  DoCmd.OpenForm stDocName, , , stLinkCriteria

Exit_Tool_manufacturing_information_Clic:
  Exit Sub

Err_Tool_manufacturing_information_Click:
  MsgBox Err.Description
  Resume Exit_Tool_manufacturing_information_Clic

End Sub

Private Sub Form_Open(Cancel As Integer)
```

```
DoCmd.Maximize

End Sub

Private Sub Tool_code_LostFocus()
DoCmd.RunCommand acCmdSaveRecord

End Sub

Private Sub Tool_component_information_Click()
On Error GoTo Err_Tool_component_information_Click

  Dim stDocName As String
  Dim stLinkCriteria As String

  stDocName = "fPU Tool components"

  stLinkCriteria = "[Tool Code]=" & "'" & Me![Tool code] & "'"
  DoCmd.OpenForm stDocName, , , stLinkCriteria

Exit_Tool_component_information_Click:
  Exit Sub

Err_Tool_component_information_Click:
  MsgBox Err.Description
  Resume Exit_Tool_component_information_Click

End Sub
Private Sub Command35_Click()
On Error GoTo Err_Command35_Click

  Dim stDocName As String
  Dim stLinkCriteria As String

  stDocName = "tTool Components"

  stLinkCriteria = "[Tool Code]=" & "'" & Me![Tool code] & "'"
  DoCmd.OpenForm stDocName, , , stLinkCriteria

Exit_Command35_Click:
  Exit Sub

Err_Command35_Click:
  MsgBox Err.Description
```

```
  Resume Exit_Command35_Click

End Sub
Private Sub Command36_Click()
On Error GoTo Err_Command36_Click

  Dim stDocName As String
  Dim stLinkCriteria As String

  stDocName = "tTool Components"
  DoCmd.OpenForm stDocName, , , stLinkCriteria

Exit_Command36_Click:
  Exit Sub

Err_Command36_Click:
  MsgBox Err.Description
  Resume Exit_Command36_Click

End Sub
Private Sub Command37_Click()
On Error GoTo Err_Command37_Click

  Dim stDocName As String
  Dim stLinkCriteria As String

  stDocName = "tTool Components1"

  stLinkCriteria = "[Component code]=" & Me![Tool code]
  DoCmd.OpenForm stDocName, , , stLinkCriteria

Exit_Command37_Click:
  Exit Sub

Err_Command37_Click:
  MsgBox Err.Description
  Resume Exit_Command37_Click

End Sub
Private Sub cmd_manufacturing_Click()
On Error GoTo Err_cmd_manufacturing_Click

  Dim stDocName As String
  Dim stLinkCriteria As String
```

```
  stDocName = "fPU Manufacturing Information (Tools)"

  stLinkCriteria = "[Tool code]=" & "'" & Me![Tool code] & "'"
  DoCmd.OpenForm stDocName, , , stLinkCriteria

Exit_cmd_manufacturing_Click:
  Exit Sub

Err_cmd_manufacturing_Click:
  MsgBox Err.Description
  Resume Exit_cmd_manufacturing_Click

End Sub
Private Sub Command49_Click()
On Error GoTo Err_Command49_Click

  Screen.PreviousControl.SetFocus
  DoCmd.DoMenuItem acFormBar, acEditMenu, 10, , acMenuVer70

Exit_Command49_Click:
  Exit Sub

Err_Command49_Click:
  MsgBox Err.Description
  Resume Exit_Command49_Click

End Sub
Private Sub Products_made_on_tool_Cmd_Click()
On Error GoTo Err_Products_made_on_tool_Cmd_Click

  Dim stDocName As String
  Dim stLinkCriteria As String

  stDocName = "fProduct Tool Relationship"

  stLinkCriteria = "[Tool code]=" & "'" & Me![Tool code] & "'"
  DoCmd.OpenForm stDocName, , , stLinkCriteria

Exit_Products_made_on_tool_Cmd_Click:
  Exit Sub

Err_Products_made_on_tool_Cmd_Click:
  MsgBox Err.Description
  Resume Exit_Products_made_on_tool_Cmd_Click
```

```
End Sub
Private Sub Moulding_machine_info_Cmd_Click()
On Error GoTo Err_Moulding_machine_info_Cmd_Click

  Dim stDocName As String
  Dim stLinkCriteria As String

  stDocName = "fPU Tool Factory Production Machines"

  stLinkCriteria = "[Tool code]=" & "'" & Me![Tool code] & "'"
  DoCmd.OpenForm stDocName, , , stLinkCriteria

Exit_Moulding_machine_info_Cmd_Click:
  Exit Sub

Err_Moulding_machine_info_Cmd_Click:
  MsgBox Err.Description
  Resume Exit_Moulding_machine_info_Cmd_Click

End Sub
Private Sub Tool_Material_Cmd_Click()
On Error GoTo Err_Tool_Material_Cmd_Click

  Dim stDocName As String
  Dim stLinkCriteria As String

  stDocName = "fPU Tool Materials"

  stLinkCriteria = "[Tool code]=" & "'" & Me![Tool code] & "'"
  DoCmd.OpenForm stDocName, , , stLinkCriteria

Exit_Tool_Material_Cmd_Click:
  Exit Sub

Err_Tool_Material_Cmd_Click:
  MsgBox Err.Description
  Resume Exit_Tool_Material_Cmd_Click

End Sub
Private Sub Mould_Base_Info_Cmd_Click()
On Error GoTo Err_Mould_Base_Info_Cmd_Click

  Dim stDocName As String
  Dim stLinkCriteria As String
```

```
  stDocName = "fSF Mould Base Types"

  stLinkCriteria = "[Mould base code]=" & Me![Mould base code]
  DoCmd.OpenForm stDocName, , , stLinkCriteria

Exit_Mould_Base_Info_Cmd_Click:
  Exit Sub

Err_Mould_Base_Info_Cmd_Click:
  MsgBox Err.Description
  Resume Exit_Mould_Base_Info_Cmd_Click

End Sub
Private Sub Command59_Click()
On Error GoTo Err_Command59_Click

  DoCmd.GoToRecord , , acNext

Exit_Command59_Click:
  Exit Sub

Err_Command59_Click:
  MsgBox Err.Description
  Resume Exit_Command59_Click

End Sub
Private Sub Tool_designers_Info_Cmd_Click()
On Error GoTo Err_Tool_designers_Info_Cmd_Click

  Dim stDocName As String
  Dim stLinkCriteria As String

  stDocName = "fSF Tool Designers"

  stLinkCriteria = "[Tool code]=" & "'" & Me![Tool code] & "'"
  DoCmd.OpenForm stDocName, , , stLinkCriteria

Exit_Tool_designers_Info_Cmd_Click:
  Exit Sub

Err_Tool_designers_Info_Cmd_Click:
  MsgBox Err.Description
  Resume Exit_Tool_designers_Info_Cmd_Click

End Sub
```

```
Private Sub Command61_Click()
On Error GoTo Err_Command61_Click

  Dim stDocName As String
  Dim stLinkCriteria As String

  stDocName = "fSF Tool Designers"

  stLinkCriteria = "[Tool code]=" & "'" & Me![Tool code] & "'"
  DoCmd.OpenForm stDocName, , , stLinkCriteria

Exit_Command61_Click:
  Exit Sub

Err_Command61_Click:
  MsgBox Err.Description
  Resume Exit_Command61_Click

End Sub
Private Sub Command62_Click()
On Error GoTo Err_Command62_Click

  Dim stDocName As String
  Dim stLinkCriteria As String

  stDocName = "f Tool Designers"

  stLinkCriteria = "[Tool code]=" & "'" & Me![Tool code] & "'"
  DoCmd.OpenForm stDocName, , , stLinkCriteria

Exit_Command62_Click:
  Exit Sub

Err_Command62_Click:
  MsgBox Err.Description
  Resume Exit_Command62_Click

End Sub
Private Sub Tool_Designers_Cmd_Click()
On Error GoTo Err_Tool_Designers_Cmd_Click

  Dim stDocName As String
  Dim stLinkCriteria As String
```

```
  stDocName = "fPU Tool Designers"

  stLinkCriteria = "[Tool code]=" & "'" & Me![Tool code] & "'"
  DoCmd.OpenForm stDocName, , , stLinkCriteria

Exit_Tool_Designers_Cmd_Click:
  Exit Sub

Err_Tool_Designers_Cmd_Click:
  MsgBox Err.Description
  Resume Exit_Tool_Designers_Cmd_Click

End Sub
Private Sub Tool_trials_information_Cmd_Click()
On Error GoTo Err_Tool_trials_information_Cmd_Click

  Dim stDocName As String
  Dim stLinkCriteria As String

  stDocName = "fSF Tool Work Request and Production Trial Records"

  stLinkCriteria = "[Tool code]=" & "'" & Me![Tool code] & "'"
  DoCmd.OpenForm stDocName, , , stLinkCriteria

Exit_Tool_trials_information_Cmd_Click:
  Exit Sub

Err_Tool_trials_information_Cmd_Click:
  MsgBox Err.Description
  Resume Exit_Tool_trials_information_Cmd_Click

End Sub
Private Sub Command66_Click()
On Error GoTo Err_Command66_Click

  Dim stDocName As String
  Dim stLinkCriteria As String

  stDocName = "fProduct Tool Relationship"

  stLinkCriteria = "[Cat #]=" & "'" & Me![Tool code] & "'"
  DoCmd.OpenForm stDocName, , , stLinkCriteria

Exit_Command66_Click:
```

```
  Exit Sub

Err_Command66_Click:
  MsgBox Err.Description
  Resume Exit_Command66_Click

End Sub
Private Sub Command67_Click()
On Error GoTo Err_Command67_Click

  Dim stDocName As String
  Dim stLinkCriteria As String

  stDocName = "fProduct Tool Relationship"

  stLinkCriteria = "[Tool code]=" & "'" & Me![Tool code] & "'"
  DoCmd.OpenForm stDocName, , , stLinkCriteria

Exit_Command67_Click:
  Exit Sub

Err_Command67_Click:
  MsgBox Err.Description
  Resume Exit_Command67_Click

End Sub
Private Sub Back_to_Main_Menu_Click()
On Error GoTo Err_Back_to_Main_Menu_Click

  Dim stDocName As String
  Dim stLinkCriteria As String

  stDocName = "Switchboard"
  DoCmd.OpenForm stDocName, , , stLinkCriteria

Exit_Back_to_Main_Menu_Click:
  Exit Sub

Err_Back_to_Main_Menu_Click:
  MsgBox Err.Description
  Resume Exit_Back_to_Main_Menu_Click

End Sub
Private Sub Solid_Work99_Click()
```

```
On Error GoTo Err_Solid_Work99_Click

  Dim stAppName As String

  stAppName = "C:\sw99\sldworks.exe"
  Call Shell(stAppName, 1)

Exit_Solid_Work99_Click:
  Exit Sub

Err_Solid_Work99_Click:
  MsgBox Err.Description
  Resume Exit_Solid_Work99_Click

End Sub
Private Sub Command71_Click()
On Error GoTo Err_Command71_Click

  Call Shell("NOTEPAD.EXE", 1)

Exit_Command71_Click:
  Exit Sub

Err_Command71_Click:
  MsgBox Err.Description
  Resume Exit_Command71_Click

End Sub
```

A.6.4 Programs for Manufacturing Information Management

```
Option Compare Database
Option Explicit

Private Sub Command0_Click()
On Error GoTo Err_Command0_Click

  Dim stDocName As String
  Dim stLinkCriteria As String

  stDocName = "f Workshop Machinery"
  DoCmd.OpenForm stDocName, , , stLinkCriteria

Exit_Command0_Click:
  Exit Sub

Err_Command0_Click:
  MsgBox Err.Description
  Resume Exit_Command0_Click

End Sub
Private Sub Command1_Click()
On Error GoTo Err_Command1_Click

  Dim stDocName As String
  Dim stLinkCriteria As String

  stDocName = "fSF Tool Work Request and Production Trial Records"
  DoCmd.OpenForm stDocName, , , stLinkCriteria

Exit_Command1_Click:
  Exit Sub

Err_Command1_Click:
  MsgBox Err.Description
  Resume Exit_Command1_Click

End Sub
Private Sub Command2_Click()
On Error GoTo Err_Command2_Click

  Dim stDocName As String
```

```
  Dim stLinkCriteria As String

  stDocName = "f Workshop (tool trials)"
  DoCmd.OpenForm stDocName, , , stLinkCriteria

Exit_Command2_Click:
  Exit Sub

Err_Command2_Click:
  MsgBox Err.Description
  Resume Exit_Command2_Click

End Sub
Private Sub Sodic_wire_cut_info_Cmd_Click()
On Error GoTo Err_Sodic_wire_cut_info_Cmd_Click

  Dim stDocName As String
  Dim stLinkCriteria As String

  stDocName = "fSF Sodic Wire Cut"
  DoCmd.OpenForm stDocName, , , stLinkCriteria

Exit_Sodic_wire_cut_info_Cmd_Click:
  Exit Sub

Err_Sodic_wire_cut_info_Cmd_Click:
  MsgBox Err.Description
  Resume Exit_Sodic_wire_cut_info_Cmd_Click

End Sub
Private Sub Factory_Production_Machinery_Cmd_Click()
On Error GoTo Err_Factory_Production_Machinery_Cmd_Click

  Dim stDocName As String
  Dim stLinkCriteria As String

  stDocName = "fFactory Machinery"
  DoCmd.OpenForm stDocName, , , stLinkCriteria

Exit_Factory_Production_Machinery_Cmd_Cl:
  Exit Sub

Err_Factory_Production_Machinery_Cmd_Click:
  MsgBox Err.Description
```

```
  Resume Exit_Factory_Production_Machinery_Cmd_Cl

End Sub
Private Sub Machine_Tools_Info_Cmd_Click()
On Error GoTo Err_Machine_Tools_Info_Cmd_Click

  Dim stDocName As String
  Dim stLinkCriteria As String

  stDocName = "fSF Machine Tools"
  DoCmd.OpenForm stDocName, , , stLinkCriteria

Exit_Machine_Tools_Info_Cmd_Click:
  Exit Sub

Err_Machine_Tools_Info_Cmd_Click:
  MsgBox Err.Description
  Resume Exit_Machine_Tools_Info_Cmd_Click

End Sub
```

A.7 Experimental Results of Investigation Genetic Parameters

PMX: Partially matched crossover.
UOX: Uniform order-based crossover
SWAP: Swap two random positions in a single route.
FLIP: Flip function to rearrange the order in single route
Loop: Generations
Pc: Possibility of crossover
Pm: Possibility of mutation
Popsize: Initial population pool size

PMX CROSSOVER+SWAP MUTATION , FLIP MUTATION
Loop=10000; Pc=0.9; Pm=0.05-0.1;
Popsize=31; Parts=32. Distance=4.0384e+003
Loop=5000; Pc=0.8; Pm=0.1;
Popsize=51; Parts=32. Distance= 3.8703e+003
Loop=10000; Pc=0.9; Pm=0.05-0.1;
Popsize=31; Parts=32. Distance=3.8730e+003
Loop=5000; Pc=0.8; Pm=0.1;
Popsize=81; Parts=32. Distance=3.6694e+003
Loop=5000; Pc=0.8; Pm=0.1;
Popsize=81; Parts=32. Distance=3.8623e+003
Loop=5000; Pc=0.8; Pm=0.1;
Popsize=81; Parts=32. Distance=3.9582e+003
PMX CROSSOVER+SWAP MUTATION Loop=5000
Loop=2000; Pc=0.9; Pm=0.05;
Popsize=61; Parts=32.
Distance= 5.3182e+003
UOX CROSSOVER+SWAP MUTATION
Loop=3000; Pc=0.9; Pm=0.1;
Popsize=61; Parts=32. Distance=3.7232e+003
PMX+ FLIP MUTATION
Loop=2000; Pc=0.7; Pm=0.15;
Popsize=61; Parts=32.
Distance=6.8509e+003
PMX +SWAP MUTATION
Loop=2000; Pc=0.7; Pm=0.15;
Popsize=61; Parts=32.
Distance= 6.2296e+003
PMX CROSSOVER+FLIP MUTATION
Loop=5000; Pc=0.9; Pm=0.1;
Popsize=61; Parts=32. Distance= 3.5682e+003
PMX CROSSOVER+FLIP MUTATION
Loop=20000; Pc=0.9; Pm=0.1;
Popsize=61; Parts=32. Distance= 3.5227e+003

Index

W

X

MIX
Papier aus verantwortungsvollen Quellen
Paper from responsible sources
FSC® C105338

If you have any concerns about our products,
you can contact us on
ProductSafety@springernature.com

In case Publisher is established outside the EU,
the EU authorized representative is:
Springer Nature Customer Service Center GmbH
Europaplatz 3, 69115 Heidelberg, Germany

Printed by Libri Plureos GmbH
in Hamburg, Germany